Model-Based Engineering for
Complex Electronic Systems

I would like to thank my wife, Caroline, and my daughter, Heather, who between them have provided me with so much love and support during the period of writing this book. While the time I spend on research and engineering is rewarding, my time with Caroline and Heather is priceless.

I would therefore like to dedicate my efforts on this book to Caroline and Heather.

Peter Wilson

I would like to thank my wife, Mary Lynn, and our three wonderful daughters, Deanna Lynn, Laura Kathryn, and Maureen Elaine, for their love, support, and patience as I spent the time working on this project along with the other dozen or so I typically have on the go. I would like to dedicate my efforts on this book to the memory of my brother, Opie Louis Mantooth, Jr (1953–1999) and his daughter, Stephanie Ann (1981–2000).

Alan Mantooth

Model-Based Engineering for Complex Electronic Systems

Peter Wilson

H. Alan Mantooth

ELSEVIER

AMSTERDAM • BOSTON • HEIDELBERG • LONDON
NEW YORK • OXFORD • PARIS • SAN DIEGO
SAN FRANCISCO • SINGAPORE • SYDNEY • TOKYO
Newnes is an imprint of Elsevier

Newnes

Newnes is an imprint of Elsevier
The Boulevard, Langford Lane, Kidlington, Oxford OX5 1GB, UK
225 Wyman Street, Waltham, MA 02451, USA

First published 2013

Notice
Knowledge and best practice in this field are constantly changing. As new research and experience broaden our understanding, changes in research methods, professional practices, or medical treatment may become necessary.

Practitioners and researchers must always rely on their own experience and knowledge in evaluating and using any information, methods, compounds, or experiments described herein. In using such information or methods they should be mindful of their own safety and the safety of others, including parties for whom they have a professional responsibility.

To the fullest extent of the law, neither the Publisher nor the authors, contributors, or editors, assume any liability for any injury and/or damage to persons or property as a matter of products liability, negligence or otherwise, or from any use or operation of any methods, products, instructions, or ideas contained in the material herein.

British Library Cataloguing-in-Publication Data
A catalogue record for this book is available from the British Library

Library of Congress Cataloging-in-Publication Data
A catalog record for this book is available from the Library of Congress

ISBN: 978-0-12-385085-0

For information on all Newnes publications
visit our website at **store.elsevier.com**

Printed and bound in the United States

13 14 15 16   10 9 8 7 6 5 4 3 2 1

  **Working together
to grow libraries in
developing countries**

www.elsevier.com • www.bookaid.org

Contents

SECTION 3: Design Methods

Foreword

Models scare engineers.

They expose, for all the world to see, any lack of understanding of the device being used or the design itself that they may have. This is why "I don't do modeling" is a common refrain among electronics engineers.

Yet they do, indeed, do modeling. Without a model, you cannot analyze a design (either on the computer or by hand) to determine if it will work correctly. Models are (and always have been) essential to the design process. Show me a designer and I'll show you a modeler.

To make matters worse, rarely, if ever, will a design fail from the aspects the engineer has considered. It more likely fails from what the engineer has ignored or didn't know about. And, to top it off, systems have become so complex that engineers have to use computers (which need models) to create the designs.

This is precisely why Peter Wilson and Alan Mantooth have written this book — to help analog and mixed-signal (AMS) engineers through this conundrum.

The "rule of thumb" that surprisingly assists is that second- and third-order effects rarely help.

If a design won't work with simple, first-order effects, it probably won't work when second- and third-order effects are included. This "saves the day".

Model-based engineering starts with the simple, first-order models with which most engineers are comfortable (commonly referred to as "behavioral models" with the *mistaken* implication that all behavioral models are inaccurate). Only when the design works with these first-order models does the engineer proceed to using more complex models. Starting with the simple first-order models helps the engineer gain confidence — in the design and the use of models. Topologies and approaches can be quickly and easily analyzed, rejected, and accepted.

This book debunks some of the myths associated with modeling, how it has been perceived as a peripheral activity, and, instead, positions it firmly at the heart of the design process.

The second contribution of this book covers the new generation of modeling tools that help with the creation of those dreaded second- and third-order models. These new tools make modeling more natural to the engineer by using a language all AMS engineers understand: circuits, diagrams, equations and blocks. This is, indeed, "AMS engineer friendly".

By bringing all these concepts together in one place, the authors make a *major* contribution to "model-based engineering" — where modeling is placed comfortably as a natural part of the design process and the AMS engineer receives the benefit of the additional insight and confidence into the system being designed.

It's really not so scary, after all.

Ian Getreu

Preface

The phrase *model-based engineering* (MBE) has been used by many people and organizations to indicate the importance of models as part of the engineering design process. Our use of the phrase for the title of this book is no different. It is not difficult to find people who will advocate for modeling to be used as a precursor to design and implementation. Conversely, it is also not too difficult to find hardware-centric designers that want as little to do with computers and modeling as they can. Fortunately, this is a bygone era and not a formula for success in modern complex designs. In point of fact, nearly all modern engineering design is performed with the aid of models, whether it be for civil, industrial, mechanical, or electrical and electronics engineering projects. Some of these engineering disciplines have substantially advanced their techniques for creating models in recent decades. This is typically reflected in the complexity of systems being designed, as, once again, necessity becomes the mother of these inventions. With all due respect to the simulation algorithm developers of the world, the single most effective way to manage growing design complexity is through modeling. Yes, when a simulation algorithm breakthrough is realized, it can have a profound effect, but these are few and far between, and, in fact, it could be argued that this has not taken place for several decades. In the scope of a given complex system design, such a breakthrough certainly cannot be relied upon. The design team must, in general, be smarter and manage that design through modeling in the context of existing solution algorithms.

The concepts in this book were conceived more than 15 years ago by the authors, but it is only recently that we can say that modeling technology has reached a point where the concepts can be taught effectively and succinctly. This book is not intended to convince engineers of a new way of thinking or profoundly new design approach. Rather, it is intended to acknowledge the model-based activity designers already engage in and provide a procedural formalism that will hopefully make them more productive. Further, it should reveal to the designer that, with investment in "sharpening this saw", many more valuable analyses are possible than are typically done today.

This book is designed to be a desk-top reference for engineers, students and researchers who intend to carry out MBE. It is also written in the style of a book that upper-level undergraduates or entry-level graduate students having taken their introductory engineering courses can utilize.

The book is organized into three sections: Fundamentals, Modeling Approaches, and Design Methods. The first section consists of three chapters that cover the basic tenets of MBE. In order to perform MBE, you have to know how to write models and how to compose them for reuse. Understanding where these models fit into the simulation flow and how they are used in design activity is important. However, many articles and texts have been written on the simulation algorithms, so we will not duplicate this information other than to summarize it for our needs. In this way, we can stay focused on the primary picture — modeling as a means for design.

The second section goes more deeply into modeling techniques to ensure the reader understands the various way behaviors can be represented. Most of these techniques are illustrated through the use of modeling tools rather than by hand coding in various languages. This keeps the modeling activity focused on the conceptual level of what is required and out of the details of any specific language syntax or semantic. This book is designed specifically to avoid the need to have to teach hardware description languages, C, or any other language in which models can be represented. Therefore, no modeling language primers are included as part of this book. However, the reader can use the included modeling tools to view what would be generated. Again, the focus is on model creation at a conceptual level using tools that will have the utility to generate these languages. In this way, the designer focuses on the utility of the model, its usefulness and requirements at a given point in the design, and in capturing that functionality efficiently. Without question, models can be written by hand and successfully employ a MBE design approach. However, the learning curve for the uninitiated is much larger, so we have chosen the simplest route to effective model based design productivity. Section 2 is predicated on the fact that even with huge libraries of models, there will remain the need to capture more models for each design. It is meant to arm the designer with these techniques while also demonstrating reuse in each chapter because *"the easiest model to build is the one you don't have to"*.

The final chapter of the book represents a case study that was performed on a complex integrated circuit between the University of Southampton and the University of Arkansas. This chip includes analog, radio frequency, and digital all on the same chip and was designed using the MBE design methods prescribed in this book.

For the practicing engineer, we hope that the book will serve as a method of self-teaching oneself the basics of MBE and, together with the online material available, give them get a quick start on checking out the methodology on a design of their own. The case studies and examples will be indispensable for the deepest learning and illustration of the MBE approach.

In the accompanying online materials, available on the companion website: http://www.modelbasedengineering.org, all of the examples given in the book are captured for ease of execution in a simulator. The modeling tools used to create the models are provided for free, so that the models can be read into the tool and a hardware description language suitable for simulation can then be generated with the push of a button.

Acknowledgments

It seems like most of my adult life has been concerned with attempting to become a good engineer. As a proud Scot, with an engineer father, it is a matter of national pride to be part of the engineering fraternity. From those first rough prototypes of electronic circuits, through to production equipment, every step of the way has been a passionate drive to make things work and make them better. For all of my professional life, my passion has been to design, innovate, and invent, and more than 20 years ago this was enhanced with the introduction to advanced modeling and simulation techniques. I can trace the moment where the "burners were really lit" to attempting to figure out a problem with a transformer, with testing providing only a limited insight. Having to understand the transformer's complex behavior, model it, simulate it, and then use that to trace the source of the problem was an epiphany that demonstrated the power of a model-based approach to engineering. For that moment I must thank John Murray at Ferranti, in Edinburgh, for the words "can you take a look at this...?".

As a junior engineer at Ferranti, working with John Murray, Frank Fisher, and Alan Abernethy, in particular, was an education, inspiration, and genuine pleasure, which I will always treasure and never forget. Moving to Analogy Inc. was a pivotal moment in my life as it was where I met my wife, Caroline, and, above everything else, this is what I am most grateful for. Professionally, it was also a formative period, working with many smart, funny, and talented individuals, including my co-author Alan Mantooth. More recently, since moving back to academia to first pursue the long-held dream of a PhD and then becoming an academic at Southampton, I have had the good fortune to work with another group of diverse, talented people, including my PhD supervisor Dr. Neil Ross, and Dr. Reuben Wilcock, in particular.

The journey of this book has deep roots, and the friendship with my co-author Alan Mantooth goes back more than 20 years. We talked for many years about why it seemed to be so difficult for engineers to be able to do what seemed to come naturally to them, and so the obvious step was to try and write down what we did to capture those techniques. The original work on the book started back in 2008, when I spent a sabbatical semester at the University of Arkansas to begin the process, and I must mention the hospitality of Alan, Mary Lynn, Deanna, Laura, and Maureen Mantooth, who invited me into their home to become "part of the family": I am sincerely grateful. As the book developed, we obtained the support, advice, and assistance of so many people along the way, both in Arkansas and also at Southampton, including Dr. Li Ke, Dr. Reuben Wilcock, Robert Rudolf, Dr. Matthew Swabey, Dr. Matt Francis, and Chip Webber. I would particularly extend my thanks to the Department of Electrical Engineering at the University of Arkansas for their generous hospitality for more than a decade — I am proud to have been a part of the Razorback family for all this time.

—Peter Wilson

In designing the scope and purpose of this book, I am sitting on the shores of Cromarty Firth near Inverness, Scotland. I am reminded of the fact that Scotland is celebrating a year-long event they call Homecoming Scotland 2009 in honor of the 250th anniversary of Robert Burns' birth, Scotland's national bard, and Scotland's contributions to golf, whisky, culture, and, most relevant to this book, inventors and innovation. As an American in Scotland people take note when I respond to questions without the local lilt in my voice. They assume I'm here on vacation, or holiday as they call it, but when they hear I'm writing a book they immediately settle in for a conversation. They are very curious as to the subject. There is a fascination with poetry and good story-telling in this land. While I must confess that most are impressed, but disappointed, that it is a technical book, I do hope that Peter, himself a Scot, and I have done them proud, and that perhaps some inspiration to invent or innovate will be facilitated through this work.

I would like to acknowledge an outstanding class of young engineers that helped work through this book and a design that became the basis for the last chapter. These young men went well beyond what students normally do in the classroom to design an ambitious chip. Their effort is most appreciated. Their names are: Shamim Ahmed, Eric Baumgardner, Tyler Bowman, Nick Chiolino, Tavis Clemmer, Zihao Gong, Erik Jakobs, Peter Killeen, Ranjan Lamichhane, John Monkus, Ashfaqur Rahman, Brad Reese, Thomas Rembert, Mahmood Saadeh, Brett Shook, and Brett Sparkman. Finally, very special thanks goes to Dr. Matt Francis, my post-doc, who helped to architect ModLyng, knows it better than anyone, and assisted me in teaching this great class.

—Alan Mantooth

Fundamentals for Model-Based Engineering

The first section of the book describes the key fundamentals for successfully accomplishing model based engineering (MBE). The first chapter introduces the topic of MBE, the alternative approaches to MBE, and what this book will teach the reader. The idea is to engage engineers from the very beginning with specific real-world examples of design, talking about how modeling is crucial to the processes involved and talking about solving real problems. We extend the scope of models from just a piece of code into all the facets of a rich entity that has a symbol, structure, equations, behavior, test circuits and validity regions, and executable specifications. Chapter 1 highlights the fact that engineers do these things already.

Chapter 2 talks about design and verification process and Chapter 3 provides a rapid tutorial on computer simulation methods typically used for analog, digital and mixed-signal electronics. These chapters are very important for understanding the context of design analysis. In deference to the late, well-known designer, Bob Pease, who was fond of giving simulators "the business", simulation is indeed a very useful tool, but cannot replace the brain. The designer must perform the design activity and use simulation as an extra tool to evaluate the design among the other tools at his disposal such as breadboarding. The biggest value in simulation is being able to quickly analyze many scenarios of electronic systems. However, it needs the guidance of the designer to design the circuitry and the approach to verifying them. Further, simulation can be used in an optimization or statistical context to evaluate many variations. And finally, simulation can be extremely valuable in troubleshooting a design when guided by the designer's insights. The major advantage of a simulation is the ability to "look inside" components within a design to enhance the designer's understanding of the behavior of that design.

Chapter 4 concludes the first section by laying down the fundamentals of modeling, modeling techniques, and describing tools and forms of model representation which will all be described in much more detail in the chapters of Section 2.

CHAPTER 1

Overview of Model-Based Engineering

1.1 Introduction

This book is intended to provide designers with an insight into effective model-based engineering (MBE). The concept behind model-based engineering is to use modeling as not merely a mechanism to perform computer simulations, but to capture specifications, clarify design intent, facilitate hierarchical designs, and be efficient at collaboration, design reuse, and verification – in other words, manage complexity. This book is aimed at engineers managing or working in design teams, including those that cross engineering disciplines, such as integrated circuits, power systems, transportation systems, medical electronics, and industrial electronics to name but a few. The unifying theme that this book brings to all of these fields is being able to create and use models effectively in the design process of complex systems.

In this initial chapter, the fundamental concepts of managing complex and hierarchical designs will be introduced with a view toward providing solutions and methods to solve these issues in the remainder of this book. This book has several learning objectives as follows:

- Create a common ground for the development of MBE by:
  - Summarizing for the reader the very basics of modeling and simulation
  - Describing the ways in which circuits and systems can be represented
- Cover the breadth of modeling techniques and methods, including some emerging methods, while demonstrating a common creation paradigm that promotes reuse and the capture of hardware behavior appropriate to the tasks at hand
- Illustrate concepts through lots of examples, which will be available online
- Describe the keys to adoption of these methods.

To begin to motivate why MBE is important, it is instructive to understand what problems engineers are facing most prevalently. A big problem for engineers in today's world is attempting to close the so-called "Design Gap".

We can see from Figure 1.1 that the complexity of design (in this case integrated circuits) is outstripping the ability of engineers to design these more complex chips in the same time. As a result, it is taking longer in real terms to design these more complex systems. This issue of complexity is a fundamental problem for modern systems designers, whether on a chip or a large industrial project, and yet we have much more computing power available to work on these problems. Clearly, the key to success is to be able to leverage our increased computing capability and use computer-aided design (CAD) tools to manage complexity and develop systems more efficiently.

One critical aspect of designing complex systems is developing models of parts of the system that enable design and analysis to proceed at an efficient pace. Modeling has become an integral part of almost every aspect of modern design, but not necessarily in the same or unified form. Some models are created for Matlab, others are created in spreadsheets, and still others are created in modeling

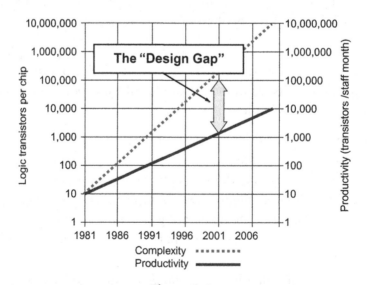

Figure 1.1:
The "Design Gap" between the increasing complexity of electronic designs and the designer's ability, on average, to produce them

languages. And yet, the number of modeling experts in each field is still relatively small, especially in the areas of mixed-signal and mixed-technology design. One reason for the development of modeling automation or productivity tools is to compensate for this lack of expertise, and provide access to the utility of hardware description language (HDL) technology without the need to become an expert programmer. This is analogous to the development of synthesis for digital systems, where, prior to the existence of automatic logic synthesis software, the synthesis process was an expert process, time-consuming and complex. When the first logic synthesis programs became generally available, the increase in productivity was huge. Being able to synthesize models from abstract descriptions is the equivalent for modeling to that of logic synthesis. This is now a capability that can be found in commercially available tools.

Moving the emphasis of the designer away from the details of the modeling language also has a profound effect on the *way* that design is undertaken. Currently, the approach is most akin to a software programming style, using text editors and then simulating to check for "syntax errors" and other "bugs" − much in the way that the compilation process is used in software. Moving to a design-based approach will have the effect of the designer being able to think about algorithms and behavior, transferring the thought process firmly back into the design domain rather than a programmer's domain. This approach itself opens up the modeling field to many more non-experts than those fortunate enough to be skilled in the syntax of one or more HDL. It also increases the productivity of the designer, even though at first glance it may appear to be another task on the heap. On the contrary, with the proper modeling tools and libraries, the modeling and documentation activity already performed by most designers is converted into a more singular approach that leverages this work and begins to close the design gap.

1.2 Multiple Facets of Modeling

We can think of a "model" as a rich object that has many facets (Figure 1.2). The conventional approach of disconnecting the implementation of a model from its specification, test-bench, and documentation is perhaps counterproductive. In contrast, we consider a model as not just the encapsulation of the behavior in a piece of HDL or code, but rather the linked set of "meta" data that fully describes

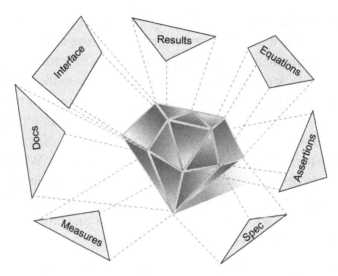

Figure 1.2:
The model is a multi-faceted object—not just code

the design in its entirety. This meta data includes documentation, test results, design context and intent, background links and references, equations, and structure. This broader view of a model will begin to merge the distinctions between model and design in a positive way, so that members of a design team are able to gain a more complete understanding of the design – both during the initial design activity and later when the design is either being enhanced or ported to a new process. The detail and data that can be obtained from simulations serve to further enhance the richness of a model. These can include specific measurements, assertions (errors or warnings familiar to digital designers), and results from advanced analyses, such as stress or sensitivity analysis.

The activities that the designer normally engages in include pre-design documentation involving requirements, functional specification synthesis, capture, refinement, and negotiation. This design documentation can be made executable with current modeling tool technology, where a combination of test benches, waveform measures, specifications, and simple models for the device being designed are captured in a single tool rather than by multiple tools and documentation methods. This has the advantage of improved productivity for the same tasks, but offers an even greater enhancement still when you consider how these specifications are directly traceable through the CAD tools from concept to realization rather than entries into a Microsoft Word document that is disconnected from the design flow.

One cannot forget that design activity is not complete until the product is firmly in manufacturing. The initial system builds are rarely defect-free. So, the design team is often re-analyzing, discussing, and debugging any remaining issues before the design is released for large-scale manufacturing. This scenario as depicted is an optimistic one considering that, in many instances, multiple design iterations are required to produce manufacture-ready results. Therefore, the design team is engaged in debugging activities to a significant degree. Given these circumstances, MBE techniques prove to provide a substantial advantage. Consider a design where all specifications are not executable, each block of the design was designed independently, and the system designer pulled it all together to create the system. The individual blocks were designed by individuals, each with their own style and notions of what rigorous design means. Even with a talented team, this will lead to issues when the debugging rounds occur. With no unified modeling, test bench, and specification methodology in place the debugging activity will proceed longer than necessary. Figure 1.3 shows where time is spent during these multiple design iterations and how MBE methods can not only save

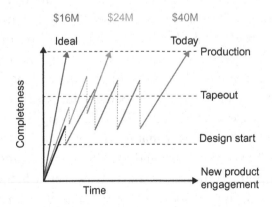

Figure 1.3:
Figure showing the design time comparison between those employing MBE-based methods and those not. Notice the relative improvement during the debugging rounds (iterations beyond the first)

time in the first iteration, but significantly decrease design debugging activity because the framework for simulation-based investigation of problems is already in place. While Figure 1.3 cannot illustrate a specific improvement in the general case, the thesis is that, starting with the initial design round, specification capture in companion with test bench definition takes no more time with a MBE approach than with traditional methods. And yet the big benefit is realized in the second and subsequent design rounds, which are focused on debugging and performance improvement/refinement. Figure 1.3 illustrates the relative improvement in time as compared to a situation where the system and its constitutive components are not modeled according to the MBE approach. The MBE approach establishes a framework so that, from the top system specifications down to the specifications of individual blocks, the design can quickly be analyzed at the level of detail required. This allows much more in-depth investigation of the issues, as well as much faster investigations. Among the benefits of this approach are: (a) reduced design cycle times, (b) shorter investigations to determine and address problems, and (c) reduced number of design cycles because of the improved framework for design analysis and verification (not illustrated in Figure 1.3, but very important).

Figure 1.3 focuses on a new design beginning from scratch, such as the first Bluetooth chipset, for example. In contrast, in our years of teaching circuit and system design we have consistently tried to set the expectations of our students that most design is "re-design". By this we mean that designing a completely new function, chip, or system that the organization has no history of designing before is much more rare than the design of circuits and systems that are being improved over previous designs, or are being undertaken for a new customer but will be based on existing designs. When a design team is performing a re-design, the benefits of MBE are naturally going to be somewhat different. The use of MBE will expedite reuse and decrease the learning curve of the new engineers engaging in the re-design activity. However, whether MBE is used or not the likelihood of design re-spins is lower going into a re-design project as compared to a project with substantially new design content. However, in spite of this, the deployment of MBE methods remains a good investment for the following reasons:

1. Designs always seem to have a longer lifetime than we ever imagine. During the lifetime of a product/design, the original design team's composition will change. New engineers will come onto the project, and others will leave it. The transfer of knowledge and the learning curve for the design is much more self-directed and efficient with the MBE approach. All of the particulars are documented, executable, and repeatable.

2. In addition to assisting the design team, the test and evaluation team is a beneficiary of the design team using MBE methods. The MBE approach actually enables the test team's preparations and activity at an earlier stage of the design cycle, thus shortening the overall design cycle again. This is because instead of design data being "thrown over the wall" at the test team once the design goes to manufacturing, the test group is able to access the same design databases once designs are frozen in order to begin developing test strategies, Automatic Test Equipment (ATE) test code, and test fixturing. This effectively moves this activity up in time, thus shortening the design cycle.

3. Product support is enhanced through the use of MBE methods for the downstream engineering teams in a production environment. The application and support engineering teams are better able to support designs and handle customer requests and issues if they have a better view into the design. The MBE approach creates that view into the design that is far more conducive for investigation, troubleshooting, and learning-much as for the new project engineer referred to in (1) above.

The motivation therefore is to have a framework to create and develop models from a *design perspective,* which implies abstracting the behavior of individual elements and systems into a form that a designer can understand and is comfortable interacting with. The remainder of this portion of the chapter will review some of the general design issues in more detail and introduce the key requirements for MBE as a technique.

1.3 Hierarchical Design

The reality of modern design is that it is generally not feasible to design a circuit or system without using some kind of hierarchy. If we consider a typical integrated

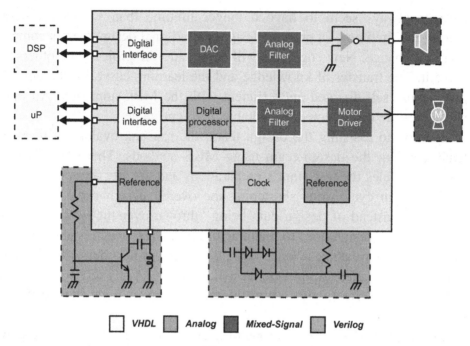

Figure 1.4:
Complex design illustrating the multiple design techniques and technologies

circuit in today's world of sensors, mobile electronics, and networks, shown in Figure 1.4, we can see that multiple technologies are often required with numerous design techniques — often spread across an entire design team.

Different blocks require design techniques appropriate for their discipline. For example, the motor driver clearly requires a different set of tools and methods than those for processor design. In most cases, the design team will consist of a number of engineers who are specialists in their own field that will work together as a team to complete the design. From a management perspective, the only way to handle this practically is to use hierarchy to implement the design and partition intelligently to assign the individual blocks appropriately.

In this context, where systems are generally mixed-signal (analog and digital content) and often mixed-technology (some interface to the "real world" — such as thermal, mechanical, or magnetic), the partitioning can often be intelligently

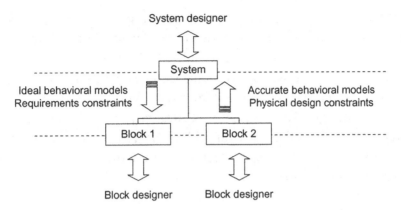

Figure 1.5:
An illustration of managing partitioning and design hierarchy

made on the boundaries, such as those given in Figure 1.4. A high-density digital section of the design, such as a processor, would be best undertaken by a digital designer using digital simulation tools. Conversely, a section of analog circuitry, such as a voltage reference, is best undertaken by an analog designer with analog tools. This is fine as far as it goes, but it has often been noted that "problems migrate to the boundaries" in such complex designs, and it is therefore critical that the mixed signal and mixed technology aspects of the design are handled in a similar smart fashion. The other aspect of partitioning that is often neglected is the hierarchy issue. The system designer or project leader may have a completely different view of the overall design than the individual designers, with different requirements and specifications to meet. It is therefore essential, that, if at all possible, the hierarchy is managed in such a way that they can observe the overall system structure and behavior, as well as the individual blocks of the design itself (Figure 1.5).

1.4 Partitioning

There are several important considerations when partitioning a design. The first is to even the spread of complexity, to ensure that the blocks in the hierarchy are as evenly matched as possible to the design team capabilities. Practically, this is important to ensure that the risk of one individual or team being late does not compromise the rest of the team unduly. Spreading the work evenly mitigates that particular risk factor. The second is to examine the boundaries and blocks intelligently and to partition generally with those boundaries in mind. For example, identifying the key blocks in certain technologies, as in Figure 1.4, helps in categorizing the individual sections of the design, and therefore assigning the designers to them becomes a simpler process. Thirdly, it helps to understand the overall system requirements and identify the critical elements in the design. This can help the design process by adding additional checks and validation to the individual block specifications that may not have been derived from the overall specification, but contribute to meeting the overall specification much easier.

1.5 Specifications

The original purpose of the development of languages such as VHDL was not actually for design, but rather as an electronic specification language. With this in mind, it is interesting to note how even with the advent of advanced HDLs, such as VHDL and Verilog, the electronic specification is still not really a reality. In practice, the use of paper specifications still exists and, in particular, the test criteria are often defined using a set of tests and limits, which are not easily defined using an HDL.

The analog and digital design communities have developed the concept of automatic test (usually implemented within an ATE) using standard techniques and languages, but these are different to the design languages used to implement the hardware itself. The digital design world has embraced the concept of "assertions" to assist in this area by providing a method by which models can be checked against a series of criteria, and if there are any deviations from the specification as defined in these assertions, then these can be identified in the model.

In general, this sophistication does not exist in a standard form for analog or mixed-signal systems. In this context, the reliance is still on simulation and post-processing to establish the integrity of the design.

1.6 Keys and Barriers to Adoption of Model-Based Engineering

In considering all of the evidence for either adopting or not adopting MBE methods in a design team, we felt it would be important to describe the barriers and keys to adoption from our experience. As the material in the book is developed, we will reinforce the keys to adoption as much as possible, culminating in our own experience with the method as chronicled in the last chapter.

The primary barrier to the adoption of any new design methodology is complacency — particularly for design teams that have been successfully moving circuits and systems out into production. The status quo can be a hard barrier to overcome because some people have an aversion to risk or do not like to change things very often. The old adage "if it ain't broke, don't fix it" is the defense of the stodgy-minded design manager. That is not to say that such decisions should be taken lightly or rushed into, but experience has also shown that too often new approaches are simply dismissed without taking heads out of the sand. In the same way that a successful professional golfer will choose to change his swing mid-career, even in the midst of enjoying tremendous success, a design team and its management is well-advised to abide by continuous improvement ideas and evaluate the entire design chain of activity to seek out ways in which processes can be streamlined without introducing unchecked risk. MBE offers that opportunity, as we will show.

Another barrier to adoption of MBE involves the perception that modeling is an offline activity that is too time-intensive to be a part of the design process. This perception may arise from an arduous modeling experience, or passed along by other engineers. Unfortunately, this perception can get locked into the mind and we as humans have trouble giving such efforts a second chance even in the presence of improved tools and methods. The fact is that many very useful models can be constructed in minutes, require very little validation, and offer no serious issues where simulation speed and convergence are concerned. This will be illustrated both through the coming chapters and the capstone chapters at the end.

The last barrier of any consequence is expense. MBE requires investment. This investment takes the form of money for tools to support MBE and, just as importantly, training for engineering staff that will execute MBE techniques. An organization that is intent on reducing CAD expenditures, designing everything with SPICE, and unwilling to give their engineers time to sharpen their saws is doomed eventually anyway. MBE is not the answer for this mindset.

On the flip side of the coin, in order to adopt MBE methods there are a few keys (some of which were alluded to above):

Investment — clearly, the organization needs to be open to adopting a method that will benefit not only the design team, but the test and application teams. In fact, it will also help the technical marketing team in winning new business. As mentioned above, the investment involves more than just money for CAD tools. It involves investment in training these various groups of engineers and engineering managers. It involves investing in the time and resources to establish and deploy the new approach internally. There never has been a serious engineering CAD tool that did not require some degree of investment in learning the tool, customizing it for use within an organization, or understanding its boundaries and limitations. It is no different with MBE methods. In fact, as it involves a change in design flow or process, its deployment requires a strategy of introduction and roll-out so that too many things do not change too quickly and bring productivity to a screeching halt. This strategy itself represents an investment or commitment that an organization needs to make in order to successfully transition to MBE.

Verifiable evidence of providing improvement — it is a fair assertion that any new approach that promises what has been said about MBE needs to be demonstrated as having real efficacy. Some of the forms that this evidence could take are:

*   First-past success in silicon integrated circuit (IC) design — said another way, faster success in silicon IC design, particularly when including all issues between design conception and delivery to market
*   Faster design time — this is related to the first bullet, but focuses more strictly on the first design cycle. The aim is to dispel the myth that design takes longer if modeling is included. This goes away over time as MBE is used more and model libraries grow
*   Reduced design cycles
*   True executable specs that are traceable and help in debugging a non-functioning system
*   Improved design collaboration, which can be measured by no mistakes at the interfaces between parts of the system
*   Demonstrating what complexities get managed and how through MBE.

Case studies — documenting the costs and benefits of MBE provides more evidence of value. The authors will be describing an experience they have been involved in on a mixed-signal integrated circuit with a company. The company wanted our team to demonstrate the MBE approach. In many ways, this is a great model of how to evaluate a new method — work with a university or two to pilot the method. If their students, who are cheaper and no risk, can do it, then it is likely that corporate engineers can. The beauty of this approach is that the company can be involved in the activity, monitor the pros and cons, and be better prepared as they choose to adopt the approach in whole or in part. Another interesting observation is that as our teams have begun to deploy the method for the second design, even with some new students, the legacy has already begun to emerge so that process improvements can be made, libraries are growing, and the new students can more easily adopt it as *the* way that things are done. This speaks the value of reuse — not just of the system models, but of the test benches, assertions, and specification capture. We submit that for those organizations that are reluctant to invest, this academic/industrial setting helps to alleviate the catch-22 where the company has to invest big dollars before having a convincing argument for MBE, but needs the convincing argument in order to invest.

Support structure — resources such as this book, which will attempt to teach the MBE method, are an important step toward creating a community of users/ adopters that share non-proprietary best practices. MBE tool companies and their associated user groups, as well as online communities to support the non-sensitive aspects of MBE, are also valuable resources once the approach is adopted.

Conclusion

This first chapter has introduced what MBE actually entails without providing all the details of implementing the approach. The rest of the book will focus on the *how* as compared to the *what* that was defined here. Table 1.1 describes succinctly the constitutive pieces of what MBE requires in the field of electrical and electronic-based technologies and design. To summarize, MBE is defined to be

Table 1.1: Model-Based Engineering Requirements Applied to Electrical and Electronic Systems

Requirements	Engineering Team Affected by MBE Adoption			
	Technical Marketing	Design Team	Test Team	Applications or Support Team
Model libraries	D	D	D	U
Modeling tools	U	U	U	U
Fault modeling	–	D	D	U
Test benches	D	D	U	U
Specification capture	D	D	U	U
Performance measures	D	D	U	U
Assertions	–	D	D	U
Stress modeling	–	D	–	U

U – indicates user, but not developer of this aspect; D – indicates developer and user of this aspect.

the use of enriched models at a variety of levels of complexity to: (a) manage the complexity of the overall design, (b) improve the ability to detect and correct design flaws, and (c) improve the overall design flow process from concept to manufacturing through reduced design, test, and debug time.

CHAPTER 2

The Design and Verification Process

Henry Royce (of Rolls-Royce fame) "Strive for perfection in everything you do. Take the best that exists and make it better. When it does not exist, design it."

In order to understand model-based engineering in the context of design, it is important to understand and define what we mean by design, particularly the major steps to be taken to achieve the goals we set.

2.1 Introduction to the Design Process

As described in Chapter 1, this book is intended to describe the techniques required for model-based engineering (MBE). It is important to keep the broader context in mind when putting a variety of techniques at the designer's disposal, and it is also necessary to be goal-oriented in order to achieve successful designs.

2.2 Validation, Verification, and Requirements

It may seem early in the process of design, even before we have defined the process itself, to introduce the concepts of validation and verification, but, in fact, these should be considered at the very first stage of the design process *so that we know where we are going*. These two concepts are often misunderstood and misused, so making some key definitions at this stage is extremely useful from a *design* perspective. If we ask the question "what are we trying to achieve?" it can be answered with the basic response "we need to make sure our design does the right thing, and does it right". This is a very useful definition of validation (doing the right thing) and verification (doing the right thing *right*) in a compact form.

The distinction between validation and verification and their joint usage is also critical for system design to understand when the requirements are to be developed. It is obviously essential to design a system that is going to "do the job" — and this is what validation is all about. The first step in any design is therefore driven by these validation tasks and ensures that whatever we do, the design is heading in the right direction from the very beginning of the design process. Keeping this in mind it is now possible to define what we mean by the design process itself.

Looking at Figure 2.1, we can see how a seemingly simple requirement from a customer can quickly become misinterpreted and poorly implemented, leading, consequently, to the customer NOT getting what they really needed in the first place (The cartoon in Figure 2.1 has been used in a large number of settings, and not attributable to anyone in particular — becoming common in use in Britain from the 1960s onwards).

From a hierarchical design perspective, the subsequent decomposition of requirements down to individual components, whether they be electronic devices or mechanical actuators, puts a stringent constraint on the designers to ensure that the requirements are managed throughout the complete hierarchy. This puts a significant constraint on the design team to ensure that overall requirements are not missed when the design is being put together.

Depending on the type of system being modeled, the verification task becomes more or less formal. For example, in the area of digital electronics there are formal techniques that can assess the test regime and indicate the coverage of tests, which not only tells the designer whether a test has been passed but also the proportion of the possible failure conditions which are actually being tested! While this is now routine in digital electronic design, it is also true to say that this is most certainly an alien concept to many designers in other fields.

2.3 The Design and Verification Process

The design and verification process can follow a number of different routes, with much analysis of different processes and options being described historically. For example, there has been much use of block diagrams, structure charts, activity charts, and so on in order to describe how the process of design is to be

Figure 2.1:
Rope swing cartoon

undertaken. In both hardware and software design, however, there has been an acceptance that, starting from an initial specification, the best approach is to define the design in increasing levels of details. This involves effectively going down into the details of the design until eventually the individual components of the design can be described in detail, and then returning back to the original high level, perhaps even of a whole system for validation.

This approach is often called the "V diagram" and is shown in general terms in Figure 2.2.

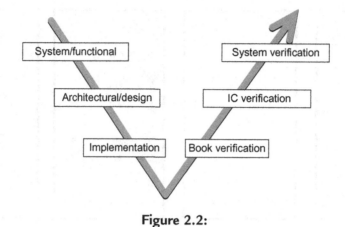

Figure 2.2:
V diagram for design

There are clearly defined stages to this diagram which show how, at each level of decomposition, there is an increasingly detailed and focused specification for the element being designed. On the upward (right-hand) half of the V, a correspondingly comprehensive verification regime until eventually, on the top right of the V, the product will meet the original specification and the goal is achieved.

For example, consider the diagram in Figure 2.2, where the classical V diagram was presented. This shows an isolated design flow, with no context of previous designs or the next iteration after the current design. Perhaps the biggest advantage of this method is the fact that the *next* iteration can take advantage of the previous or current ones to inform decisions and get to the final design faster. Consider the scenario where the first version of a product has been completed and the next starts almost immediately. The result will be a progression where the verification of version 1 has been completed and the design for version 2 starts from a much more informed position. This is illustrated in Figure 2.3, although superficially showing the same steps, where the version 1 design is in place.

This leads to the "W diagram" shown in Figure 2.4 where repeated cycles of design iterations take place and gradually refine the design until the specification is achieved.

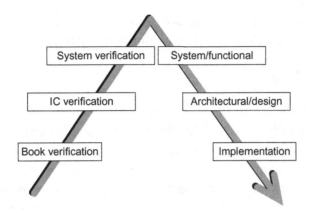

Figure 2.3:
The second iteration in a design process

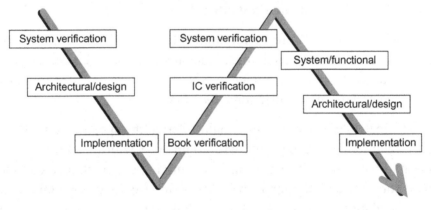

Figure 2.4:
Developing the W diagram

2.4 System/Functional Level: Executable Specification

Proceeding to the next iteration of the design activity involves revisiting the system level and adding additional information to our models to address any shortcomings that were identified as part of the design and verification process the first time. At this phase we capture a model so the specification is self-consistent and makes sense. The model will essentially be a top-level system model with inputs and outputs; however, the model itself will consist of an algorithm, not necessarily a structure of the real hardware. Looking at the V diagram, it is also clear that, in conjunction with the model, the process requires the development of a coherent test plan and criteria for acceptance that will be consistent with that plan.

A key element of the result of the initial system design is that whatever model is produced essentially encapsulates the system specification. This is a crucial point in model-based engineering and is where the concept of model-based engineering differs from conventional "paper"-based approaches. In a paper approach, the specification document is the primary source of requirements. In contrast, in a true model-based engineering scenario, there needs to be an executable specification in such a way that the model entirely captures the requirements. This is a non-trivial task and puts the onus on the system level designer to properly and completely capture the requirements in the model and fundamentally be able to demonstrate that this is the case.

When considering the capture of requirements in a model, it is important to establish what aspects of a requirements document need to be defined and how these can be tested to ensure completeness and accuracy. In order to understand this key concept, we can consider some basic examples.

If I want to create a calculator, how should I define the specification? For example, many requirements documents will commence with an initial word-based description of what the customer wants. This might be something along the lines of "a hardware device is required that allows the user to press numeric keys to facilitate calculations such as $1 + 1 = 2$". This is an interesting statement that can give some basic information, but clearly leaves a lot of room for ambiguity (ambiguity being the enemy for *any* specification, never mind executable). Looking at the statement more closely, we can identify attributes such as "a hardware device is required that allows the user to press numeric keys...". This defines that this is a hardware product requiring a certain number of keys to enter the numbers and functions. But perhaps this could actually be defined as a software application running on a standard hardware platform such as a mobile phone? If we define that this is to be a standard electronic calculator, then the requirements do not define the display format — which could be liquid crystal display (LCD), flat panel, or some other means.

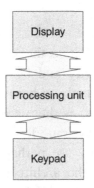

Figure 2.5:
Simple electronic calculator hardware block diagram

The obvious next step in the definition of a more precise specification would be to define a hardware block diagram perhaps of the form shown in Figure 2.5.

Within this block diagram, it is clear that specific functionality can be defined explicitly for individual blocks and that these can be tested individually. However, whatever approach is taken, the overall system requirements must be met.

The second aspect of the specification is the rules that define correct operation. This has several factors. The first issue is what numeric inputs are allowed. Is this to be integer based, fixed point, floating point, or some other? Are exponents to be included? What about decimal points? The second issue is what are the specific calculations that are to be implemented? Is this a simple add/subtract/multiply/divide calculator with just some basic arithmetic calculations, is it a more complex scientific calculator, or could it be a graphical function calculator? These aspects are crucial to the functionality and ultimately the quality of the product.

At this point there is an obvious issue to be resolved by the system level designer: at what point does the specification execute and what aspects of the product are implemented in the executable specification? For example, is it essential to include the buttons and display or simply the calculation engine?

2.5 Architectural Level

The architectural level model is where the specific sub-system blocks are put in place, with accurate connectivity and behavior. This model may be more or less abstract, but correlates directly to hardware to be built. This is an interesting stage of the design process, as there are many tools which do an excellent job of encapsulating a system level description, such as Matlab (from Mathworks). However, they are perhaps less efficient at capturing the *architecture* itself. If we define the architecture as consisting of the correct inputs, outputs, and parameters, then HDLs would be more useful for these functions.

The other related issue is the detailed structure of a design (whether hardware or software). Given the complexity of modern designs, it is commonplace to partition designs into multiple sections, blocks, and routines. This partitioning naturally leads to a hierarchical design, which is, again, well suited to an HDL approach.

As far as the function of the architectural level design is concerned, this may be very simple (transfer function level or equivalent high level behavioral) or begin to encapsulate details of behavior that make it more complex. The important thing is that the fundamental aspects of the model are incorporated into the model. For example, an amplifier model may not have all the details of offset, power, delays, slew rate, or noise at the architectural level, but it is important that the model at the very least has the correct connections and some gain so that the signal path can be tested.

The signal path test is an important evaluation for any system design — a sanity check that all the basic blocks are in place and connected correctly. For example, consider the simple design shown in Figure 2.6, which illustrates a simple transceiver down-conversion channel with several basic blocks.

Each of these blocks could be modeled with large amounts of detail, but, in order to ensure that all the signal names are correct, and are connected correctly, a very simple model could be used that models the power amplifier, filter, and automatic gain control (AGC) amplifier as unity gain blocks. This would ensure that the connectivity to the analog-to-digital converter is consistent and correct.

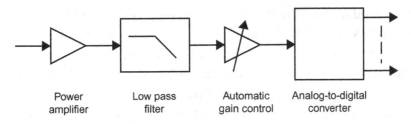

Power amplifier | Low pass filter | Automatic gain control | Analog-to-digital converter

Figure 2.6:
Diagram of a simple partial down-conversion channel

2.6 Implementation Level

The implementation level model is where the detailed hardware design is crystallized. The model at this level is defining exactly what is going to be fabricated. The initial architectural model has the correct connections and structure. Therefore, the focus is much more on the design details.

Initially, the task for the designer is to ensure that the correct functionality is implemented. The conventional approach is to use fairly abstract models, with a limited amount of detail in place to establish fundamental design parameters, such as gain or bandwidth, are consistent with the specification. The related task for the designer is to ensure that the specification is also consistent and correct. This is an extremely valuable use of the model to validate the design and its specification prior to any detailed modeling or design being undertaken.

As the model is implemented in ever-increasing levels of detail, more analysis can be undertaken, and more detailed analysis can be carried out to ensure that the design will meet the requirements of the design.

2.7 Model-Based Engineering — A Winning Approach

In this chapter we have introduced the important concepts of the design and verification process, and now we need to know the mechanisms by which we can accomplish that.

This is an introduction to how the ideas behind model-based engineering will make the process *work*. The V diagram is a way of communicating the activities that take place regardless of the approach. Our conjecture is that MBE is the way to undertake the same activities, except that the approach is more rigorous and efficient than doing it by hand. In Chapter 1 we presented this approach in general terms, and we have now related it directly to the design process. Going forward we will now delve into the specific techniques required to accomplish these tasks, and, finally, apply these techniques to specific design examples to illustrate exactly how these make a positive impact.

CHAPTER 3

Design Analysis

3.1 Introduction

In colleges and universities the world over, young engineering students are taught the basics of design analysis. So much of our curricula are oriented toward teaching the analytical methods necessary for students to gain an understanding of the fundamental principles that govern physical systems. In the electronics world, this translates into Kirchhoff's voltage and current laws (i.e., KVL and KCL), energy conservation, and the governing equations of the basic elementary devices in electrical circuits. Unfortunately, academics are missing the opportunity to begin driving home the idea of model-based engineering when instructing students on analysis techniques and capabilities. The situation is compounded when computer-based analysis tools are brought into the classroom. These tools are the embodiment of model-based engineering (MBE), but they are often relegated to a minimalist treatment because students are expected to acquire the knowledge needed to use the tools mostly on their own. Some level of self-instruction is certainly acceptable, but not at the expense of teaching the essential theoretical underpinnings models, modeling methods, and their role in the design process.

What are really needed are instructional modules that become an integral part of the circuits, electronics, and control systems courses taught at the undergraduate level. In this way students learn that these analytical techniques, whether done by hand or on a computer, are all essentially based on models. The students need the instruction on the basic theories of analysis and computation being utilized by the tools, and on the boundaries between solution algorithms and models. This chapter begins to reveal these ideas for circuit simulation, which, with the proper courseware, could be taught as modules in the aforementioned undergraduate courses. In addition to all that has been mentioned to this point, MBE is also a useful perspective toward design that provides the benefits of executable specifications through design exploration and efficient management of the verification of complex systems.

This chapter is organized into sections regarding various important forms of design analysis. The first is what we will term as "manual" analysis, or devoid of any automation. In this section classic hand calculations, prototyping, and other traditional forms of analysis are described as they relate to MBE techniques. The next section is a fairly in-depth description of computer simulation as it relates to physical systems, electronics in particular, but has been proven equally applicable to multi-discipline systems [6]. Finally, the ever-important, and yet fleeting, skill of troubleshooting is described along with how MBE methods enhance a design team's ability to deduce problems — a task for which modeling and simulation can be indispensible, particularly in integrated circuit design.

3.2 Manual Analysis

As alluded to earlier, one way to think about MBE is as a perspective on design that embraces the models as the specifications and tools for evaluating proposed design alternatives. From the earliest days in engineering schools, students are taught the methods of hand analysis independent of discipline (i.e., mechanical, electrical, civil, chemical, industrial, etc.). The next section of this chapter describes hand analysis of electric circuits as a model-based analytical approach.

3.2.1 Hand Calculations

Nodal and mesh analyses are analytical techniques for electric circuits that were derived based on Kirchhoff's current and voltage laws (KCL and KVL), respectively. In a first circuits course, the governing equations of the passive elements used in circuits for resistors, capacitors, inductors, and now memristors are re-introduced to the students. Their initial exposure was most likely a physics course. These equations are not typically described to students as the behavioral representations of the elements themselves, which is a missed opportunity to begin a distributed treatment of MBE in the curriculum as new devices are introduced. For it is a fact that Ohm's law for a resistor is merely a behavioral model for how a resistor behaves. It doesn't account for temperature variations or other nonlinearities that may be present in some resistive materials. The same holds true for the first order equations for the other elements.

Is it any different once the students begin to study nonlinear devices such as diodes and transistors? No, it is not. The equations first introduced as the governing equations for diodes, MOSFETs, and bipolar transistors *a la* Sedra and Smith [7] are also simply *behavioral* models of these devices. Modern devices have completely different expressions to describe their behavior in circuit simulation, as noted by Liu [8], Getreu [9], and Cressler [10] for MOSFETs, bipolar, and HBT devices, repectively. As the authors belong to that group of academics that teach these courses, we are keenly aware of the fact that these "models" are introduced as the governing equations to be used for circuit design and yet are often not treated as models at all. However, there is a useful instructional point to be made about this. Even though the governing equations are much-simplified versions of the physics-based equations one might find implemented in a transistor model in a circuit simulator, circuit designers all agree that these relatively simple expressions, and the linear small-signal models and expressions derived from them, are very useful design tools. They allow the designer (analog circuits designer in this case) to make assumptions about the regions of operation of devices and approximations to the real behavior, all for the purpose of determining if a proposed circuit can be expected to achieve the design objectives. This is all performed with models. It is also the case that the equations and models used to describe many non-linear devices start with a relatively simple description for basic calculations and become progressively more complex.

Let us consider the example of a MOSFET circuit, such as the common-source amplifier shown in Figure 3.1 and taught in electronics courses. The MOSFET drain-source current can be expressed by the square-law equation:

$$i_{M1} = \frac{\mu_n C_{OX}}{2} \left(\frac{W}{L}\right)(v_{GS} - V_T)^2(1 + \lambda v_{DS}) \qquad (3.1)$$

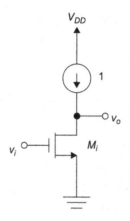

Figure 3.1:
Common-source amplifier often used as a gain stage in multi-stage amplifiers

which is often simplified by dropping the output resistance term (i.e., $\lambda = 0$) when actually used to compute the current, thus leaving

$$i_{M1} = \frac{\mu_n C_{OX}}{2} \left(\frac{W}{L}\right)(v_{GS} - V_T)^2 \tag{3.2}$$

The low-frequency, small-signal model that corresponds to this equation for the saturation region of a MOSFET is given in Figure 3.2. In the case of a common-source amplifier, the back-gate bias-dependent current generator is absent. Also, the higher-frequency version of this model would include capacitances. As this stage is often used as a low-frequency gain stage, the only important elements are the g_m current generator and the output resistance of the transistor. Once the output resistance of the current source r_{cs}, also implemented with transistors, is accounted for, the voltage gain of this circuit becomes:

$$\frac{v_o}{v_i} = -g_m(r_o||r_{cs}) \tag{3.3}$$

where g_m and r_o are the partial derivatives of (3.2). Specifically, $g_m = \partial I_{M1}/\partial v_{GS}$ and $r_o = 1/(\partial I_{M1}/\partial v_{DS}) = 1/\lambda$.

This example serves to illustrate that hand calculations performed using simple equations and simple circuits to build up a more complicated circuit performance *is* how analog design is performed. These simple equations are models.

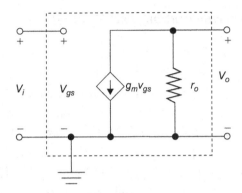

Figure 3.2:
Small-signal circuit of the common-source amplifier

They are not even all that accurate because they do not need to be for the task at hand. The simplifying assumptions made (i.e., using these simple models) are then validated and circuit performance verified using computer simulation where more accurate models, too unwieldy to be used in the design activity, are employed. Also, through the use of the simulator and these more accurate models, second- and third-order effects can be investigated to determine their effect on the performance of the overall circuit or system.

Experience in teaching this approach has shown that the most unsettling aspect is the fact that we are not using ultra-accurate models for our design activity. The design approach of: (1) make simplifying assumptions, (2) rigorously apply the approximations that result from those assumptions, (3) perform your design activity, (4) validate your original assumptions, and (5) verify the final performance of your circuit, seems to unsettle many because the result can be off by as much as 50%. However, if the circuit does not work under simplified scenarios, it certainly will not work once noise, nonlinearities, distortion, and other nonidealities are accounted for. The logical conclusion that follows here is that MBE is precisely what we do today. We just do not often refer to it by that name. By more fully embracing MBE we can generalize our approach to design. As a result, design productivity will improve, troubleshooting will be made much more straightforward, and complex systems can be managed much more effectively in terms of verification and coverage before fabrication.

3.2.2 Emulation, Experimentation, and Prototyping

Another form of design analysis that has been used for decades is to actually build a rapid prototype of the system under design to validate some concepts, as well as uncover the "real-world" effects that some components will have on the overall system [11]. This sometimes takes the form of a combination of hardware and software, thus using software emulation of the hardware. Field Programmable Gate Arrays (FPGA) are an example of an emulation of digital hardware where the Very High Speed ASIC Hardware Description Language (VHDL) or Verilog code that has been written for subsequent synthesis can be downloaded into a FPGA platform so that the code can be executed with other hardware in the system.

It is often very instructive to build a board-level prototype as part of the verification process. The value of this activity is based on abstraction. Each component on a board is a black box, which is the moral equivalent of a behavioral model with all nonidealities present. As a result, once the components are populated on the board, several "what-if" scenarios from a system perspective can be investigated. A small set of alternatives can be investigated, but this soon becomes impractical. Nonetheless, this prototyping provides extremely valuable insight into making the entire system work − even if it is to ultimately be integrated into a single chip.

The prototyping activity, at first glance, would seem to be totally removed from any MBE methodology. However, those adept at rapid prototyping and building quality prototypes prior to design fabrication are only one step removed from formulating the models of the constitutive components that would be able to represent the key nonidealities at a behavioral level, so that such analyses could subsequently be performed in simulation. The advantage of making this transition is the fact that now a large set of "what-if" scenarios can be investigated. As tolerances on the specifications now are executable in a simulator, they can be more definitively determined (or validated) for the individual blocks by simulating many cases to find out when the system begins to fall out of specification.

Such prototyping-modeling activity might appear to be far too labor-intensive for designers to contemplate. However, there are many alternatives that make it quite attractive: (1) a second spin of silicon is worth how much?, (2) the fact that not every single design starts from ground zero and that as you perform this activity more and more, the library of models grows and forms the basis for a tremendous amount of reuse, and (3) troubleshooting is made far easier, as will be discussed later in the chapter.

3.3 Computer Simulation

Clearly, for MBE, this book would not be complete without a suitable coverage of computer simulation algorithms typically used for lumped physical systems. In this way, this book serves as a core book for the field guides that will accompany it. A treatment of simulation algorithms in those guides will not be necessary, nor will a detailed description of many basic modeling principles as they are covered in this book.

Simulation affords the designer the possibility to interact with either simple abstract models or very detailed ones that may be characterized to actual experimental results. In this way, a vast array of performance characteristics, statistical behavior, and reliability assessments can be evaluated prior to fabrication, thus increasing the chances of a successful design. In the event that a design is not completely functional, simulation once again proves to be a very effective tool in diagnosing any issues and a means of verifying the proposed corrections in the context in which the design error or flaw was discovered. It is important to repeat the fact that an MBE approach must not be evaluated solely on the merits of its first time use, but rather the fact that the model libraries and techniques are cumulative and become less and less of an incremental cost as designs proceed. Having said this, in many cases, a gradual rolling out of MBE into the design process mitigates any substantial up-front costs and pays for itself through the adoption process. This discussion is elaborated more fully in Chapter 13 when we talk about successfully adopting MBE.

The next two subsections talk about algorithms for analog, digital, and mixed-signal systems, and practical issues that have to be dealt with when developing models for nonlinear solvers, respectively.

3.3.1 Simulation Algorithms

Before embarking on a detailed description of simulation, it is important to understand the relationship between modeling and simulation. In essence, the simulator is a black box process with the models as the data source for that process. The purpose of an analog simulator in the context of analog electronic system design, for example, is to solve nonlinear ordinary differential equations that describe the behavior of the system. The model is the container within which the governing equations of an element are described. Generally, each system is described using a circuit diagram (schematic) or list of models, often called a netlist. This term is a throwback to the days of integrated circuit design, where each connection ("net") was listed literally with the individual component connections so that a computer operator could correctly connect the signals on an Integrated Circuit (IC) layout. A more sophisticated term often used in modern design is "structural description", which is more cumbersome, but, nevertheless, more accurate. The structural description can often be generated from the circuit diagram by computer-aided design (CAD) software automatically, or written by a designer using a hardware description language, which is becoming more common, especially for complex designs with substantial digital content. The real issue is not whether a "behavioral" or "structural" description is used to describe a system, but how this is translated into the equations or elements to be solved by the simulator. Each individual element in an analog circuit diagram, a transistor, for example, is represented by an underlying set of equations in the final problem to be solved by the simulator. An equivalent also exists in the digital domain, where, ultimately, a circuit may decompose to individual logic gates or equations. The simulator interprets the list of individual models and for analog systems constructs a matrix of equations or a set of logical expressions for the complete design. This is then solved by the simulator, whether analog or digital. In this chapter, we will introduce the key concepts in analog and digital simulation, and provide some insight into how these processes operate, some common pitfalls, and a few methods to potentially make simulations faster, more robust, and more accurate.

3.3.1.1 Continuous and Discrete Time Simulation

The details of the simulation processes will be discussed later in this chapter, but there is an important aspect of the model definition to be considered first.

Electronic circuits fall into two main areas: analog and digital. While it is true that all electronic systems can, ultimately, be described as analog circuits, for example a *digital* logic inverter can also be described using individual *analog* transistors, it is useful to consider methods for simulating using an analog or digital approach. The final stage of integration is to consider how these two very different methods can be combined to efficiently and accurately deliver a system model that a mixed-signal designer can use to design more effectively. In order to help understand this process, the analog and digital approaches will first be considered individually, and then the combination of the two will be viewed as a true mixed-signal approach.

Analog simulation using digital computers is based on approximating the continuous-time behavior of an electronic circuit, solving nonlinear differential equations using iterative techniques such as Newton-Raphson, and using some form of a predictor-corrector method to estimate time steps between approximation points during a transient analysis. The individual connections in the circuit are solved using network analysis techniques such as KVL or KCL with the complete circuit stored in a large matrix.

Digital simulation with digital computers relies on an event-based approach. Rather than solve differential equations, events are scheduled at certain points in time, with discrete changes in level. Resolution of multiple events and connections is achieved using logical methods.

If the two approaches are compared it is clear that in the analog simulation approach, although the time axis is discretized, if the time steps are small enough it is effectively continuous. At each time step, the analog simulation requires the complete system to be solved. This can be time-consuming and the simulation time increases on the order of 1.2 ($O(n)^{1.2}$) with the size of the system. In contrast, the digital variables only need to change if an event occurs, leading to much faster simulations (typically 1000 times faster than analog). Using logical variable resolution rather than numerical techniques, as in the analog case, is also generally faster and simpler to implement.

Developments in simulation technology combine the two techniques into a form of *mixed-signal* simulation that can take advantage of the relatively high accuracy of the analog simulation and the high speed of the digital simulation.

Regardless of the techniques used, it is useful to understand at least the basic concepts used in each approach to improve the understanding of the mixed-signal simulation process as a whole.

3.3.1.2 Analog Simulation Techniques

Analog simulation is a method whereby electric circuits are modeled using behavioral blocks representing individual components, such as resistors, capacitors, transistors, and diodes, and a set of network equations formulated from the connectivity of these components are solved to determine the function of the circuit. Vlach and Singhal [1], Pillage [2], and Kielkowski [3] are good books for explaining circuit simulation. However, it is important to review the key concepts involved as they are common to almost every commercial simulator available on the market today.

Overview of Analog Simulation

The analog simulation process begins with the network listing, or netlist, that articulates the various components in the circuit however many instances of each there may be and how they are interconnected. Figure 3.3 shows a flowchart that describes an overview of this process starting from the netlist, proceeding to the simulation process, which, in turn, produces results such as voltage and current as functions of time and/or frequency, and the post-processing tools that may be used to derive other quantities from these (e.g., power dissipation).

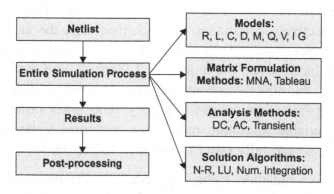

Figure 3.3:
Flowchart indicating an overview of the simulation process

The simulation process includes accessing and instantiating the various component models used, stamping these models into a matrix, and performing the requested analyses. These analyses require the use of specific solution algorithms for linear and nonlinear systems.

Descending into the simulation process a bit further, the flowchart in Figure 3.4 shows how the netlist is used to form the system matrix of equations. The order of this matrix is determined by the number of nodes in the circuit – not the number of elements in the netlist. This matrix is formulated through a process known as matrix stamping. Each model has a matrix element "stamp" that is used to formulate the system matrices quite efficiently based on circuit connectivity. For example, a two-terminal device such as a resistor, capacitor, diode, or inductor will impact four entries in the system matrix associated with the rows and columns that correspond to the nodes to which the component is connected. The remainder of Figure 3.4 indicates a very important context for models in the simulation process as it gets executed once the system matrices are formulated. The system matrix has to have its values repeatedly updated and solved iteratively in the nonlinear case to achieve a solution. The model's function in this flow is to accept the values of the independent variables that are predicted by the simulation algorithms (e.g., node voltages), compute the dependent variables according to the governing equations of the model (e.g., Ohm's law), and return these values to the system

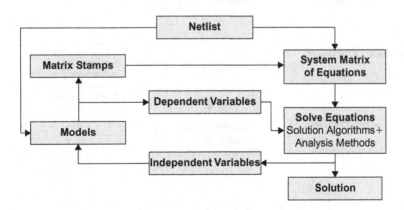

Figure 3.4:
Relationship of algorithms, models, and matrix formulation in analog simulation

matrix for the next iteration. This process continues for all components in the matrix. This leads to a definition of a model in the context of simulation: *models take independent simulation variables as inputs, compute the dependent variables according to their governing equations or behavior, and return those to the system matrix for determining if a solution has been achieved.*

The next step into the depths of analog simulation takes us to the solution algorithms themselves. From a mathematical standpoint, the transient solution of nonlinear differential algebraic equations (DAEs) is the most general algorithm required. All other solutions, such as the nonlinear algebraic equations that must be solved when performing a DC analysis or the series of linear algebraic equations being solved for an AC analysis, are subsets of this one. This is depicted in Figure 3.5 where the arrows on the left-hand side indicate the transient, nonlinear algebraic, and linear processes, respectively.

Referring to Figure 3.5, the nonlinear differential, algebraic equations are transformed into nonlinear algebraic equations through numerical integration. Numerical integration techniques (e.g., Gear or trapezoidal) are first employed to transform the given system of DAEs into a series of nonlinear algebraic equations. An iterative technique, typically Newton-Raphson, is used to transform the nonlinear algebraic equations into a series of linear algebraic

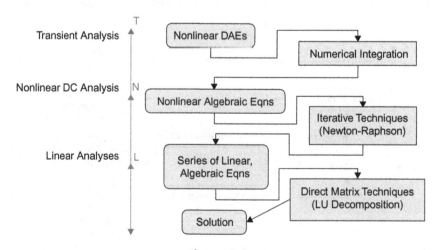

Figure 3.5:
Solution of nonlinear differential algebraic equations (DAEs)

equations. The system of linear equations can now be solved using direct matrix techniques (such as LU [Lower-Upper] decomposition).

The precise way in which these equations are solved is best described bottom-up starting from the solution of linear systems, then nonlinear algebraic systems and then the transient solution of nonlinear DAEs.

Matrix Formulation and Solving Linear Equations

The first step in the simulation of an electric circuit is to construct a set of equations that describe the elements in the circuit and their connections. Taking the circuit example shown in Figure 3.6, the nodal equations for the circuit can be obtained using KCL and summing the currents at nodes corresponding to those labeled as v_1, v_2, v_3, and v_4. The circuit itself consists of two current sources (I_1 and I_9) with seven resistors (R_2-R_8). While modified nodal analysis (MNA) will produce the correct matrix, it is not the most computer-efficient algorithm for matrix formulation in software. These nodal equations can be explicitly realized using items known as *nodal admittance equation stamps* of the elements in the circuit.

While delving into these low-level issues may appear a digression from our mission of MBE, some details of matrix stamping are given here to illustrate how the system of equations is created and to reveal to the reader yet another role that models play in the simulation process. These matrix stamps are generated by formulating the model's governing equations with current as the dependent variable and voltage as the independent variable when possible. The admittance is then entered into the matrix in the rows and columns associated with where

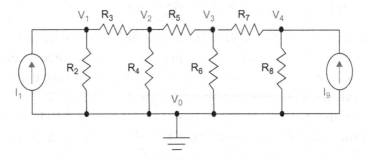

Figure 3.6:
Simple linear network to illustrate matrix formulation and linear circuit analysis

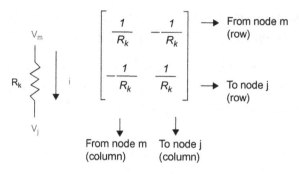

Figure 3.7:
Arbitrary resistance in a circuit and its matrix element stamp

the device is connected in the circuit. Examples will best illustrate this process, but a more exhaustive treatment can be found in [2].

To illustrate matrix stamps, we shall begin with the resistor. The governing equation of a resistor is Ohm's law and is given for this arrangement in Eq. (3.1). Figure 3.7 shows an arbitrary resistor R_k connected between nodes m and j with a current i flowing through the resistor.

$$i = \frac{v_m - v_j}{R_k} \tag{3.4}$$

The admittance $1/R_k$ is stamped into the nodal admittance matrix $\mathbf{Y}$ in the system formulation $\mathbf{Y v} = \mathbf{J}$ in the locations indicated by rows m and j and columns m and j, respectively. Note that the formulation of Eq. (3.1) is cast as the dependent variable (current) as a function of the independent variables (node voltages) leaving it in admittance (i.e., $1/R_k$) form.

In an analogous fashion, the matrix stamp for the ideal current source is given in Figure 3.8, where the entries are stamped into the source vector $\mathbf{J}$ as the element is a sourced current. The notational convention being employed is that the positive direction of current flow through the element is from node m to node j, while the positive voltage drop is also from node m to node j. Maintaining this consistency ensures that Kirchhoff's laws are obeyed.

Returning to our example circuit of Figure 3.6, we begin by stamping out the current source I_1 in Eq. (3.5) and then proceeding to stamp out the resistors in

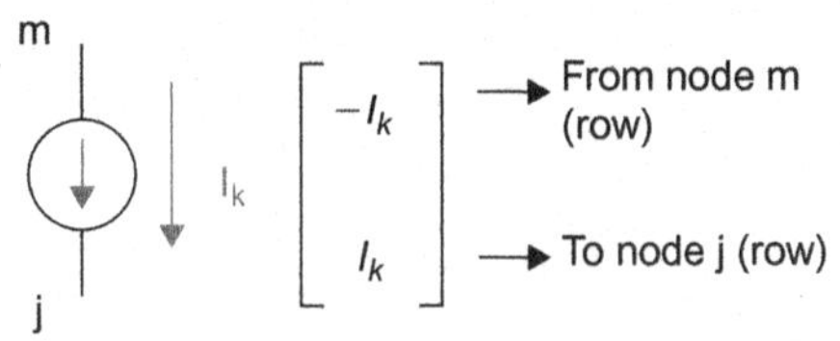

Figure 3.8:
Arbitrary ideal current source in a circuit connected from nodes *m* to *j* and its matrix element stamp for the source vector J

succession in Eqs. (3.6)–(3.8). The final stamp added is that of I_9 in the source vector in (3.8).

$$
\begin{bmatrix}
0 & 0 & 0 & 0 & 0 \\
0 & 0 & 0 & 0 & 0 \\
0 & 0 & 0 & 0 & 0 \\
0 & 0 & 0 & 0 & 0 \\
0 & 0 & 0 & 0 & 0
\end{bmatrix}
\begin{bmatrix}
v_0 \\ v_1 \\ v_2 \\ v_3 \\ v_4
\end{bmatrix}
=
\begin{bmatrix}
-I_1 \\ I_1 \\ 0 \\ 0 \\ 0
\end{bmatrix}
\qquad (3.5)
$$

$$
\begin{bmatrix}
\dfrac{1}{R_2} & -\dfrac{1}{R_2} & 0 & 0 & 0 \\
-\dfrac{1}{R_2} & \dfrac{1}{R_2} & 0 & 0 & 0 \\
0 & 0 & 0 & 0 & 0 \\
0 & 0 & 0 & 0 & 0 \\
0 & 0 & 0 & 0 & 0
\end{bmatrix}
\begin{bmatrix}
v_0 \\ v_1 \\ v_2 \\ v_3 \\ v_4
\end{bmatrix}
=
\begin{bmatrix}
-I_1 \\ I_1 \\ 0 \\ 0 \\ 0
\end{bmatrix}
\qquad (3.6)
$$

$$
\begin{bmatrix}
\dfrac{1}{R_2} & -\dfrac{1}{R_2} & 0 & 0 & 0 \\
-\dfrac{1}{R_2} & \left(\dfrac{1}{R_2}+\dfrac{1}{R_3}\right) & -\dfrac{1}{R_3} & 0 & 0 \\
0 & -\dfrac{1}{R_3} & \dfrac{1}{R_3} & 0 & 0 \\
0 & 0 & 0 & 0 & 0 \\
0 & 0 & 0 & 0 & 0
\end{bmatrix}
\begin{bmatrix}
v_0 \\ v_1 \\ v_2 \\ v_3 \\ v_4
\end{bmatrix}
=
\begin{bmatrix}
-I_1 \\ I_1 \\ 0 \\ 0 \\ 0
\end{bmatrix}
\qquad (3.7)
$$

$$\begin{bmatrix} \left(\frac{1}{R_2}+\frac{1}{R_4}+\frac{1}{R_6}+\frac{1}{R_8}\right) & -\frac{1}{R_2} & -\frac{1}{R_4} & -\frac{1}{R_6} & -\frac{1}{R_8} \\ -\frac{1}{R_2} & \left(\frac{1}{R_2}+\frac{1}{R_3}\right) & -\frac{1}{R_3} & 0 & 0 \\ -\frac{1}{R_4} & -\frac{1}{R_3} & \left(\frac{1}{R_3}+\frac{1}{R_4}+\frac{1}{R_5}\right) & -\frac{1}{R_5} & 0 \\ -\frac{1}{R_6} & 0 & -\frac{1}{R_5} & \left(\frac{1}{R_5}+\frac{1}{R_6}+\frac{1}{R_7}\right) & -\frac{1}{R_7} \\ -\frac{1}{R_8} & 0 & 0 & -\frac{1}{R_7} & \left(\frac{1}{R_7}+\frac{1}{R_8}\right) \end{bmatrix} \vec{v} = \begin{bmatrix} -I_1 - I_9 \\ I_1 \\ 0 \\ 0 \\ I_9 \end{bmatrix}$$

(3.8)

Systematically following the stamping procedure, as illustrated by this example, produces a system of equations that are in indefinite form because each row and column sums to zero. This means that we do not have an independent set of equations. To address this issue, one simply removes one row and column. In reality, the ground node is never explicitly stamped and this leaves us with the elimination of row 1 and column 1 in our example, as shown in Eq. (3.9).

$$\begin{bmatrix} \left(\frac{1}{R_2}+\frac{1}{R_3}\right) & -\frac{1}{R_3} & 0 & 0 \\ -\frac{1}{R_3} & \left(\frac{1}{R_3}+\frac{1}{R_4}+\frac{1}{R_5}\right) & -\frac{1}{R_5} & 0 \\ 0 & -\frac{1}{R_5} & \left(\frac{1}{R_5}+\frac{1}{R_6}+\frac{1}{R_7}\right) & -\frac{1}{R_7} \\ 0 & 0 & -\frac{1}{R_7} & \left(\frac{1}{R_7}+\frac{1}{R_8}\right) \end{bmatrix} \vec{v} = \begin{bmatrix} I_1 \\ 0 \\ 0 \\ I_9 \end{bmatrix}$$ (3,9)

The method of equation formulation is basically the same regardless of whether the system contains linear or nonlinear elements. The resulting equations can then be rewritten in a matrix form as illustrated above and then solved using standard methods for solving simultaneous equations such as Gaussian elimination (described in good numerical analysis textbooks, such as [4] or [5]). This is effectively how simulators such as SPICE calculate operating point solutions for circuits, although, in practice, a more efficient approach of the Gaussian elimination method known as LU decomposition [2,4,5] is applied. In the LU decomposition method, the nodal admittance matrix Y is factored into a lower triangular matrix L and an upper triangular matrix U. The reader is directed to [2] and [4] for an overview of how these optimized techniques perform in general, and [5] provides details of how the algorithms are implemented in the C programming language.

Solving Nonlinear Algebraic Equations

It is, of course, not realistic to restrict the model equations to a linear form, so a method for solving nonlinear equations must be used. In fact, repeated use of a linear solution method can be used to solve nonlinear algebraic equations, as shown in Figure 3.9. Common nonlinear solution techniques are the Newton-Raphson and Katzenelson iterative methods. The loop in Figure 3.9 contains the linear DC analysis steps discussed in the previous section. That is, matrices are formed and then solved through LU decomposition. The nonlinear DC analysis indicated here updates the nodal admittance matrix entries as the nonlinear elements are re-linearized at successive iterations. Thus, a very efficient linear DC analysis directly affects the efficiency of the nonlinear analysis.

The Newton-Raphson method is commonly used in commercial simulators, such as SPICE, and applies an iterative technique to approach the exact

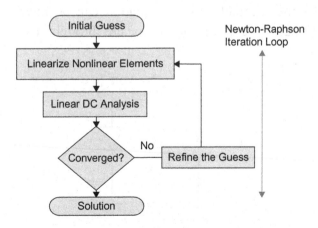

Figure 3.9:
Solution of nonlinear algebraic equations using iterative techniques

solution for a nonlinear equation. The basic Newton-Raphson formula is given in Eq. (3.10) where $F(X) = 0$ is the equation to be solved and $F'(X)$ is its derivative. X_n is the value of the solution X during the current iteration and X_{n+1} is the value of X at the next iteration.

$$X_{n+1} = X_n - \frac{F(X_n)}{F'(X_n)} \tag{3.10}$$

This method requires the formula to be iterated until the value for $F(X)$ approaches zero. It is necessary to specify the value of target error, and when $F(X)$ falls below this value, convergence is achieved. It is important to notice that if the derivative $F'(X_n)$, or slope, of the function $F(X_n)$ tends to zero, then Eq. (3.10) will approach infinity and therefore not converge to a useful solution. To illustrate how the method operates in practice, a nonlinear example of a diode connected to a current source is shown in Figure 3.10.

The equation for the example nonlinear diode model used is given by Eq. (3.11). The exponent of the ideal diode equation, $40V_D$, is based on the thermal voltage (V_T) being approximately 25mV and the ideality factor (n) being 1. As the exponent is $V_D/(nV_T)$, this gives the resulting approximate exponent of $40V_D$.

$$i_d = 10^{-12}(e^{40V_d} - 1) \tag{3.11}$$

Using Eq. (3.11), the complete circuit equation can be rewritten in the form shown in Eq. (3.12), where all the terms are brought onto one side (this

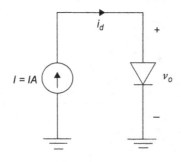

Figure 3.10:
Diode circuit to illustrate the solution of nonlinear equations

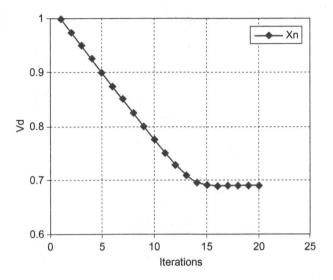

Figure 3.11:
Graph of the diode voltage $V_d(X_n)$ showing convergence to 0.69 V

becomes $F(X_n)$ in Eq. (3.10)) and differentiated to provide the slope for the Newton-Raphson equation $F'(X_n)$ as shown in Eq. (3.13).

$$F(V_d) = 10^{-12}(e^{40V_d} - 1) - i_d \tag{3.12}$$

$$F'(V_d) = 40 \times 10^{-12} e^{40V_d} \tag{3.13}$$

Figure 3.11 shows how the values of diode voltage $V_d(X_n)$ converge to a solution value of 0.6908 V within 20 iterations from an initial starting point of 1 V.

Integration Methods and the Solution of Nonlinear DAEs

Introduction

In order to establish the time-domain response of a network, the differential equations for the system variables need to be integrated numerically. We will describe the basics of some of these integration methods and also how the nonlinear DAEs are solved according to Figure 3.12. In order to perform a transient analysis, the algorithm needs a starting point solution. This starting point can be derived via a nonlinear DC solution of the circuit for the response at time zero, or specified as initial conditions or some combination thereof. Then, analogous

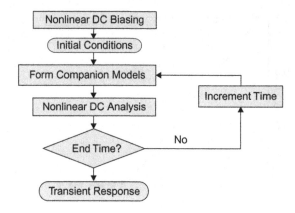

Figure 3.12:
Time-domain (i.e., transient) analysis flowchart with nonlinear DC analysis from
Figure 3.9 as a procedure used within this flow in two places

to how matrix element stamps are used to formulate the system matrices instead of direct application of Kirchhoff's Current Law, entities known as *companion models* are formed based on the selected integration algorithm. This then allows subsequent stamping of the elements present in the companion models to form the system matrices. In this way, numerical integration is performed far differently than a brute force implementation of the integration algorithm. It is performed through the use of companion models being stamped into the system matrix, which is mathematically equivalent. We shall begin with a description of the integration algorithms and then proceed to describe this interesting transient solution method.

The methods for numerical integration consist of two main types, the Linear Multi-Step (LMS) and Runge–Kutta methods. The LMS techniques contain methods such as forward and backward Euler, and variations on the trapezoidal method. In the descriptions of all the integration methods, the following basic notation is used. The variable to be found is x with respect to time (t) and so the equation to be integrated is:

$$x' = f(x, t) \tag{3.14}$$

which can be rewritten as:

$$x = x(a) + \int_a^b f(x, t)dt \tag{3.15}$$

The value of x at the current time, t_n, is the current solution and is denoted by x_n. The initial condition is defined by:

$$x_0 = x(a) \tag{3.16}$$

The solution at a time after a small step forward, t_{n+1}, is denoted by x_{n+1} and the resulting time difference between the current and next solutions is called the time step (h), and is given by Eq. (3.17).

$$h = t_{n+1} - t_n \tag{3.17}$$

Forward Euler Integration The forward Euler method uses the knowledge of the current point (x_n), the time step (h), and the derivative (x'), and approximates the slope between the current and next point (x_{n+1}) using:

$$x_n' = \frac{x_{n+1} - x_n}{h} \tag{3.18}$$

Rearranging Eq. (3.18) provides an expression for the next value of x as given by:

$$x_{n+1} = x_n + x_n' h \tag{3.19}$$

This is the simplest and probably least accurate method for calculating the integral of x'. Forward Euler is known as an *explicit* integration algorithm because it is a function of known quantities (i.e., past and current values). Figure 3.13 illustrates how the current value of x is used at time t to approximate the slope.

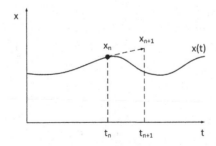

Figure 3.13:
Forward Euler integration illustration where the current value of x at t is used to approximate the slope

Backward Euler Integration The backward Euler method is similar to the forward Euler approach except that it uses the next point x_{n+1} as the point for calculating the derivative:

$$x'_{n+1} = f(x, t) \tag{3.20}$$

And similar to the forward Euler, the next value, x_{n+1}, is estimated using:

$$x_{n+1} = x_n + hx'_{n+1} \tag{3.21}$$

This procedure is then iterated until x_{n+1} converges onto a solution. The integration approach is illustrated in Figure 3.14. Backward Euler, trapezoidal, and Gear integration methods are known as *implicit* integration methods because the value being determined is a function of other unknown variable(s) at that same point in time (e.g., $v(t + \Delta t)$ depends on $i(t + \Delta t)$).

Trapezoidal Integration Trapezoidal integration uses a combination of the derivatives at the current and next simulation points to calculate the integral (Figure 3.15). The average value is taken as shown by:

$$x_{n+1} = x_n + \frac{h}{2}(x'_{n+1} + x'_n) \tag{3.22}$$

Gear Integration Higher-order methods, such as Gear's method, use several preceding time points to better predict the value at time $t + \Delta t$. A polynomial of some order is fit to the solution at preceding time points and either extrapolated

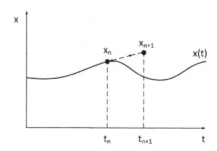

Figure 3.14:
Backward Euler integration illustration where the next value of x at $t + \Delta t$ is used to approximate the slope

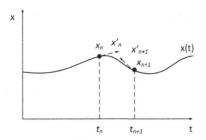

Figure 3.15:
Trapezoidal integration illustration where both the current and the next value of *x* are used to approximate the slope

(explicit) or interpolated (implicit) to the present time point [12]. Gear's method is one of the most popular methods available in circuit simulators. For instance, Gear order two through six is available in the Saber simulator [13].

The Gear method is nearly as accurate as the trapezoidal approach; however, it is more stable and therefore does not have a tendency to oscillate. If we consider the second-order case, then the equation for the next step is as follows:

$$y(t_{n+1}) = \left(\frac{4}{3}\right) \times y(t_n) - \left(\frac{1}{3}\right) \times y(t_{n-1}) + \left(\frac{2}{3}\right) \times stepsize \times \frac{d}{dt}(y(t_{n+1})) \quad (3.23)$$

Runge-Kutta Integration One problem with the previously described LMS methods is the poor accuracy that can result unless a very small time step (*h*) is used. This can be a severe penalty in time domain simulation, so alternative techniques are sometimes used, such as the Runge-Kutta method [4]. The basis of this technique is to take a trial step and use the information to provide a more accurate estimate of the average slope over the complete time step. The Runge-Kutta method may use different orders, which define the number of trial steps to be taken. If the second-order Runge-Kutta is used to illustrate the method, the equations can be derived easily. The first step is to calculate an intermediate point, k_1:

$$k_1 = hf(x_n, t_n) \quad (3.24)$$

And use this point to calculate a second intermediate point, k_2:

$$k_2 = hf(x_n + k_1, t_n + h) \tag{3.25}$$

The next value is then obtained by using the intermediate points in:

$$x_{n+1} = x_n + \frac{1}{2}(k_1 + k_2) \tag{3.26}$$

Hosking, Joe, Joyce, and Turner [4] have compared the methods of integration and show that even a second-order Runge-Kutta integration of a simple equation can be over 25 times more accurate than the corresponding backward Euler approach. However, it is often much slower in practice than a Gear or trapezoidal method, and so, in practical circuit simulators, either a Gear or trapezoidal approach is generally employed.

Time Domain Simulation

With the methods described previously in this section, it is possible to carry out time domain (often referred to as "transient") analysis simulations. The flowchart in Figure 3.12 shows how this can be carried out. As described earlier, this flow begins with an initial solution point and then evolves the solution over time by performing numerical integration via companion models to transform the nonlinear DAE system of equations into a system of nonlinear algebraic equations that are solved iteratively according to the methods given in Figure 3.9. It is worth noting that the time step, h or Δt, is usually allowed to be variable in circuit simulators. This is important particularly in nonlinear and switching systems to obtain efficient use of the available computing power where the time step can be decreased to improve accuracy on fast switching waveforms and then increased where the waveform becomes relatively slowly changing.

To understand the formation of companion models we shall take the example of forming the companion model of a capacitor. The equation for a capacitor is:

$$i = C\frac{dv}{dt} \tag{3.27}$$

In seeking to find a solution at $t + \Delta t$ we integrate to obtain:

$$\int_{t}^{t+\Delta t} \frac{dv}{d\tau} d\tau = \frac{1}{C} \int_{t}^{t+\Delta t} i(\tau) d\tau \tag{3.28}$$

$$v(t + \Delta t) - v(t) = \frac{1}{C} \int_{t}^{t+\Delta t} i(\tau) d\tau \tag{3.29}$$

$$v(t + \Delta t) = v(t) + \frac{1}{C} \int_{t}^{t+\Delta t} i(\tau) d\tau \tag{3.30}$$

The integral on the right-hand side can now be approximated by one of several formulas depending on which integration method is chosen (e.g., backward Euler, forward Euler, trapezoidal, etc.). Performing this yields

$$\int_{t}^{t+\Delta t} i(\tau) d\tau \approx \left\{ \begin{array}{l} \Delta t \cdot i(t) \\ \Delta t \cdot i(t + \Delta t) \\ \dfrac{\Delta t}{2} \cdot (i(t) + i(t + \Delta t)) \end{array} \right. \tag{3.31}$$

where the top equation is forward Euler, the middle is backward Euler, and the bottom is trapezoidal integration. Therefore, the voltage across the capacitor at the next time step can be written as:

$$v(t + \Delta t) = v(t) + \frac{1}{C} \cdot \left\{ \begin{array}{l} \Delta t \cdot i(t) \\ \Delta t \cdot i(t + \Delta t) \\ \dfrac{\Delta t}{2} \cdot (i(t) + i(t + \Delta t)) \end{array} \right. \tag{3.32}$$

Here is the key point surrounding these equations — they each have an equivalent circuit representation known as a companion model. For the forward Euler case, the capacitor element is replaced with a voltage source as shown in Figure 3.16. To start with the original capacitor, we seek the current through and voltage across the capacitor at time $t + \Delta t$. Eq. (3.32) expresses this where the voltage at $t + \Delta t$ is equal to the expression in the figure, which is only a

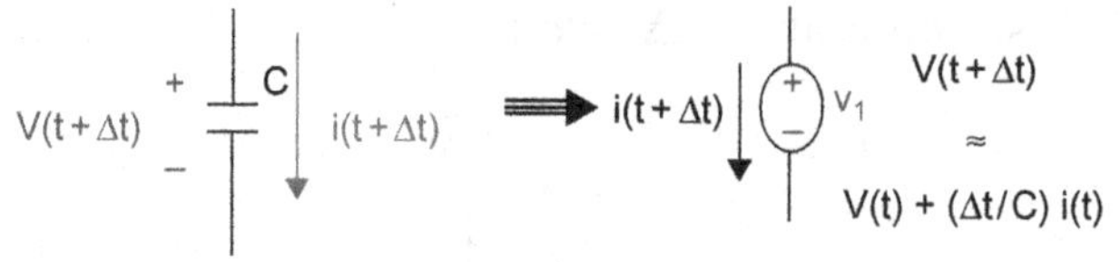

Figure 3.16:
The companion model of a capacitor based upon the forward Euler integration
method. The capacitor is replaced by a voltage source in the matrix formulation.
The value of the voltage source is updated at each time step

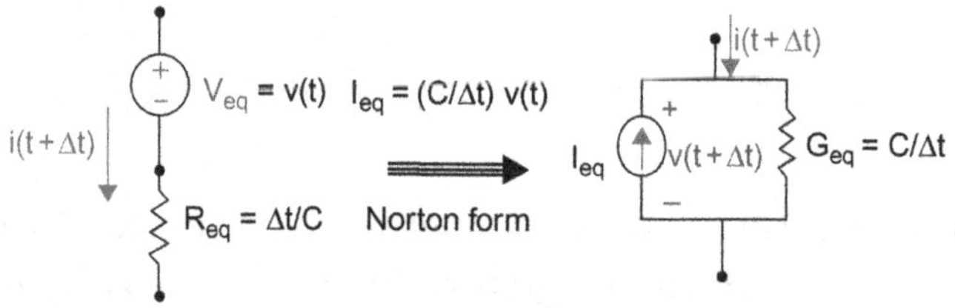

Figure 3.17:
The companion model of a capacitor based upon the backward Euler
integration method

function of past values $v(t)$, $i(t)$, C and the time step Δt. Thus, the nodal admittance matrix has a voltage source stamped in place of the capacitor and is updated according to (3.32) to perform integration.

If backward Euler integration is selected, then the companion model is derived from the middle expression in Eq. (3.32) as shown in Figure 3.17. Note that the Norton form of the companion model is preferred for the nodal admittance matrix, as you would expect. The Thevenin form on the left of Figure 3.17 shows an equivalent resistance that reflects the instantaneous relationship between $i(t + \Delta t)$ and $v(t + \Delta t)$ that will be present in all implicit integration algorithms. It also shows an offset voltage due to the past value of the voltage. These two elements are then converted into Norton form, which yields a current source in parallel with a conductance. What happens if Δt becomes zero? As one can see from the equations, the conductance value becomes infinity and the simulator will issue the infamous "time step too small" error.

To complete the companion model description, Figure 3.18 shows the companion models if trapezoidal integration is used for the capacitor. More companion

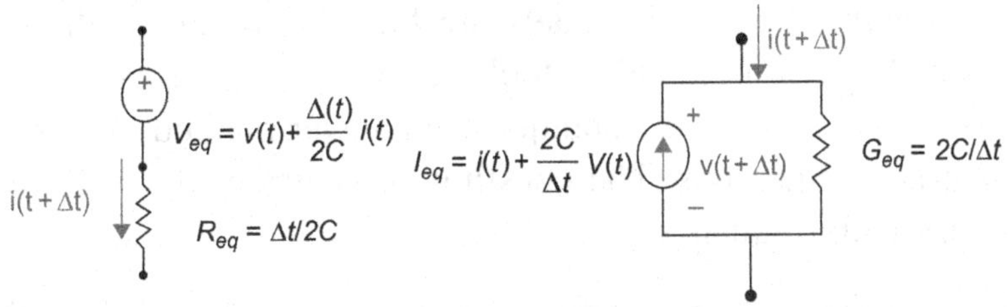

Figure 3.18:
The companion model of a capacitor based upon the trapezoidal
integration method

models of circuit elements are described in [2], including nonlinear capacitive
and inductive elements. The approach is completely general.

Small-Signal AC Analysis

Small-signal AC analysis is a specialized analysis for linear systems. While we
can use linear elements only to create a linear system, in analog circuit design,
we are designing circuitry to have a linear response over a finite operating
regime using nonlinear (i.e., transistors and diodes) and linear elements. In order
to accomplish this linear analysis, simulators utilize the matrix element stamping
method described earlier to create the system of linear equations. So even though
nonlinear elements are present in the original netlist, the linearization about the
DC operating point creates a new linear system that will simulate really fast. The
stimulus to such systems is often a swept frequency source in order to produce
the frequency response (magnitude and phase). As the system is now linear, the
magnitude applied to the system is not important, whereas a transient analysis of
the original nonlinear circuit would require a stimulus that more accurately
mimics a small-signal input (say, 10 mV, or so). This would ensure that the nonlinear circuit remained in its "linear" operating region about the bias point.
Therefore, in the AC analysis it is common to set the magnitude to 1 and the
phase to 0 degrees. By doing so, the output values are, indeed, the ratio of the
output to the input and represent the gain (magnitude and phase) directly.

Different simulators accomplish the linearization process from the original nonlinear circuit differently. One way is that the small-signal models are stamped
because they consist of only linear elements and their values are determined by

the operating conditions. These models are built-in similar to the companion models described previously. Other tools derive the linearization mathematically based on finite difference techniques. Whatever method is used, the result is a linear matrix of equations. The constitutive values of the nodal admittance matrix will vary with frequency.

The key aspect of the AC (often called a frequency) analysis is that the models must be considered from a complex variable point of view. This is often best done using Laplace notation ("s" domain) to describe the individual equations. For example, a simple first-order filter model can be described using a basic Laplace equation shown in Eq. (3.33)

$$output = \left(\frac{1}{1 + s\tau}\right) input \tag{3.33}$$

where τ is the time constant of the filter, which, of course, determines the cut-off frequency of the filter. The AC analysis carries out a frequency sweep across a specified range of frequency values, for example from 1 Hz to 1 MHz, and at each individual point the linear solution is calculated. This is similar to the DC analysis, except that in the AC analysis, the variables are complex (with magnitude and phase). When the variables are examined after the analysis has been completed, we can observe the real or imaginary parts of the variables. It more common to plot the gain of the system or element using a Bode plot (Figure 3.19). Figure 3.7 shows the magnitude portion of the Bode plot for the filter. The response is the magnitude in dB (a log function) versus the frequency (on a log axis).

The AC analysis is therefore ideal for calculating the frequency response of linear analog elements such as filters; however, it is not applicable to digital systems. Even when nonlinear systems are linearized, prior to the linear AC analysis being carried out, the simulation is restricted to that single operating point for validity. In some mixed-signal simulators, this analysis is not even supported to avoid this possible confusion.

3.3.1.3 Digital Simulation

Introduction

In contrast to the continuous time of analog simulation, the first key difference with any "digital" simulation is that the time axis is considered to be discrete.

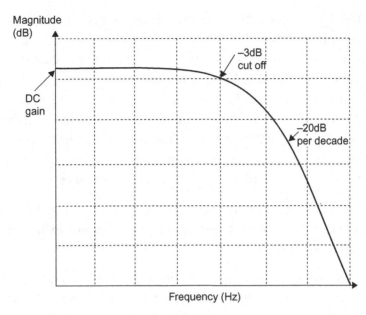

Figure 3.19:
Bode plot of low-pass filter example

This means that there is a finite resolution, the smallest time step that can occur, limiting the accuracy of the variables being simulated. The second important difference is that the simulator is based on a series of discrete "events" rather than solving equations. This is fundamentally different and effectively means that simulations can generally be carried out much faster than analog equivalents. The basic element in digital simulation is the concept of the logic gate. The logic gate contains a Boolean expression of its behavior that is solved when and only when an input signal changes (an "event"). Therefore, from a digital system perspective, there are logic expressions inside models to be executed – using logical methods, signals to connect between models, and finite delays from events (usually input to output in a model).

Level of Abstraction and Logic Resolution

In addition to the discrete nature of the time axis, the types of signals are often reduced to enumerated types rather than a complete scale of values in order to simplify logic calculations. For example, a simple logical Boolean type may consist of two values – "true" and "false". These could be represented using the enumerated type values "1" and "0", respectively, or, perhaps, "H" and "L". Therefore, when representing these values in a simulator we need to consider how to display signals graphically for a designer's viewpoint and also how to resolve signal conflicts when more than one signal is assigned concurrently. An important aspect of the "digital" nature of such signals is the ability to hold two values at one absolute value of time, and this is inherent in the basic nature of digital simulation. For example, representing an "analog" or real world electrical signal in a digital simulator would require the translation of real voltages into an equivalent logic context as in Figure 3.20.

This is interesting because clearly there are threshold and rise time or fall time issues to consider in practice, but the logic signal does not exhibit any of those details—it is either 1 or 0.

It is possible to add an extra state to the logic, "X", which is often called "don't care", but which should probably be more accurately defined as a "don't know" value. This can be used to provide a more realistic value during particularly longer transitions, such as that in Figure 3.21.

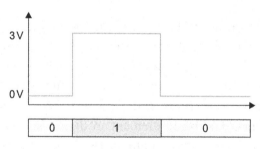

Figure 3.20:
Idealized digital signal representation

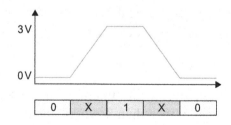

Figure 3.21:
Addition of "don't know" X states

An interesting problem with logical resolution is what happens when multiple signals of different values are connected to the same node or "net". How is this to be resolved in practice? Consider the simple example of two logical gates connected to the same net in a system in Figure 3.22. What is the value on the net?

This is dependent on a logical resolution table in practice, which simply looks at the list of connected signals and, depending on their values, defines the appropriate output. For example, in the example in Figure 3.22, if both signals are "0" then the output will be "0", and if both are "1" then the output will be "1". This is obvious so far, but what if one is "1" and the other is "0"? In that case we must *define* the state of the output, and this could be defined as "X" (i.e., don't know) in this simple logical example. The obvious question is how can we incorporate another state such that we can avoid this problem in reality, just like a tristate driver output? The answer is to add one further possible value – high impedance "Z". We can therefore modify our example and incorporate a more sophisticated logical resolution to the problem, as shown in Figure 3.23.

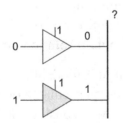

Figure 3.22:
Logical resolution when a net is being driven by two gates toward conflicting values

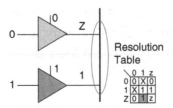

Figure 3.23:
Logical conflict resolution employing "Z" and "X"

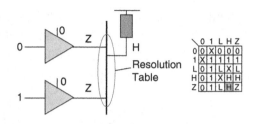

Figure 3.24:
Logical conflict resolution with pull-ups and pull-downs

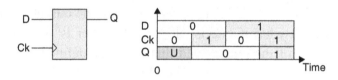

Figure 3.25:
Initialization of memory elements

The final sophistication to be added is the difference between a strong and a weak value. In the strong case, the signal is being actively driven, but in the weak one the signal may be pulled up or down by a resistor (or equivalent). We can represent this using a "pull-high" or "pull-low" option in our logical definition to further extend the logic resolution table (Figure 3.24).

Timing is also a key issue in digital modeling; therefore, we need to consider the possibility of incorrect values, particularly on initialization of memory elements (such as flip-flops). If we consider a simple D-type flip-flop in Figure 3.25, it is clear that the initial value of Q needs to be set to something, otherwise the overall simulation values could be unknown.

This leads to a final possible value of a generic logical enumerated type of "U" − un-initialized. The result of this development of logical enumerated types is the standard definitions used in many logic simulators for VHDL and Verilog. For example, the IEEE standard logic definition is given below:

```
type std_ulogic is
( 'U', - uninitialised
'X', - forcing unknown
'0', - forcing 0
```

```
'1', - forcing 1
'Z', - high impedance
'W', - weak unknown
'L', - weak 0
'H', - weak 1
'-' - don't care
);
```

This shows the basic types of logic level described in this section, and the resulting resolution table is given in Figure 3.26.

With the basic logic expressions and resolution in place, the second aspect of digital simulation can be considered — timing. With instantaneous calculation of logic, it is vital that accurate delays of those expressions are implemented in models. One such implementation is inertial delay, which is summarized in Figure 3.27.

One key aspect of the inertial delay is that a fixed delay results regardless of the loading and connections to the gate. The inertial delay also limits the size of the shortest pulse able to pass through the gate. This models the real world situation regarding glitches, where very short spikes are "damped out" by the real capacitance on a system. This is illustrated in Figure 3.28.

The delays can be made asymmetrical, which is reflected in Figure 3.29. This is helpful for modeling driver transition times accurately in line with real devices.

	U	X	0	1	Z	W	L	H	-
U	U	U	U	U	U	U	U	U	U
X	U	X	X	X	X	X	X	X	X
0	U	X	0	X	0	0	0	0	X
1	U	X	X	1	1	1	1	1	X
Z	U	X	0	1	Z	W	L	H	X
W	U	X	0	1	W	W	W	W	X
L	U	X	0	1	L	W	L	W	X
H	U	X	0	1	H	W	W	H	X
-	U	X	X	X	X	X	X	X	X

Figure 3.26:
The std_ulogic logic resolution table

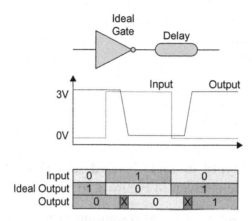

Figure 3.27:
Depiction of inertial delay

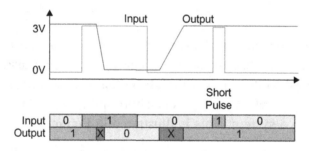

Figure 3.28:
Inertial delay — short pulse cancellation

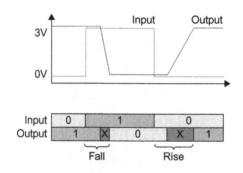

Figure 3.29:
Asymmetric rise and fall times in delay modeling

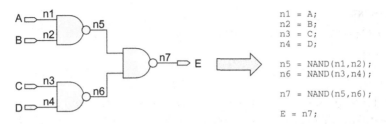

Figure 3.30:
Compiler-driven simulation where everything is compiled to code for execution

Types of Digital Simulators

There are two main types of digital simulator: compiler- and event-driven. Compiler-driven simulators have the characteristic feature that the model is translated into code models that are compiled — so every model effectively becomes its own simulator. Clearly, these can be extremely fast, but they tend to model only simple delays and are often restricted to synchronous circuits owing to the poor simulation capability to handle tight feedback loops. The other common type of digital simulator is the event-driven simulator, where the models are read in to a library and commands interpreted to run simulations. These are usually slower but very powerful, and able to handle multiple libraries and types of delay models. If we consider the compiler-driven simulator first and wish to simulate a circuit, then it must be mapped onto equivalent sequential code, as shown in Figure 3.30.

It is worth noting that even in this simple example, the code generation must be mapped in the correct order from input to output (Figure 3.31).

The resulting code is such that it can then be compiled using a standard compiler and run as a standalone program as listed as follows.

```
for (time = 0; time < end; time++) {
new[1] = A;
new[2] = B;
new[3] = C;
new[4] = D;
new[5] = NAND(old[1],old[2]);
new[6] = NAND(old[3],old[4]);
```

```
new[7] = NAND(old[5],old[6]);
E = old[7];
old = new;
}
```

Delays can be implemented with simple or more sophisticated values as required. The key problem with such systems is the inability to handle asynchronous feedback, and, in such systems, this feedback must be broken and registers inserted to provide an artificial barrier to instant feedback (Figure 3.32).

Event-driven simulators are much more common in practical use and are a more general solution for arbitrary circuits. They are based on the idea that few gates are switching at any one time; nevertheless, they are usually much slower than compiler-driven simulators.

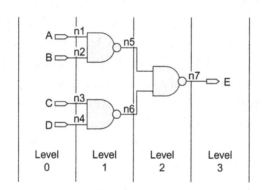

Figure 3.31:
Compiler-driven simulation illustrating coding order

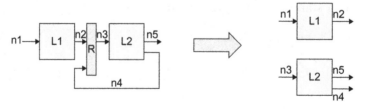

Figure 3.32:
Feedback being broken in order to process the simulated values in a compiler-driven scenario

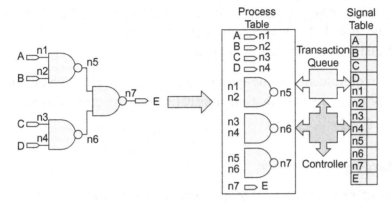

Figure 3.33:
Event-based simulator approach

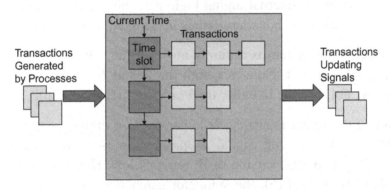

Figure 3.34:
Transaction queue in an event-driven simulator

The key concept in event-based simulators is the idea of an "event queue", which represents an ordering of events and logical expressions to be carried out when they are needed (Figure 3.33).

In contrast to the compiler-driven simulator, the event-based simulator has a queue of transactions. This allows each individual gate or signal to be considered in parallel as a process and the individual events to be taken in turn (Figure 3.34). The heart of this technique is managing the transaction queue, which can prove to be a bottleneck in the overall efficiency of the simulation.

Transactions can have any delay associated with them — there's no restriction on the delay model. Each transaction is added to the transaction queue at a time slot equal to the current time plus the transaction's delay.

The critical requirements of the queue are that transactions can be added to the queue at any time slot and that all transactions at the current time slot can be fetched as a group.

3.3.1.4 Mixed-Signal Simulation

In the case of mixed-signal systems, there is a mixture of continuous analog variables and digital events (Figure 3.35). The models and simulator need to be able to represent the boundaries and transitions between these different domains effectively and efficiently. Consider the following example where there is an analog input, internal digital logic gate, and analog output. This simple example illustrates the boundary issues in a mixed-signal system.

The first stage of the system is a purely analog filter where the input A is filtered to give the signal B. Note that both the A and B signals are analog and continuous (i.e. not discrete in either time or value) (Figure 3.36).

At the point B, the signal is analog, but changes to a digital input to the buffer (logic gate). As the analog signal B progresses (see time steps 1−4 in Figure 3.37) B_d, the digital version of B, remains low. However, when the gate input threshold is crossed (5), the simulator calculates the exact crossing time and schedules an event (6) on the digital signal B_d.

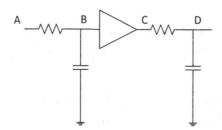

Figure 3.35:
Mixed-signal circuit for illustrating the boundaries between analog and digital simulation

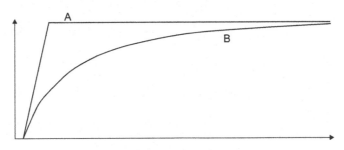

Figure 3.36:
Mixed-signal example — A input to B

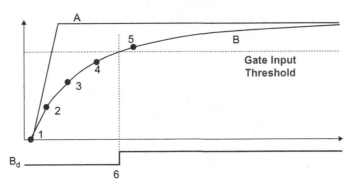

Figure 3.37:
Mixed-signal example — analog to digital transition

When the digital signal B_d event is scheduled, the digital model schedules another event at a delay time of t_{plh} (low to high transition) later. This schedules the digital output signal C_d to change state (number 8 in Figure 3.38). Notice that the analog signal B continues (7).

When the digital event C_d is scheduled, the analog signal C is triggered to change — this requires a time step (9) at the transition point (8) and the output signal C is initiated (Figure 3.39). The system is now back into the analog domain where the analog solver behavior applies.

Thus far, we have discussed a potential method for connecting between the analog and digital simulators, but there are, in fact, several methods that could be employed. The simplest and earliest approach was to use a technique called

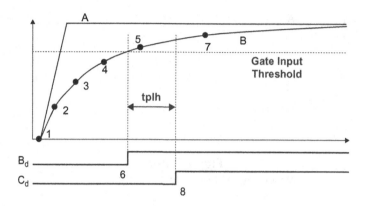

Figure 3.38:
Mixed-signal example — digital to digital transition

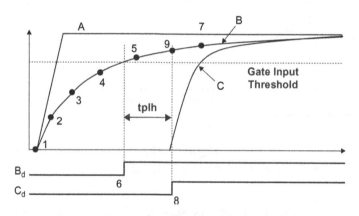

Figure 3.39:
Mixed-signal example — digital to analog transition

"lockstep" (Figure 3.40) which simply linked the analog and digital time steps together. However, this is extremely slow and hardly ever used in practice.

A development of this approach is to use a periodic synchronization (Figure 3.41), which is more efficient, but still can introduce errors into the mixed-signal boundary.

Perhaps the most efficient type of approach is to allow both the digital and analog simulators to be able to operate at their optimal speeds and then

Figure 3.40:
Mixed-signal time step synchronization (i.e., lockstep) between the analog and digital simulator engines

Figure 3.41:
Mixed-signal periodic time step synchronization between the analog and digital simulator engines

Figure 3.42:
Mixed-signal time step synchronization between the analog and digital simulator engines when performed as necessary

synchronize when values at the analog–digital interface need to be made self-consistent. This is highlighted in Figure 3.42.

3.3.4 Practical Issues

3.3.4.1 Speed

As we have seen so far in our descriptions of both analog and digital simulation, a common theme is complexity. Whether it is nonlinear equations or complex logical expressions, the simulation engines have a lot of work to do in order to achieve an accurate solution. In general, the digital simulation will be much faster for an equivalent structure, for fairly obvious reasons. As we have seen, the digital simulation of a single gate will require a direct solution of a logic equation, whereas the analog simulation needs to carry out LU decomposition for a linear matrix and a Newton-Raphson iterative loop for nonlinear equations. However, it is not simply a case of analog = slow and digital = fast, although that is a good working assumption. There are some specific issues that we need to consider in some detail to truly understand what is going on. We can break these down into some specific aspects to consider in some detail.

Initial Conditions

If we reconsider our simple diode example from earlier in this chapter, with the schematic shown in Figure 3.10, we can see that in the specific case of an initial starting point of 1 V the model converges to a solution value of 0.6908 V within 20 iterations. Some interesting questions arise at this point. Why did we choose 1 V? What is the "best" starting point? It is a useful assumption to make that "zeros" are particularly bad values to use in many initial condition settings. This can lead to awkward matrix calculations, with division by zero being an obvious problem with convergence — more on this particular issue later. If we are considering speed, then we obviously need to get to the correct solution as quickly as possible, and non-convergence does not help us achieve that.

For example, if we consider the simple example of the diode again and run a number of different simulations of the same model using different initial conditions, we can measure the number of iterations to achieve the specified error value. This is particularly interesting when we observe the results of running a simulation using a "pure" Newton-Raphson simple code model and compare the results. The target tolerance was 1.0 e-3 and the number of iterations for each starting voltage is given in Table 3.1. The final solution value was 0.70814 V for this circuit. The results are interesting as they seem to indicate in the first instance that the closer to the final solution, the better — with a reduced number of iterations as we start off closer to the correct final solution. However, the large number of iterations at 0.6 V seems to go against this principle at first glance. This is explained by the fact that the simple implementation of the Newton-Raphson for this illustration takes the solution at 0.6 V, and the tangent approximation leads to a next guess of approximately 2.5 V, which then subsequently requires the algorithm to "crawl" back down the exponential relationship in 73 more iterations. A second interesting aspect is that at 0.0 V,

**Table 3.1: Comparison of Iterations versus Starting
Voltage for Diode Circuit**

Starting Voltage (V)	Number of Iterations
5.0	174
3.0	94
2.0	54
1.0	14
0.9	10
0.8	6
0.7	2
0.6	73
0.0	Did not converge

there is no solution. This is again due to the "ideal" Newton-Raphson algorithm relying on a finite value of $F(x)$ to start the algorithm.

This initial condition is vital for all the simulation analyses that we carry out as we need to start from some operating point for both time domain and AC analyses. The solution we obtain can dictate how fast and accurate the other analyses will be.

This simplistic model of the Newton-Raphson method is not what is used in most circuit simulators, which take a more electronic "device centric" approach. This recognizes that many circuits are, in fact, based on semiconductor devices, which have an intrinsically exponential current−voltage characteristic. As a result, a more sophisticated error-checking mechanism is required. Rather than a simple check for an absolute error on the nodal voltages, the checking is carried out in two stages.

The first stage is to check the difference between the current and past nodal voltages and compare this to two tolerance parameters (using SPICE notation) VNTOL and RELTOL. This equation is described in Eq. (3.34):

$$|V(n) - V(n-1)| < RELTOL * V(n) + VNTOL \qquad (3.34)$$

where the default value for RELTOL is 0.0001 and VNTOL is 1.0 e-6 (usually in SPICE).

In addition, the simulator may often check the individual branch currents (SPICE does this as well) in a similar fashion to ensure that the often larger relative changes in current are satisfactory due to the exponential nature of the equations. This is done using the RELTOL parameter as before and also ABSTOL, which evaluates the branch current tolerances. This is shown in Eq. (3.35):

$$|I(n) - I(n-1)| < RELTOL * I(n) + ABSTOL \qquad (3.35)$$

In this case, ABSTOL often has a default value of 1.0 e-12 (1 pA), again to reflect the context of the basic device simulators for small semiconductor devices. Different settings of these parameters (or their equivalents) will have a profound effect on the speed of simulation by directly affecting the number of iterations and the basic stability of the simulation. Clearly, they are also directly linked to simulation accuracy, which we will consider in more detail in the following section.

3.3.4.2 Accuracy

As we have seen in the previous section, accuracy is not defined in an analog simulation by one single parameter. In addition, the methods of solution for analog and digital systems are so different that accuracy has to be considered in two completely different ways. For the analog part of a simulation we have absolute and relative error bounds, whereas the digital simulation is limited by the basic model representation and timing resolution. In addition, even in an integrated mixed-signal simulation, the error bounds are defined and implemented separately.

This is not just a subtle difference, but can have a major effect on the behavior we measure in the circuit as a whole, as the *cumulative* effects can become extremely significant over time. For example, consider the case of a simple ring oscillator, where there are both analog and digital elements in the circuit. If we take this example, with the circuit shown in Figure 3.43, then there are several sources of error that can be introduced depending on how we model not only the individual models, but also the interface between the analog and digital domains.

If we model the transistors correctly and carefully with the detailed technology models of a particular silicon process, then we can say with

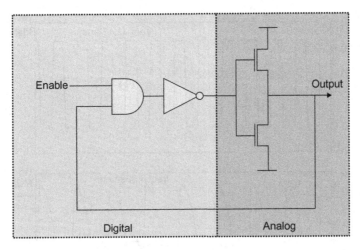

Figure 3.43:
Mixed-signal example — ring oscillator

confidence that the output waveform will look very similar to that obtained in practice. Using this example, when we simulate using ideal logic and a 0.35 μm process, we find that the output frequency of the oscillator is approximately 340 MHz. Clearly, this is not realistic, as the logic gates will have delays and their own rise and fall characteristics, and when these are incorporated into the design, the result is a more realistic 12 MHz. This assumes that the interface between the logic gates is not affected by capacitive loading and other nonlinear behavior, but when these effects are considered the results are radically different. In this example, adding nonlinear driving interface models based on MOS transistors gives a much slower response around 8 MHz—a 30% reduction. The model also changes the waveform shape, as shown in Figure 3.44. Clearly, the difference from a designer's perspective is vital to the ultimate design.

3.3.4.3 Convergence

"Failure to Converge"

The number one problem for most designers using simulation is when the simulation "fails to converge". The term convergence is one that is often misunderstood and is a frustrating problem in many cases with apparently no obvious

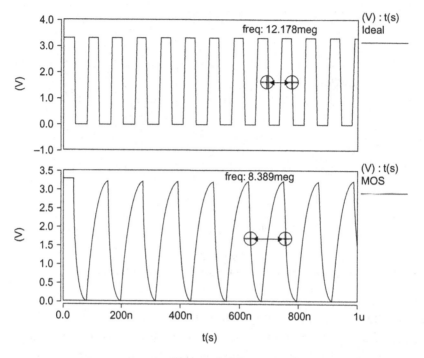

Figure 3.44:
The effect of mixed-signal interface models on performance

solution. It is worth discussing this topic from the viewpoint of gaining an understanding of the common causes of non-convergence in a simulation and some basic techniques to minimize the chances of this happening. The first stage is to define what convergence actually means. In this context, what we mean by "fail to converge" is that the simulation has failed to find a solution *within the specified number of iterations*. Notice the use of italics, as the model may be fine, but the initial conditions mean that it will take a larger number of iterations to find a solution within the error criteria defined by the designer. Consider the diode example from earlier in this chapter, and notice that for initial conditions within 0.2 V of the target value, the number of iterations was less than 20. However, for an initial starting voltage of 5 V, the number of iterations required was 174. Now, if our maximum allowed number of iterations happens to be 173, then this simulation will "fail to converge", but if it is 175 then the simulation will be successful. Therefore, perhaps a more correct way to express this failure condition is "failed to converge *within the specified number of iterations*".

With this in mind, is it simply a matter of increasing the number of iterations until a solution is found? In practice this is not the case, as there are some other sources of "non-convergence" that we need to consider. Furthermore, another possible (related) suggestion is to simply change the error criteria (which is clearly linked to the number of iterations permissible). This is an acceptable method to obtain *a* solution and can be backed off to establish a stable set of simulation parameters that provide an adequately accurate solution.

Easing the Path to Convergence with GMIN

The most common approach in many simulators to improve the chances of a simulation converging, especially if it contains nonlinear elements, such as semiconductor devices, is to add finite conductances to branches in the model. This is a specific response to the Newton-Raphson approach of numerical solution, where the dependence on a ratio of the previous term and conductance value means that for highly nonlinear systems it is easy for large slopes in the I-V characteristic to lead to huge changes in the calculated voltages and currents, iteration to iteration.

A common method of reducing the sensitivity of the solver to individual conductances, therefore, is to add artificial conductance branches to each individual branch in the model as a whole, and then set them to a value that doesn't affect the basic behavior of the system or circuit significantly, but does allow an easier path for iteration to a solution. For example, in the diode example examined previously, to aid convergence, an extra resistor is added across the terminals of the model, with a conductance value often referred to as GMIN (this is used in SPICE). This value can be assigned globally across a circuit to aid convergence.

This type of model modification can often explain why a simplified "ideal" model of a diode or a nonlinear switch may exhibit much inferior convergence behavior than a more complex equation-based equivalent. In fact, often the simplest models can cause the most severe convergence problems as a result—with a particularly prone model being a piecewise linear model of a diode, which can often cause significant problems, especially in switching circuits.

Of course, the "time step too small" issue that was mentioned earlier in the section on companion models and transient simulation is yet another case of where adding conductance such as GMIN in parallel with nodes would hide the fact that the G_{eq} of the companion model was blowing up at small Δt.

Discontinuities

The task of model creation can be easy or difficult, but in anything other than the simplest of models, there lurks the most common pitfall to face the model designer − the discontinuity. This is a particularly insidious form of model deficiency, as it may not be apparent as a problem except in very specific or very rare conditions. Therefore, the model may be operating correctly for some time until a seemingly random set of conditions can instigate a failure of the simulation to converge.

The source of many discontinuities can be traced to the approximations that are made to replicate the behavior of real devices in models. For example, we may use a number of equations to represent device behavior, with transitions between the regions to these different equations. It is vital to ensure that the transition from one region to another is smooth, with no breaks. Otherwise, it can become very difficult or even impossible for the simulator to converge to a correct solution in a reasonable time, or even at all.

Discontinuities can fall into two basic types, the first being a break in the characteristic from one value to another. This is illustrated in Figure 3.45, where the equation can be seen to change from one region to another at an input voltage of V1, and the solver may jump from one side of this clear discontinuity to the other, and, depending on the size of the discontinuity and also whether the solution is close to the discontinuity, perhaps fail to converge.

The second form of discontinuity occurs when there is a change in the *first derivative*. This is more subtle, but can also lead to failure to converge. In the Newton-Raphson technique, the approach is to use the previously solved values to predict the next solution point. However, Newton-Raphson not only relies

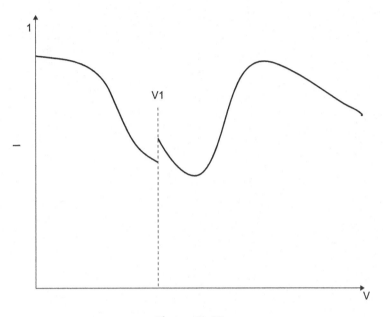

Figure 3.45:
Discontinuity in an equation

on the previous values, but the *slope* from the previous solution to predict where the next one is going to be. As a result, if the slope changes radically, then this can become difficult to solve. This is illustrated in Figure 3.46, where a characteristic is simple to solve, and then when a steep slope occurs the predictor may tend to undershoot initially and not be able to react quickly enough to find a solution within the required number of iterations, leading to a failure of convergence.

A particular problem can occur with ideal models, with zero slope (Figure 3.47). Traversing the zero slope area is easy, as the output value is the same each time (such as an ideal voltage source). However, when a transition occurs, then clearly there is no slope for the Newton-Raphson algorithm to work with and then the problem of excessive iterations may result.

The solution to this particular problem is to use a similar technique to the GMIN approach described previously, and add a finite slope to the characteristic in the zero slope region (Figure 3.48). This aids the solver in identifying a

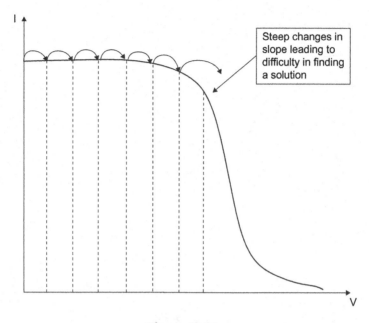

Figure 3.46:
Steep changes in slope make convergence difficult

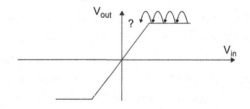

Figure 3.47:
Zero slope and transition causing non-convergence

suitable trajectory for the algorithm to iterate to a sensible solution in a reasonable time.

Summary

The basic issue with any modeling approach is ensuring that convergence can be reached to a sensible accurate solution in a reasonable time. From a modeling perspective, it is useful to understand the mechanisms behind the potential

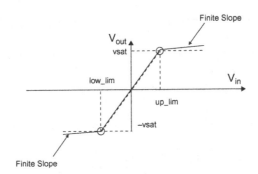

Figure 3.48:
Adding finite slopes to aid convergence

convergence issues and at least minimize the problems at source. This also gives greater flexibility for the designer when the model is in use in a particular system or circuit, enabling model parameters (such as GMIN, for example) to be used to improve convergence rather than excessive use of simulation parameters. The good news is that over the past 20 years much attention has been given to convergence issues, and better algorithms and modeling practices have emerged that have helped to address these problems.

Conclusion

In this chapter we have introduced the key concepts in analog, digital, and mixed-signal simulation. It is clear that there are more details than can be covered in a brief overview, and the reader is directed to the references for further detailed information about simulation algorithms and techniques. The main message to take into the remainder of this book is the fundamental difference between the analog and digital simulation engines, and how this can affect the implementation of models in either domain. This is exacerbated when the models are combined into a mixed-signal form, and the designer now has choices as to how to partition the model and implement from a mixed-signal perspective, rather than just "bolting together" two separate analog and digital pieces. In the next chapter, we will introduce the concept of modeling, and begin to illustrate how we can achieve mixed-signal models with an appropriate use of analog or digital modeling techniques for the application in question.

References and Further Reading

[1] J. Vlach, K. Singhal, Computer Methods for Circuit Analysis and Design, second ed., Van Nostrand Rheinhold, 1994.

[2] L. Pillage, R.A. Rohrer, C. Visweswariah, Electronic Circuit and System Simulation Methods, McGraw-Hill, 1995.

[3] R. Kielkowski, Inside SPICE, Overcoming the Obstacles of Circuit Simulation, McGraw Hill, 1994.

[4] R.J. Hosking, S. Joe, D.C. Joyce, J.C. Turner, First Steps in Numerical Analysis, second ed., Arnold, 1996.

[5] W.H. Press, S.A. Teukolsky, W.T. Vetterling, B.P. Flannery, Numerical Recipes in C, second ed., Cambridge University Press, 1992.

[6] P.R. Wilson, N.J. Ross, A.D. Brown, A. Rushton, Multiple domain behavioural modeling using VHDL-AMS, IEEE Int. Symp. Circuits Syst., ISCAS, 2004.

[7] A.S. Sedra, K.C. Smith, Microelectronic Circuits, sixth ed., Oxford University Press, 2009.

[8] W. Liu, MOSFET Models for SPICE Simulation Including BSIM3v3 and BSIM4, John Wiley, 2001.

[9] I. Getreu, Modeling the Bipolar Transistor, (First edition published by Tetronix in 1974), Published by the Author, 2009.

[10] J. Cressler, Fabrication of SiGe HBT BiCMOS Technology, CRC Press, 2008.

[11] R. Pease, Troubleshooting Analog Circuits (EDN Series for Design Engineers), Newnes, 1991.

[12] C.W. Gear, Simultaneous numerical solution of differential-algebraic equations, IEEE Trans. Circuit Theory 18(1) (1971) 89−95.

[13] The Saber Simulator, Synopsys, Inc., <http://www.synopsys.com/>.

CHAPTER 4

Modeling of Systems

*As Captain Jack Sparrow says in the Pirates of the Caribbean films, "the hull, keel and sails are what a Ship needs, but what a Ship **is**, is freedom". In a similar way it is important to think beyond the parts of a model and how the model is put together to understand what a model **is**. For example, a model consists of connections, parameters, behavior, and structure, but what the model is really is the vehicle by which we can accomplish design. This is a crucial concept for model-based engineering as we use models to do **design**, not modeling for the sake of it.*

4.1 Modeling in the Context of Design

As has been described in Chapter 1, this book is intended to describe the techniques required for model-based engineering (MBE). Rather than just presenting a series of unconnected ideas in a more formalized mechanism it is important to keep the broader context in mind when putting a variety of techniques at the designer's disposal. Therefore, in the context of this book, in Chapter 1 we presented a rationale for MBE with some initial thoughts as to why it is important and necessary. In Chapter 2 we described design verification approaches and value, as this is a big part of what the models are used for. In Chapter 3, we went on to define some of the fundamental *simulation* methods commonly used in engineering applications. In this chapter, we now develop those ideas further into the realm of modeling itself. Also, in keeping with the thought that the model is the vehicle for design, the most important concept is how models can be used to accomplish design objectives.

As was discussed in Chapter 1, the real issue for designers today is not generally one of the detailed modeling of individual devices or components, as this can often be accomplished using specific point tools or methods, but rather the management of large systems. This is evidenced with the problem of the increasing complexity of systems, leading directly to the "design gap" in many technologies where the pace of change and sheer scale of the systems are driving the productivity of designers and required delivery apart. It is also important to recognize that although there are a number of different approaches to design across various engineering disciplines, there are some fundamentals that apply across them. It is often the case that modeling techniques used in one discipline may find a use in another; it just may not be initially apparent that this could be the case. For example, an engineer modeling a *thermal* network may not be aware that a simple way to model the system could be to use an *electrical* simulator. Even in the same discipline there are often differing attitudes and preferences as to the best approach to do engineering design — *using modeling*. On one hand, some engineers are very comfortable writing code, and so using a language-based approach is perfectly natural to them, but, on the other hand, some engineers rely on graphics and diagrams to establish design ideas. This dichotomy also strongly implies that, just like in design, there is no "one size fits all" approach to modeling.

This chapter therefore completes the fundamentals section of this book by pushing further outward to the management of complex systems through modeling. The concept of hierarchy will also be discussed from a systems perspective in detail, and the options for expressing elements in the system will also be reviewed in detail. While individual techniques are to be introduced, it must be stressed that ultimately it is the designer's choice as to which will work best for them. These techniques can be viewed like the potential constituent parts of a "ship" in the sense that there may be a warehouse to choose from, but the final design will consist of the pieces that the designer chooses to use. In addition, there will also be an introduction to some of the commonly used software tools that are used in modeling for design. The ship analogy is also useful here as these can be seen as the box of tools from which the designer can select the appropriate one for them to use.

4.2 Modeling Hierarchy

4.2.1 Hierarchy Concepts

Often the biggest question for a designer or a design team is "where to start?". It is a glib statement to simply state that we need to use a "top down" approach, as the reality is often much more complex than this. Often, some model elements are already available at a device or component level, in one format or another, as well as some outline information of the overall system structure. It is also true to say that in many cases the initial specification is not entirely fixed or truly representative of what the system is *actually* required to be.

It is therefore essential that whichever technique is used to handle the hierarchical modeling of a system needs to have inherent flexibility, so that as changes are made as the design evolves into its final state, those changes feed through into the constituent parts of the design.

If we revisit the diagram from Chapter 1 that illustrates the information transfer relationships in hierarchical design, this can be seen to be the case on closer inspection. It is incumbent on the system designer to ensure that the requirements for each sub-block are transmitted correctly to the individual section designers in addition to any constraints for that section of the design. Also, there is a mechanism by which the *physical* constraints are fed back to the system designer (e.g., this may be a performance limitation of the technology being used) so that the overall partitioning can be refined or revised. In particular, at all levels of the design there must be an awareness of the limitations of the design and where improvements must be made to ensure conformance to the overall specification (Figure 4.1).

Whether the system is being designed by a single designer or a large design team there are clearly different requirements at each level of the hierarchy. For example, at the highest level of the hierarchy there are overall requirements that must be met by the design, and these can be investigated by the "system designer". At first glance that can be an easy thing to do, as the highest level models tend to be very abstract. However, one issue is that relatively small

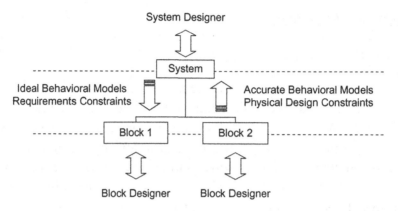

Figure 4.1:
Hierarchical design from a systems perspective

changes at this level can have a fundamental effect on a very large number of individuals at more detailed levels of abstraction. For example, the system designer may decide that a technology is not appropriate to meet the overall requirements, and the effect of this would be felt by everyone else on the design team.

4.2.2 Partitioning

One of the key tasks for the system designer at the higher levels of abstraction, in addition to establishing the requirements to be met, is the partitioning of the system into separate sub-blocks. In partitioning it is essential that the overall requirements are divided and assigned to the relevant parts of the design. This is a critical step from a number of perspectives in that the individual requirements must be managed to ensure that the overall system requirements are met, requirements are self-consistent, individual sections of the design can be verified, and problematic interfaces are properly defined. (A learned modeling expert, Ian Getreu, author of *Modeling the Bipolar Transistor*, is fond of saying "problems migrate to interfaces." It is vital that, in the process of partitioning, the interfaces are well-defined and modeled to ensure that all subsequent verification is valid.)

The process of deciding how to partition the design is, by nature, a complex and design-dependent process. However, the very fact of capturing a hierarchical model enables the basic connectivity and parameter subdivision to take place. Many potential errors and issues can be identified by simply defining a "skeleton" of the model architecture without any behavior.

A useful technique commonly employed on many system designs is to assign nominal behavior to all the main blocks in the system and then check the signal paths throughout (from inputs to outputs). This is a "sanity check" that, at the very least, the paths exist so that when the complete model behavior is put in place, the basic infrastructure can be relied upon. Also, if the system does not work at this level, then there are serious architectural and requirements issues to address before worrying about design partitioning.

4.3 Fundamentals of Modeling

Having discussed some of the *requirements* for a model, it is useful to define some of the important concepts that encapsulate what we mean in general *by* a model. There is often a discrepancy between different disciplines (such as electronic and mechanical engineering, for example) in terminology used; however, in most cases the basic fundamentals are actually surprisingly consistent. In this section of the book we will introduce some of those important concepts and establish some common ground that can be a framework for some of the advanced modeling techniques to be applied later in this book.

4.3.1 Definition of a Model

Generally, in simulation a model *represents* a physical element. The physical elements could be as simple as a resistor in electronic systems or as complex as an aircraft. In every case, it can be said that, in general, *any* model consists of a core set of behavior (expressed using equations of some form), internal structure, and the external interface consisting of both the connection points to other elements in the circuit or system and parameters that characterize the expressions that comprise the behavior of the physical element. This is illustrated in Figure 4.2.

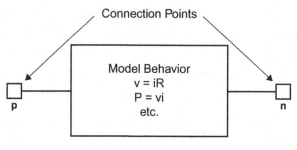

Figure 4.2:
General concept of a model

4.3.2 Representing Model Variables

There are generally four basic ways of representing variables, which can be broadly categorized into *parameters*, *procedurals*, *simultaneous*, and *events*. *Parameters* are variables that are defined once at the beginning of a simulation and do not subsequently change. An example of this is the resistance value of a resistor, for example 100 Ω. Analog-type variables consist of *procedural* and *simultaneous* values. Procedural variables can be thought of as analogous to variables in a sequential programming language, such as C. The values of the variables are calculated sequentially in the order the equations appear at each time step or iteration of a simulation. Simultaneous variables, however, are solved concurrently using matrix solution methods. The final type of variable is *event-based* and this is used for the digital part of a model, such as digital logic or sampled data systems. Each of these fundamental types of variable will be discussed in detail in this chapter. As in any areas where there are multiple forms of definition, the possibilities for confusion are endless. For example, even using the term *variable* is fraught with difficulty as it has a specific meaning in many contexts, such as in many hardware description languages (HDLs). We are attempting to maintain neutrality in our definitions, thus avoiding references to particular HDLs.

4.3.3 Representing Model Behavior

Representing behavior depends largely on the context of the system. For example, electronic circuits tend to fall into two categories, analog and digital, whereas mechanical systems may require lumped or distributed elements to be represented. Fundamentally, however, model behavior can be categorized as being either of a continuous or discrete nature.

Often, continuous models are described using nonlinear differential algebraic equations, and these provide the simultaneous equations fundamental to the matrix to be solved by a simulator. Parameters and sequential variables may use sequential assignments, including *if-then-else*-type expressions, but, ultimately, they will feed their values into the appropriate simultaneous equations to be solved. Event-based equations are not included in the analog solver directly, but generally use a discrete logic solver called an event queue. Logic methods are used to handle events, changes, scheduling, and conflict resolution. Mixed-signal simulation requires the transfer of information across the boundary elements of the analog and digital systems, respectively, as well as the solution of the analog and digital parts of the model. In addition to the basic partition between analog and digital, analog expressions can be represented using a variety of different techniques in a model. These can be loosely considered as parameter, sequential, or simultaneous equations.

Parameter equations are defined in the header or beginning of a model and are usually evaluated once at the beginning of a simulation or when a parameter to a model is changed during a pause in simulation. These are often used to calculate internal constant values for use in the other equations of the model.

Sequential equations are evaluated strictly in order and, as such, the values of variables are dependent on the order of evaluation. This can be an efficient method of handling sections of equations and minimizing the number of simultaneous equations. Different model languages handle this type of expression in different ways, and this can be a very effective way of model reuse by encapsulating the functionality in procedures or functions that can be applied to a variety of models.

Simultaneous equations define the pure architecture of the model and are an essential part of the basic analog solution mechanism, as we have seen in the previous chapter on simulation. Clearly, by minimizing the number of simultaneous equations, it is possible to greatly reduce the computational effort required to obtain a solution; however, care must be taken to ensure that enough equations exist to completely describe the behavior of the model with respect to the connection points. This will be discussed later in this chapter in more detail.

4.3.4 Representing Model Structure

A second attribute that any model has is structure. Certainly, the most degenerate form of structure is a single element or branch. However, as most engineers that have ever created or used a macromodel know, the structure of the model is how the various behaviors within the model relate to one another. A schematic diagram of an electronic circuit is a perfect analogy. The way in which the transistors are interconnected to one another determines the function of the circuit. A NAND versus a NOR gate or a differential amplifier versus a folded-cascode amplifier are key examples of this. As we will see in subsequent sections of this chapter, the structure and the behavior of a model work together to determine its overall functionality. This functionality is accessed through the external interface of parameters and connections described next.

4.3.5 Analog Connections

There are two general types of connection used in analog systems, *signal flow* and *conserved energy*. Signal flow is the classical control system definition, where the signal has a single value only. An example is shown in Figure 4.3(a), with a simple gain block. The model has an input pin *in* and an output pin *out*. In this type of model, there is no concept of loading or impedance, so the output will always be the input times the gain. In contrast, the

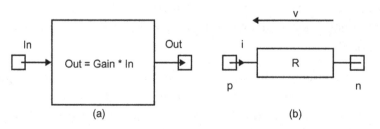

Figure 4.3:
Signal flow and conserved energy model examples

conserved energy connections have two values, which are the *through* and *across* variables (these will be described in more detail later in this chapter). This apparent contradiction is best explained using a simple example of an electrical resistor, shown in Figure 4.3(b). At the connection pin *p*, there is both a potential difference between the pin and ground (voltage), and a current through the pin. This becomes important when models are connected together in a circuit as the conserved energy model equations are dependent on impedances and loading. This is another concept in modeling that will be discussed in more detail later in this chapter.

4.3.6 Discrete Connections

In contrast to the analog connections, which are essentially based on variables of a real number type, as we have seen in the review of digital simulation, event-based digital models use enumerated or number types with discrete time steps and sometimes discrete levels (as in the case of digital logic). In digital models, therefore, there is no concept of energy conservation, and, as such, the connection points will have a direction defined as IN, OUT, or INOUT. The connection type must also define the resolution table for when these models are connected together arbitrarily.

In addition to the discrete nature of the time axis, in order to simplify logic calculations, the types of signals are often reduced to enumerated types rather than a complete scale of values. For example, a simple logical Boolean type may consist of two values — "true" and "false". These could be represented using the enumerated type values "1" and "0", respectively, or perhaps "H" and "L". When representing these values in a simulator, therefore, we need to consider how to display signals graphically for a designer's viewpoint and also how to resolve signal conflicts when more than one signal is assigned concurrently. An important aspect of the "digital" nature of such signals is the ability to hold two values at one absolute value of time, and this is inherent in the basic nature of digital simulation. For example, representing an "analog" or real world electrical signal in a digital simulator would require the translation of real voltages into an equivalent logic context as in Figure 4.4.

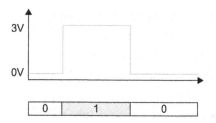

Figure 4.4:
Ideal case of digital signal representation

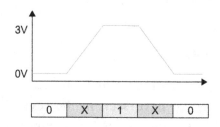

Figure 4.5:
The addition of "don't know" X states to an ideal digital representation

This is interesting because clearly there are threshold and rise-time or fall-time issues to consider in practice, but the logic signal does not exhibit any of those details; it is either 1 or 0.

It is possible to add an extra state to the logic, "X", which is often called "don't care", but which should probably be more accurately defined as a "don't know" value. This can be used to provide a more realistic value during particularly longer transitions such as that in Figure 4.5.

An interesting problem with logical resolution is what happens when multiple signals of different values are connected to the same "net". How is this to be resolved in practice? Consider the simple example of two logical gates connected to the same net in a system in Figure 4.6. What is the value on the net?

This is dependent in practice on a *logical resolution table*, which simply looks at the list of connected signals and, depending on their values, defines the appropriate output. For example, in the example in Figure 4.7, if both signals are "0" then the output will be "0", and if both are "1", then the output will be "1". This

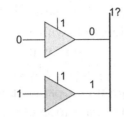

Figure 4.6:
Logical resolution example

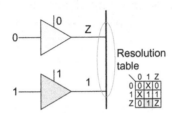

Figure 4.7:
Logical resolution with the addition of "Z" and "X"

is obvious so far, but what if one is "1" and the other is "0"? In that case we must *define* the state of the output, and this could be defined as "X" (i.e., don't know) in this simple logical example. The obvious question is how can we incorporate another state such that we can avoid this problem in reality, just like a tri-state driver output? The answer is to add one further possible value – high impedance "Z". We can therefore modify our example and incorporate a more sophisticated logical resolution to the problem as shown in Figure 4.7.

4.3.7 Generic Versus Component Models

A model can be defined in two forms, one of which is a *generic* model that is the general behavior of a physical type of part, for example an operational amplifier or MOSFET model. The other type is the *component* or characterized model. This is where the model parameters of a generic model are characterized such that the model behavior matches that of a specific physical part. Using the example of an operational amplifier, the generic model may be a simple macromodel with the specific component model defined with parameters for the real part (e.g., 741 op amp).

To develop this distinction further, a *generic* operational amplifier model would often consist of a gain function with perhaps a couple of equations to define the frequency response, a section of the model to allow power supply limiting, input impedance, and output impedance. Each of these aspects of the model would then have an appropriate parameter (e.g., the gain) so that the engineer could specify how much gain to assign to the model. This is in contrast to a *component* model, where a real device, for example the ubiquitous 741 operational amplifier, has been characterized in detail, with the parameters for gain, input impedance, output impedance, etc., being defined and fixed. When this *component* model is used, the engineer knows that its basic behavior is consistent with a real part.

4.3.8 Models and Effects

When models become more complex, we then have choices about how we implement the behavior: whether to use a structural-based strategy or a more equation-oriented one. However, there is a more subtle distinction that we can make in the definition of constitutive model elements which can define such an element as a model or an effect. The term "effect" is one that we will use frequently, so it will be introduced here in the context of describing models in general. In modeling, we can consider an element that stands on its own as a "model"— with connection points, parameters, structure, and behavior. We can instantiate it in a simulation and it will exhibit the behavior we assign to it. When we macromodel in the traditional fashion, we instantiate models together to create a hierarchical model of something else. However, we can also create and utilize an element called an "effect", where this may not be a complete standalone model, but does exhibit some specific aspect of behavior than can be used *in* a model. This is an interesting concept, as an effect may consist of a structural topology, parameterized equations, and connections, like a model. However, it is not necessarily complete in its own right in that it does not represent a physical device, but rather an effect that the device contains. The key difference is in parameters and ports. Parameters for an effect are really just constants because there is not a container, or wrapper, around the effect to allow parameter passing. Connection points in effects are merely variables, but are used just like electrical and other types of physical ports in models. However, again, they are not part of a model wrapper and become variables in the ultimately generated model. For example, a distinguishing characteristic that an effect may possess is a connection type that is a pure real number variable. As such, the effect may not stand on its own as a model *per se*. The advantage of considering effects in this way is that we can build a complex topology in a model, including multiple effects, very quickly, rather than having to take all the constituent parts and build them into a new model each time the effect is required. Effects can be hierarchical. Effects will be contained in a library for reuse much like full-fledged models. This unified way of modeling removes all burden from the designer to have to deal with these definitions. They simply compose models by instantiating other models and/or effects, generate code and stay focused on design activities, such as analysis and verification.

Examples of effects might include the specific behavior of a section of a semiconductor device that has no meaning outside the device. Another example is an environmental effect, such as temperature or radiation, which is then simple to add to an existing model to include the extra behavior associated with that effect. A third example might be a mechanical model of a beam which has the basic intrinsic mechanical behavior (the "model"), but which may have a thermally sensitive aspect — which would be implemented as an "effect". An example of an effect that has variables as its connection points is the comparator effect. Figure 4.8 illustrates several variations on comparator effects, including one that is purely variable at its connection points to those that are hybrids of variable connections and actual physical ports.

This aspect of modeling where the behavior is divided more subtly into models and effects will be discussed in much more detail with examples, later in this book.

4.3.9 Conservation of Energy

Conserved energy models can be considered to encompass the physical models required to represent the "real world" and, as such, include electrical, thermal, magnetic, mechanical, and other physical "domains". The single most

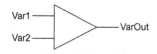

(a) Variables In and Out:
If (Var1>Var2) VarOut = 1 else Varout=0

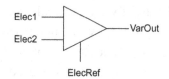

(b) Electrical In, Electrical Reference and Variable Out:
If (Elec1>ELc2) VarOut = 1 else Varout=0

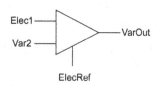

(c) Electrical In, Variable Reference and Variable Out:
If (Elec1>Var2) VarOut = 1 else Varout=0

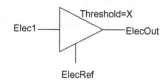

(d) Electrical In, Parameter Threshold and Elec Out:
If (Elec1> Threshold) ElecOut = 1 else ElecOut=0

Figure 4.8:
(a) Variables In and Out. (b) Electrical In, Electrical Reference and Variable Out.
(c) Electrical In, Variable Reference and Variable Out. (d) Electrical In, Parameter
Threshold and Elec Out

important aspect of these models is that in any system, energy must be conserved. This is a fundamental law of physics and states that in an isolated system, the total amount of energy must remain constant, and neither be created nor destroyed. In a similar way, we must (or at least are well-advised to) ensure that the models we create in continuous time simulators conform to this law.

We can illustrate this with a simple electrical example of a capacitor connected to a resistor as in Figure 4.9.

In this case we know that the voltage across both the capacitor and the resistor will be a voltage V. We also know that at any time the two fundamental equations of the individual elements will hold true:

$$V = I_R \cdot R \tag{4.1}$$

$$I_C = C\frac{dV}{dt} \tag{4.2}$$

Obviously, in this circuit, the voltage is the same across both elements; however, owing to Kirchoff's current law, we can also say that $I_R = -I_C$. We also

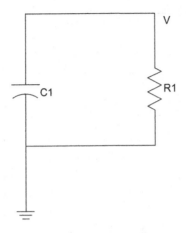

Figure 4.9:
Circuit example to illustrate conservation of energy

know that the energy stored in a capacitor is ½CV². But how do we calculate the energy consumed in the resistor? The answer in many cases is that we need to calculate the integral of the power over the period of operation:

$$Energy = \int Power \cdot dt = \int V \cdot I \cdot dt \qquad (4.3)$$

Now we have enough information to establish whether energy is conserved in this system. If we apply an initial voltage to the capacitor of 1 V, then from the equation of stored energy in a capacitor we can calculate:

$$E = \frac{1}{2}CV^2 = 0.5 * 1 * 1^2 = 0.5\ J \qquad (4.4)$$

If we run a simulation of this simple circuit over time, we can calculate the power dissipated in the resistor (which is not really lost, but rather translated into a different form of energy − in this case heat) by taking the instantaneous product of voltage and current. We can then integrate this to establish the energy consumed by the resistor, and compare this with the energy remaining in the capacitor. If we look at the resulting energy graph over time, we can see that as the energy stored on the capacitor decreases, it is consumed by the resistor in equal measure (Figure 4.10).

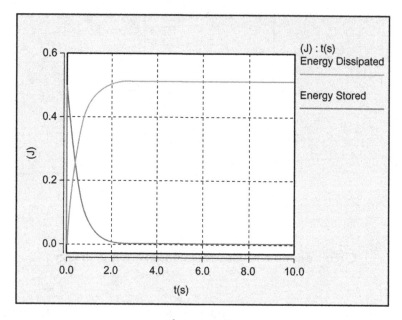

Figure 4.10:
Energy balance in a closed electrical system

This simple example is not as trivial as it first appears in that it can be easy to make a system that does not adequately model this conservation of energy correctly. The second point is that the energy is not "lost" from this system when it is consumed in the resistor – in fact, it is dissipated as heat, in other words it is converted into another physical form of energy.

This is an extremely important aspect when considering energy conservation in any model or simulation of a physical system, as it is very easy to make a mistake and end up with a physically unrealistic system that does not represent reality. While this concept is clearly important in the interconnections between blocks, how is this ensured *inside* models and the resulting equations? That is where the concept of a *branch* comes in to play, and this will be described in the next section.

4.3.10 Branches

The relationship between connection points and equations in conserved energy systems is defined with the concept of a *branch*. The "branch" is the

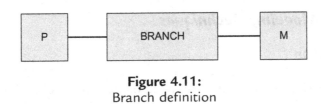

Figure 4.11:
Branch definition

definition of the variables that ensures that energy is conserved in the analog model when a simulation is carried out. This somewhat abstract concept is best illustrated using a simple example (Figure 4.11). Consider two analog connection points p and m. How can we represent a "branch" that defines the electrical behavior in terms of current and voltage such that the branch is equivalent to a resistor?

If we consider this simple case, what do we mean by the "branch"? It is the method by which the current from pins p to m and the voltage across pins p to m are defined with an equation. In electrical terms this is, of course, Ohm's law for a resistor ($V = I \cdot R$). Each of the pins (p and m) has a voltage and current associated with it, and there is, therefore, a voltage difference across the branch $v(p) - v(m)$; the current through the branch must be equal to the current through pin p and this, in turn, must be the same as the current through pin m. In general terms the variable *through* the branch is called the "through variable" and the variable representing the difference across the branch is called the "across variable". In this example, the across variable is the voltage (potential difference) and the through variable is the current.

We can therefore define the "branch" in a model using the following items:

- connection pins
- across variable definition
- through variable definition
- branch equation.

Using these four elements enables a complete branch definition, and this can, in turn, be used in a more complex model and will ensure the conservation of energy when the system is solved in the simulator.

4.4 Specific Modeling Techniques

4.4.1 Introduction

There are different methods of creating component models of real devices ranging from highly abstract models using simple, approximate equations to complex structural models including detailed physical effects. Behavioral modeling can take the form of direct implementation of model equations using HDLs or so-called macromodeling using building blocks (gain, derivative, combinations, etc.) to provide the correct behavior. Adding more structural elements to a behavioral model can be taken to the level of building as close a representation of the actual physical device as possible. Generally, adding more structural elements and the associated governing equations to the model leads to a more detailed model.

4.4.2 Behavioral Modeling Using HDLs

Implementing behavioral models directly using languages is rapidly becoming the technique of choice in a wide variety of design situations. It is already prevalent in high-density ASIC design (VHDL and Verilog are used in this case for digital designs), but the same techniques are becoming widely used in mixed-signal electronic design. The advantages of this type of approach are that a direct implementation of the equations in a model allows direct control of the accuracy and features in the model, with an efficient and fast simulation. In this case, accuracy does not necessarily get sacrificed at the expense of faster simulations, as that depends entirely on the accuracy of the behavioral equations.

As we will see later in this chapter, using HDLs in a language-based design flow can often be useful when a "programming" approach is taken.

4.4.3 Behavioral Modeling Using Macromodeling

Macromodeling has historically been used with SPICE-based simulators for one simple reason: the original SPICE simulator contained a fixed number of pre-compiled primitive models, requiring any new model to be built up from these basic primitives. The technique has been successfully applied in a wide variety of cases through necessity. Modern SPICE simulators, such as PSpice, do allow limited equation-based entry for "pseudo" behavioral modeling. The approach can provide similar improvements in simulation speed depending on the equations, but may give slightly less efficient and robust models than a HDL. This is because the underlying macromodel may not have the same stability as a tuned direct equation model, and extra components are required to achieve convergence (such as the ubiquitous large resistances to ground in SPICE). HDLs also generally provide convergence aids, which help nonlinear circuits in particular.

4.4.4 Structure in Behavioral Modeling

It is often the case that, despite the behavioral model being accurate and fast, a full physical model of the device is required. This is where adding more structure to the model becomes necessary. One example of this is using a library of effects to capture the behavior of an operational amplifier. These effects can be wired together to create a more detailed model than a simple gain block with output limiting. Another example of this would be an operational amplifier where, after a behavioral simulation demonstrates the basic performance is acceptable, the transistor-level design needs testing to ensure that the individual transistor parameters (length, width, technology) will meet the designer's requirements. This type of simulation can be much slower than behavioral methods and, in general, judicious choice of which type of model to use in different circumstances is required for optimum efficiency.

In the digital domain this might well take the form of a gate level structural description, such as the full adder shown in Figure 4.12. This approach can be very effective when a library of standard parts exists and can greatly speed up design re-use.

4.4.5 Signal Flow Models

Signal flow models, as the name suggests, are implemented such that the signal values propagate through the system in one direction. This is in contrast to the conserved energy approach used in most circuit simulators. The most common application for signal flow modeling is in high-level modeling of systems such as control systems. In general, we can consider signal flow models to be analog in nature; however, they are not always solved using an analog circuit simulator,

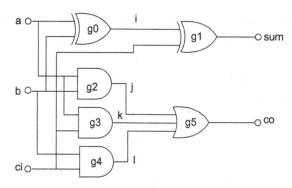

Figure 4.12:
Digital structural model example

as there are dedicated system simulators available that will simulate this kind of model, with the most commonly used being Matlab/Simulink.

The key aspects of signal flow models are that they have number-based connection pins (usually of a real number type), the connection pins have no units associated with them (although this is not always the case), the connections have a direction specified (input or output), and the model equations are defined to produce the input to output characteristics.

As signal flow models are often used to describe control systems, typical models consist of linear elements such as gain, filters, Laplace functions and also nonlinear elements such as backlash, hysteresis or dead-band. If we consider a simple gain element shown in Figure 4.13 as a first example, then how do we model this using a signal flow approach?

The way to consider any model is to think about each of the elements we have introduced already: connections, parameters, structure, and equations. In this case we have an input X and an output Y. In a signal flow model these are defined as having a real number base type and direction IN and OUT, respectively. The structure of a simple model such as this is not particularly important, as this behavior can be described with a single equation; however, in more complex models we may need to think about how we divide the model into different sections to make the overall modeling process easier and more efficient. In this example, the structure is a single model with one equation. The parameter K defines the gain of the block, and this is also defined as a real

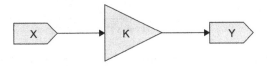

Figure 4.13:
Signal flow gain block

number in this case (although we could use an integer if we wished). The final step is to define the model equation and, in this case, a simple gain block allows us to define the expression in a single equation:

$$Y = X \cdot K \tag{4.5}$$

In this type of model, we can use the connection names directly, as we shall see this is not so straightforward for conserved analog models, but for signal flow leads to simple and easy-to-understand models.

Extending this type of approach to nonlinear or other more complex mathematical functions is essentially a function of the equation, and we can express our equations using whatever mathematical functions are available in the libraries associated with the simulator being used. In many cases this is based on the Institute of Electrical and Electronics Engineers (IEEE) *Math* library, which defines a range of mathematical functions for the user, such as trigonometric functions, log functions, etc. For example, we could define a logarithmic function based on the standard log function available in the *Math* library, if available in a simulator, by expressing the equation using the correct function name:

$$Y = \log(X) \cdot K \tag{4.6}$$

Clearly, the syntax and usage will depend on the final simulator to be used, but these issues, especially portability and compatibility, will be covered in much more detail later in this book.

The other main use of signal flow models is to describe systems using a Laplace or "s-domain" approach. This allows complex system analysis to be carried out, using the AC (frequency) analysis defined in the previous chapter. If we consider the simple example used previously of an ideal low-pass filter with the Laplace equation given in Eq. (4.7), how can this be modeled using a signal flow approach?

$$Y = \frac{K}{1 + s\tau} X \qquad (4.7)$$

where X is the input and Y the output as before, with K being the basic (DC) gain and τ the time constant of the filter. The "s" term is the Laplace operator, which can be represented using the derivative operator in a model definition. It also helps to describe the equation graphically to illustrate how this model can be put together. If we rearrange Eq. (4.7), we can see that the output Y is dependent not only on the input X, but also on a scaled version of the differentiated output Y.

$$Y = KX - s\tau Y \qquad (4.8)$$

This can be more easily seen using a simple schematic showing the different parts of the resulting model (Figure 4.14).

We now have a choice of how this can be modeled. We could take a structural approach and represent this equation using blocks for gain, difference, and Laplace operator, or we could implement the equation directly in a model. This may depend on the eventual simulator platform, but both options are generally available to designers. This type of equation implementation is a common tool to not only simplify the modeling process, but also the understanding of the model in question.

Care must be taken with these system level models as assumptions are often made about the appropriate usage of such models in different contexts. Consider the signal flow Laplace model of the filter shown in Figure 4.14.

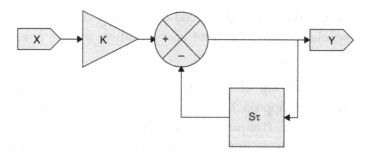

Figure 4.14:
Signal flow Laplace filter schematic

In this case, the Laplace nature of the model implies that it is *linear*. This is a fundamentally important concept as it means that a frequency response analysis can be carried out. As has been described in Chapter 3 this is a linear analysis that calculates the gain from inputs to outputs using a linear model, across a range of specific frequencies, resulting in a Bode plot, as shown in Figure 4.15.

It is easy to assume that *any* high-level model can be described using this approach, but care is required, as the instant a nonlinear element is placed into the design, then the linear analysis will only be valid for a *specific* operating point of the design. For example, if there is a gain term and it is nonlinear, if the gain term saturates, then the effective gain could be very small and the Bode plot would indicate almost no gain across the entire frequency range; however, it would actually be indicating a potentially unique situation, whereas under most *normal* conditions, the filter would operate as normal.

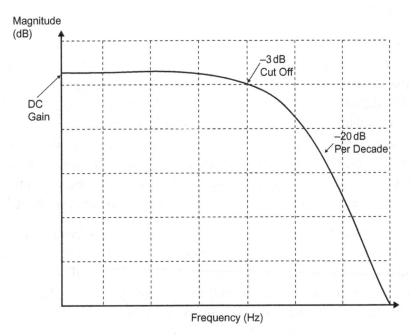

Figure 4.15:
Linear response—magnitude Bode plot

4.4.6 Analog Conserved Models

As we have seen previously in this chapter, analog models in a circuit network must satisfy the basic laws of energy conservation. This is clearly in contrast to the signal flow models which do not. Conserved analog models will generally use "branches" to define the basic relationship between the fundamental model equations and the connection points to enable this to take place. This is not to say that we can *only* use branch equations because we can use a mixture of branch equations, parameter expressions, sequential equations, and even non-conserved expressions in the model. However, the fundamental equations of the model need to be assigned to enough branches to ensure that all conserved analog connections have a connection to at least one branch. If this is not the case, then the simulator will be unable to find a solution. If we extend outwards from these continuous-time models, into the discrete domain, then a completely different form of modeling is required.

4.4.7 Discrete Models

Digital models, as we have seen in the previous chapter, must be based on a standard logic definition to enable appropriate resolution tables to be applied to connections to the same net, and also appropriate logical expressions to be carried out. Digital models will have an event-based connection with a suitable type definition (standard logic, real, Boolean logic) to characterize their detailed behavior. The connections will have a direction assignment similar to the signal flow, with IN and OUT, but also another type INOUT. In most simulators, the IN and OUT directions can *only* be used as inputs or outputs, respectively, and so, if a connection is required for both, then it needs to be of type INOUT.

The basic variable type used in logic models to express basic behavior is a "signal". This is a digital variable that can have its value defined in the event or transaction queue described in Chapter 3. Multiple signals can be assigned at the same time slot, so it is important to note this and not make the assumption that the behavior is defined in a sequential manner.

Digital logic model behavior can be defined in a number of ways, with the most common fundamental approaches being logic gate definitions and also logic expressions. Simulators often have standard libraries of simple logic elements that can be instantiated to construct basic gate level models using hierarchy. More generic models can be constructed using logical expressions. For example, consider the example of the full adder shown in Figure 4.12; this can be represented using logical expressions as follows (in VHDL syntax):

```
sum <= a xor b xor ci;
co <= (a and b) or (a and ci) or (b and ci);
```

In a similar fashion to the sequential and simultaneous options in analog models, in digital models there can be sequential and concurrent expressions. This is inherent in the concept of a "process" which is a standalone block of digital behavior that operates concurrently with other digital "processes". Signals can be shared and used in multiple processes, and so care must be taken to ensure conflicts do not arise. Another potential confusing aspect of this is that inside the process the code may appear to be sequential in nature (with "if" statements, "for", and other types of loop, for example). The model is usually defined in a concurrent manner, especially if "signals" are the basic type of variable being used. Sequential code can also be used, although this is often moved into procedures or functions in libraries, making it clear what part of the model is sequential and concurrent; however, it is possible to mix the two types of approach in a single digital model.

The final aspect to note in digital models is that often, ultimately, the model is being used for synthesis to hardware. This is important as the method of modeling can influence the final hardware directly. An important aspect of this is the use of latches and flip-flops. For example, the use of a "case"-type statement in a digital model will deliver an inferred latch in the synthesized logic, and, in practice, the designer should take great care to manage the sensitivity lists in processes and decision statements to ensure inadvertent latch insertion does not occur where possible.

4.4.8 Event-Based Models

An interesting specific type of model that seems to offer a hybrid of the analog and digital domains is the concept of event-based models. These models run in a digital simulator, so they use events and have discrete time steps; however, the values of the connections are numeric (real or integer) types. A classic example of this is sampled data system ("Z" domain) models. The obvious advantage for this type of model is speed, with the potential issue of poorer accuracy being a specific drawback. In many respects, the technique can be likened to the signal flow models, but with discrete connections. If we take the example we used to illustrate the signal flow framework, a simple gain block, how can this be represented in the event-based domain? The input and output are real types as for the signal flow model; however, they are discrete, rather than continuous, in time. This means, in practice, that the model will be run in the digital simulator and, as we have seen, this can be very efficient as an event will only be triggered inside the model when an event occurs on the input. The basic equation is still the same, except that there is no simultaneous equation to be added to the matrix in the analog simulator, thus increasing the model efficiency.

Clearly, the potential is there for extremely fast simulation and, as in all of these trade-offs in modeling, there is a potential reduction in accuracy at the expense of significant speed increases. The other issue to consider is the time resolution, as discussed in the previous chapter, which this can have an important effect on the overall simulation accuracy, as the events can only propagate as fast as the time resolution will allow. This can also affect the behavior when tightly coupled loops are involved.

With the definition of specific continuous time and discrete time modeling techniques it looks like we can model almost any type of system; however, there is an issue that we need to consider: What happens when we mix these types of model together in the *same model*? To address this important point, we must now consider: What happens when we do mix these types of models and, in particular, what happens when we attempt to connect them together?

4.4.9 Mixed-Signal Boundaries

4.4.9.1 Analog to Digital

The interface from the analog domain to the digital domain can be implemented in one of two ways. The first way, which is consistent with a hardware-oriented approach, is to consider a basic sampling approach. In this technique, a digital signal is used to "flag" the time at which a snapshot can be taken of an analog variable and then convert it into its event-based or digital equivalent. The conversion of an analog real number into a digital equivalent will be covered in detail later in this chapter, but the important aspect of this approach is that the sampling is controlled by a digital event.

The second common approach is to generate a digital flag, depending on the condition of the analog variable itself. A typical example is to set a digital flag when the variable crosses a threshold value, as shown in Figure 4.16.

When the input analog voltage crosses the threshold value, if the input is greater than the threshold, then the output is set to "1"; otherwise, the output goes to "0". As we have discussed in the previous chapter on mixed-signal simulation, this is very efficient as the event does not directly affect the behavior of the analog simulator. One note of caution is that some simulators will not automatically insert an analog time step exactly on the threshold point, so, in that case, the exact time of the threshold crossing may be interpolated (Figure 4.17).

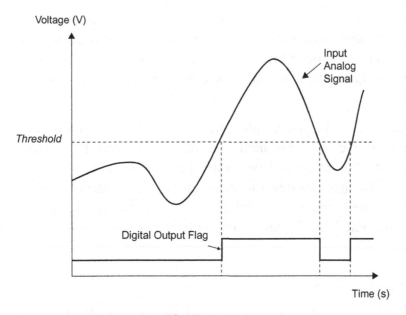

Figure 4.16:
Analog to digital threshold crossing

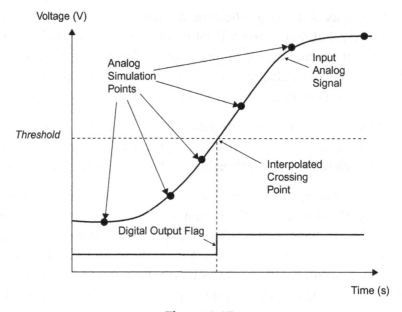

Figure 4.17:
Interpolated threshold crossing

This basic approach is useful for simple interface conditions, but often a more sophisticated understanding of the crossing is necessary. Many simulation languages also offer the ability to determine the direction of the crossing (rising or falling) with specific functions or parameters to the crossing function.

In addition, simply having an ideal threshold function can lead to oscillation, particularly if the crossing model is tightly coupled in a feedback loop. Therefore, it is normal to have some kind of hysteresis on the threshold function so that the crossing threshold on the rising signal is slightly higher than the nominal value and slightly lower for falling signals. This difference is illustrated in Figure 4.18 with the same analog input as used previously.

4.4.9.2 Digital to Analog

The digital to analog boundary in general is simple to implement on a basic level, where the analog value is set to the equivalent of the digital input when

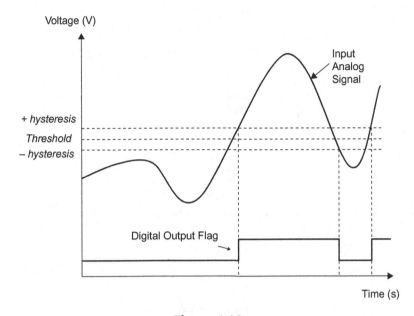

Figure 4.18:
Threshold crossing with hysteresis

the event occurs on the signal. This is simple in principle; however, there are some issues in practice that need to be considered carefully. The first is how analog values transition from one value to the next. In a digital simulation, being an event-based approach, values change instantaneously from one value to another; however, in an analog simulation, there is always a transition from one time step to the next. Therefore, in the digital to analog interface the most common method of ensuring this takes place in a controlled manner is to specify a transition time and define the rate of change from one value to the next. This is particularly the case for logic inputs and a rail to rail (V_{SS}/V_{DD}) transition of the analog equivalent variable.

Most HDLs provide some form of transition time for the analog variable to change in, and this provides a basic model for the analog to digital boundary. As most real life logic transitions are asymmetric, this has to be able to be defined with discrete rise and fall time values as shown in Figure 4.19.

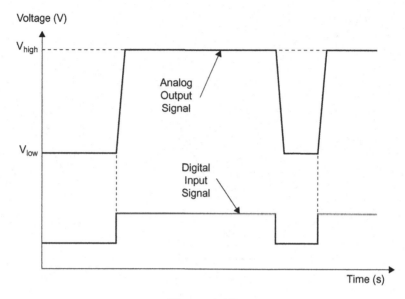

Figure 4.19:
Digital to analog transition

4.5 Forms of Representation

There is a plethora of modeling representations available to the designer, and, in many respects, the ideal representation does not exist and will probably *never* exist for everyone. Systems designers often use the commercial tool Matlab for capturing and simulating generic systems, and sometimes even quite detailed modeling effects in electrical or mechanical systems. In contrast, circuit designers in electronics may use a variant of the SPICE simulator containing primitive circuit elements, or power engineers may use the EMTP program [5], which requires a significant programming effort. The key point is that it is almost never the case that one single point tool (simulator generally) or even modeling language will be exclusively used for a system design.

In practice, a number of tools will be used, and it is the job of the design engineer to ensure that they understand the approaches and limitations of each one. In this section the various *types* of representation will be discussed, but the *specific* details will not be, as that is really the subject of specific textbooks available in each area.

4.5.1 HDLs

HDLs began to appear in the late 1980s with the original purpose being execut-able specifications of integrated circuit hardware. Digital design, in particular, has been pushing designers to ever-increasing levels of complexity. The mas-sive breakthrough of digital synthesis, whereby the physical design could be obtained automatically from a model written in a specific HDL, was one of the significant factors driving *designers* towards using a language-based approach, rather than a more typical schematic design methodology. Test is a second driver that came from the digital design community. This was, again, a result of the increased complexity in many systems, and the need for automatic test equipment to be able to not only complete functional testing (i.e., to see whether the design operated correctly), but, in particular, to be able to identify specific faults, which leads to the ability to establish fault coverage figures for a design, originating from the high-level HDL model.

The two main HDLs in current use are Verilog and VHDL. It is beyond the scope of this book to describe these languages in detail here; it is well-established that while the syntax may be different in almost every respect both HDLs are able to accomplish the same design goals and are generally inter-changeable for most software tools available today.

It is useful to look at some basic examples of a model written in both Verilog and VHDL in order to observe the similarities and differences for each. In this case we use a 4-bit counter, which is a digital electronic example of the kind of element (albeit very simple) that VHDL and Verilog are used by designers to encapsulate.

An important fundamental of the concept of HDLs is that they are designed to model *hardware*. A common misconception is to look at the model that *appears* to look like code, and to assume that it is dealt with in the same way. In fact, most HDL code is actually considered to be *concurrent*, whereas most programming code (in C for example) is, in fact, purely sequential (Figures 4.20, 4.21).

Both VHDL and Verilog are clearly targeted at the digital electronic design field. However, there has been an extension of these HDLs into the analog domain with the development of analog and mixed-signal extensions (-AMS)

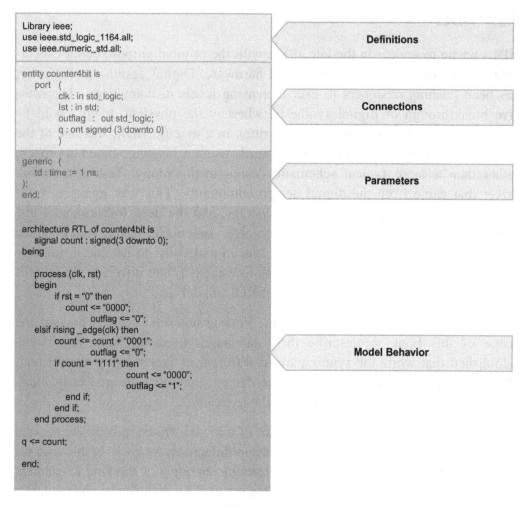

```
Library ieee;
use ieee.std_logic_1164.all;
use ieee.numeric_std.all;

entity counter4bit is
    port  (
            clk : in std_logic;
            lst : in std;
            outflag  :  out std_logic;
            q : ont signed (3 downto 0)
            )

generic  (
    td : time := 1 ns;
);
end;


architecture RTL of counter4bit is
    signal count : signed(3 downto 0);
being

    process (clk, rst)
    begin
        if rst = "0" then
            count <= "0000";
                outflag <= "0";
        elsif rising _edge(clk) then
            count <= count + "0001";
                outflag <= "0";
        if count = "1111" then
                        count <= "0000";
                        outflag <= "1";
            end if;
        end if;
    end process;

q <= count;

end;
```

Definitions

Connections

Parameters

Model Behavior

Figure 4.20:
VHDL code example of a 4-bit counter

to both languages. In fact, for Verilog there are two extensions, Verilog-A and Verilog-AMS, which have different functions. Verilog-A is designed essentially for purely analog modeling of devices, such as operational amplifiers, whereas Verilog-AMS is intended for more mixed-signal applications where the design contains digital and analog elements. The following code is a VHDL-AMS description of a resistor.

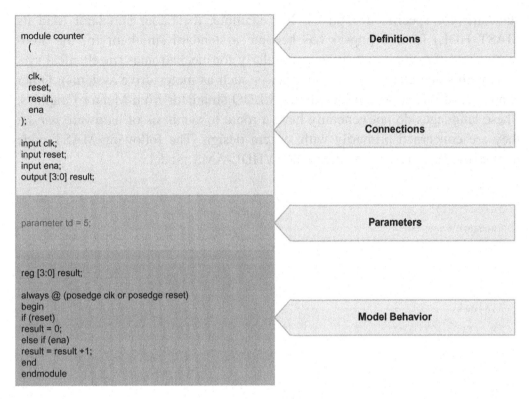

Figure 4.21:
Verilog code example of a 4-bit counter

```
entity resistor is
generic (rnom : real : = 0.0); - resistance
port (terminal p, m : electrical);
end entity resistor;
architecture beh of resistor is
quantity v across i through p to m;
begin
i = = v/rnom;
end architecture beh;
```

In addition to the Verilog and VHDL languages, and AMS variants, there are a number of proprietary languages that have evolved and become *de facto* standards for specific disciplines. For example, the Saber simulator (and its MAST HDL) from Synopsys has become a standard simulator for the automotive and aerospace industries, particularly for mechatronics (mechanical systems with some electrical content, e.g., − such as motor drive systems). Other examples of this approach include the ELDO simulator from Mentor Graphics. These languages do not generally have a route to synthesis of hardware *per se*; they are concerned primarily with system design. The following MAST code models the same resistor as the earlier VHDL-AMS model.

```
template res p m = rnom
electrical p,m
number rnom = 100;
{
var i i
val v v
values{
v = v(p) − v(m)
}
equations{
i(p->m) + = i
i: i = v/rnom
}
}
```

4.5.2 C and System-C

For those who have been in the modeling field for many years, it is interesting to note that, as in many technological fields, fashions come and go, and often "new" ideas are actually something that has been used previously in a different way. This is the case with programming languages. Before the commercial simulation tools became widely available and, in many respects, could be considered commodities, it was often the case that to solve a modeling problem design engineers would write small programs in C or FORTRAN to achieve their goals. This has never really changed over the last 20 years, although it may be that scripting languages, such as Python or

Tcl/Tk, have certainly become much more popular, as well as proprietary languages, including Visual Basic and a variety of object-oriented variations (.NET, C#).

When this is combined with the increased awareness of HDL design, where the HDLs themselves have more than a passing similarity to programming languages such as C, the step for designers to simply writing their own programs is a short one. In addition, the increasing prevalence of embedded systems, where the need exists to include software algorithms in conjunction with the hardware design, leads to a desire to "blur" the boundaries between the models as a result.

This has led to an effort to develop a platform-agnostic simulation library in C++, which can be considered under the generic "System-C" umbrella, which includes primitives, simulation techniques, and limited data-handling routines. The key advantages of this type of approach are obvious for embedded systems, and it is clearly a very flexible approach for those from a *programming background*. Again, the point must be stressed that, in many cases, there is no right or wrong approach, but simply a matter of preference.

System-C does have the clear advantage of being based on C++, which means that, in general terms, the designer can experiment with algorithms very effectively. However, the same limitations exist for *hardware*-oriented design, where the mechanisms within System-C are effectively no different from any other HDL simulator (Figure 4.22).

4.5.3 System Level Modeling: Matlab

The title of this subsection could, in many respects, be seen to be confusing and mixed up. You could quite rightly say that Matlab® is a software tool and therefore not a language; however, it is also true to say that for many system designers the underlying representation of models in Matlab is a *de facto* modeling language in its own right.

In almost every aspect the one common denominator across these system modeling languages is that there is generally a disconnect between the model and the hardware itself. This is not entirely the case, but while HDLs can be

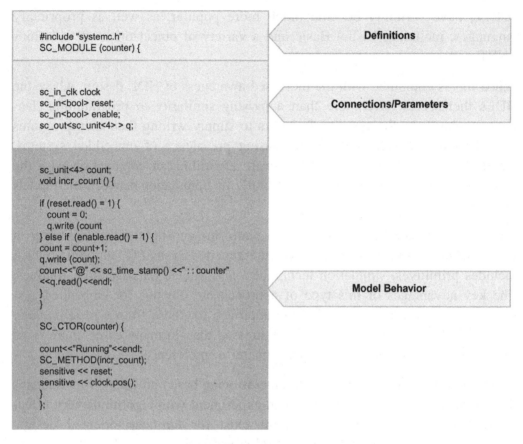

```
#include "systemc.h"
SC_MODULE (counter) {


sc_in_clk clock
sc_in<bool> reset;
sc_in<bool> enable;
sc_out<sc_unit<4> > q;


sc_unit<4> count;
void incr_count () {

if (reset.read() = 1) {
   count = 0;
   q.write (count
} else if  (enable.read() = 1) {
count = count+1;
q.write (count);
count<<"@" << sc_time_stamp() <<" : : counter"
<<q.read()<<endl;
}
}

SC_CTOR(counter) {

count<<"Running"<<endl;
SC_METHOD(incr_count);
sensitive << reset;
sensitive << clock.pos();
}
};
```

Definitions

Connections/Parameters

Model Behavior

Figure 4.22:
SystemC example of a 4-bit counter

synthesized directly into hardware, in many cases high-level system models are used for design and modeling, not for design synthesis. There are specific cases in Matlab where individual elements can be synthesized, but what this means in practice is a translation into a HDL and then a conventional synthesis procedure.

Taking Matlab as the first system modeling approach, the most important thing to remember with Matlab is that it started as a mechanism for encapsulating algorithms in a format of libraries that could easily be re used. The libraries were essentially focused on mathematics and algorithms, so when the available functions are investigated this becomes apparent. Over the years, the libraries and supporting mechanisms have been extended to a much broader range of applications from mechanical systems, hydraulics, advanced power systems, and electronics. There has also been a shift in the usage of Matlab from a mathematical modeling engine using textual input to a graphical approach called Simulink. This graphical representation approach will be discussed later in this chapter, so, for now, we will concentrate on the textual approach.

For example, the RC example we looked at previously could be defined in terms of a mathematical function that defines the basic equation for the differential equation network as defined here:

$$\frac{V}{R} = -C\frac{dV}{dt} \tag{4.9}$$

Rearranging this equation gives an expression in terms of the voltage differential:

$$\frac{dV}{dt} = \frac{-V}{RC} \tag{4.10}$$

(where, of course, there is an initial condition voltage on the capacitor which can be defined as Vinit). This equation can be implemented mathematically in Matlab using a simple function:

```
function vdot = eq1(t,v,r,c)
vdot = -v/(r*c);
```

And then a solver model can be used to calculate the relevant voltages and energy in the system:

```
r = 1.0;
c = 1.0;
vinit = 1;
tmax = 10;
n = 100;
deltat = tmax/n;
tspan = [0:1:n];
options = odeset('RelTol',1e-2);
```

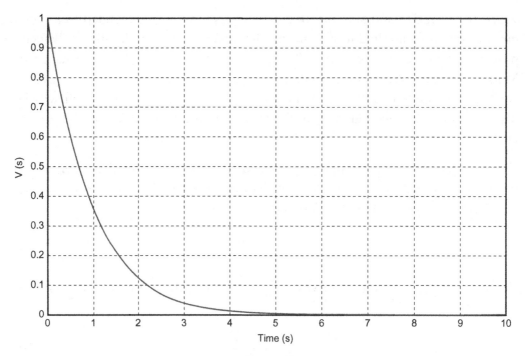

Figure 4.23:
Discharge voltage in RC circuit modeled in Matlab

```
[t1,v] = ode23(@eq1,tspan*deltat,vinit,options,r,c)
v2 = v.*v;
ecc = 0.5*c*v2;
i = v/r;
pr = i.*v;
er = cumtrapz(t1,pr);
plot(t1,er);
grid
```

Using this model we can predict the discharge curve of the voltage as before (Figure 4.23).

In more recent years, the use of Matlab has exploded owing to the addition of a graphical front end to this powerful mathematical engine called Simulink. This has enabled complex system level models to be implemented and simulated in Matlab.

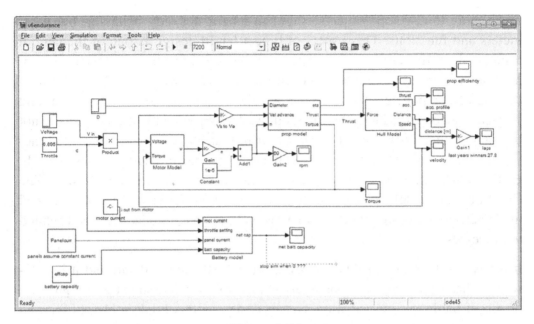

Figure 4.24:
System model example in Matlab/Simulink

A typical system example is shown in Figure 4.24.

This model consists of recognizable standard blocks, such as adders, gain, and constants, and also more complex user-defined hierarchical functional blocks. This combination of extensive libraries and hierarchy provides a very useful tool for system designers to quickly put together a high-level behavioral model of a system and understand its overall function.

4.5.4 System Level Modeling: UML

UML or the "Unified Modeling Language" has developed an increased following in the system design world over recent years as an attempt to provide a unified approach for modeling systems that would encompass both hardware and software elements of the design process. UML is really a collection of methods that make sense in all aspects of systems development and can be seen in an "object-oriented" perspective. In modern software design, rather than code being written in an old-fashioned functional manner, the recent trend has been much more to develop "objects" that can inherit not only behavior, but parameters and functions. This aspect of object-based design is one of the central aspects of UML-based design. The next related, important aspect is that of how objects are implemented, and, in the case of UML, this is done through the use of components. Components are the major elements in a system, and they can be connected together or "deployed". Objects and components are not the same thing. They are subtly different in that objects can encompass functionality and communication, as well as being instantiated in a system. In UML collaboration between objects is handled through message passing. The final form of model description in UML is that of activity. This is described using activity diagrams where decisions and actions are described and related to processes.

It is beyond the scope of this book to discuss UML in any detail, but it is noteworthy that while there are potential benefits to its employment at a high level of mixed hardware and software system design as this book is clearly focused on relatively hardware specific applications, UML is of much less relevance than, say, in a large-scale software engineering project.

4.6 Modeling Tools

4.6.1 Bottom-Up Tools

As many hardware design engineers start from circuits and schematics, it is often a natural starting point for understanding a design to begin at the lowest, most detailed level of description—circuit level. This is still a "behavioral" model in the sense that even transistor models are just that — behavioral. However, it is clear that this is probably the lowest practical common denominator for designers.

The obvious disadvantage of such an approach regardless of technology, is the generally slow rate of simulation. With often thousands of transistors, in many cases operating in highly nonlinear regions, or employed in switching circuits, large-scale time domain simulation is time-consuming and slow. The benefit is relative accuracy and the generally high level of confidence that the results represent something approaching reality.

Over the last decade or so, there has been some effort in the research community to develop software that can extract behavior from these low-level models, and provide a higher level of abstraction suitable for faster behavioral simulation. Examples of this include the Ascend software from the University of Arkansas, USA [1], the AMG tool from Grenoble Electrical Engineering Laboratory, France [2], several research endeavors involving tools for particular classes of circuits [3]–[7], and Orora's frequency domain modeling tool, Arana [8].

It is safe to say that there has not been much significant progress from a designer's perspective, however, as very few commercial bottom-up modeling tools are readily available. Further research is required in some cases and technology transfer is needed in others to get advances into the field.

4.6.2 Top-Down Modeling Tools

Model synthesis is a "holy grail" of system designers, and one which has effectively been realized for the digital electronics community. Register Transfer Logic (RTL) synthesis has been in common use by digital engineers for nearly 20 years and has provided a revolution in productivity for this group of designers. In conjunction with automatic place and route software, the time required to turn a relatively high-level design description written in a HDL into silicon-ready layout, has reduced dramatically to potentially a few minutes for standard designs.

The same cannot be said for any other field, however. Some specific point tools exist for filter synthesis where there is a mathematical relationship that can be extended easily into specific technological domains; however, a general purpose system synthesis methodology does not exist. It probably never will, as the development of wider and wider scopes of engineering systems encompassing numerous technologies means that it becomes a divergent problem to solve.

4.6.3 Graphical Modeling

In contrast to model synthesis, graphical modeling tools have been around almost as long as graphical interfaces to computer systems in one form or another. It can be argued that a schematic capture tool is one form of model capture using a graphical interface, especially when blocks are representative of behavioral functions. It could be said that Matlab/Simulink falls neatly into this category of graphical modeling, where blocks and functions are used to represent complex functions at a component or system level.

Of perhaps more pertinence to this book are those graphical tools which capture design concepts and output a model that can be simulated in at least one simulator. These tools generally fall into two main categories. The first category is a tool that is for general purpose model creation. The second is a tool that is focused on characterizing a model to a specific part or component. This categorization is consistent with the two types of models introduced previously: generic and component. Component models are generic models that have been characterized to a specific set of data. This data may have originated from a data sheet or experimental measurements. The Synopsys Model Architect™ toolset falls into the component modeling tool category. Generic models are those that represent an entire class of circuits, such as an operational amplifier, or even a transistor model, such as BSIM3. Generic models for analog, digital, and mixed-signal devices can be created. One example of a digital generic model could be a finite state machine (FSM). In models such as this, a model can easily be synthesized from a state machine captured graphically. There are a number of generic FSM modeling tools which can do this [9−12]. A more general case exists where arbitrary behavior must be implemented in a model. For an arbitrary model a more comprehensive approach is necessary. This is where the ModLyng™ software is applicable. This modeling software allows the capture of equations, data, curves, variables, and expressions in a graphical form, from which models in a number of languages can be synthesized. This software will be used extensively later in this book.

4.7 Future Proofing

One of the perennial issues for design engineers working with real systems is that of product lifetime. From one extreme, the aerospace industry, where designs may be flying in aircraft for decades to commercial products with a much shorter lifetime, the issue for designers is how to maximize their productivity and effectiveness. This is true whether it is one design lasting a lifetime, or numerous designs incrementally improving over several years. *Future proofing* becomes an important issue, which is the insulation of the designer to changing platforms and technology.

4.7.1 Common Frameworks

Clearly, the framework within which a designer operates is critical to the long-term success of a design. If the designer has to repeatedly change and update the framework, then productivity will be reduced radically. Whichever approach is used, longevity is vitally important and the ability to adapt to change equally so.

4.7.2 Libraries

One of the most powerful weapons in the designer's armory is that of libraries. When a model or component is created and validated, the designer knows that this could be re-used, even it if requires rework. This is a huge benefit to productivity, and the issue of libraries has become vital for engineers to succeed using whichever software platform or technology chosen. Libraries occur in a variety of locations. Sometimes there will be basic libraries provided by technology vendors (e.g., integrated circuit fabrication companies or chip suppliers), intellectual property provided by companies specializing in specific technology, vendor-specific libraries to help designers use their software tools, and internal libraries at a corporate, design group, or individual level.

There is no "right or wrong" approach to library development, other than whatever elements are used, are fit for *your* purpose and can be maintained over the period required to be useful. In our experience, the most important aspect is for the designer to have control over the libraries they work with and have an ability to extend and organize them in the best way for *them*.

4.7.3 Standards

Language standards have been essential in the development of synthesizable models in the digital electronics field and, to a lesser extent, in other domains. Generally, a *de facto* standard emerges in most fields, and the official standardization process follows on behind this at a much slower pace. Examples of the ability of language standards to provide common frameworks for engineers include digital VHDL (IEEE Standard 1076) [13] and mixed signal extensions to VHDL (IEEE Standard 1076.1) [14].

4.7.4 Language Independence

In recent years the plethora of simulation tools has led to designers wishing to "mix and match" their models using a variety of languages and methods. Electronic design automation (EDA) vendors have been releasing progressively more flexible simulators that can accommodate this flexibility. To this extent, there is no real issue other than the rate of this progress by the EDA companies.

However, it is dangerous to assume that just because a model exists, it is actually any good. Lack of understanding of different languages can lead to false hopes of a model's promised quality, and language independence often leads to "lowest common denominator" models with limited usefulness. Having said that, a "language neutral" approach is often very productive, but leads neatly into what we think is the way forward for modeling and model-based engineering in practice–graphical modeling.

4.7.5 Graphical Representation

All engineers tend to describe behavior using diagrams. Whether it is simple schematics and block diagrams or graphs, or state machines, more often than not we resort to a pictorial representation. With this in mind, it is perhaps not surprising that there has always been a drive to develop design software that is essentially graphical in nature.

One of the anomalies in this has been the rise of multiple domain hardware description languages, such as VHDL-AMS, which has seen quite a bit of "programming-style" hardware engineering of multiple domain systems. While this has been popular in the academic community in particular (often well-versed in mathematical and programming techniques) it has had a less favorable reception in the wider design community. For practicing engineers it is all about getting the job done efficiently and effectively, and, in most cases, a graphical representation is the way to go.

New software tools, such as ModLyng, seem to offer an opportunity for design engineers to accomplish this graphical style of modeling and thereby provide an effective increase in understanding of their design, and, ultimately, better and more efficient execution.

Conclusion

In this chapter, we have reviewed some of the fundamental concepts in modeling from a theoretical and practical perspective. Some of the specific techniques will be the subject of a more detailed analysis later in this book; however, the important types of modeling (structural or behavioral) have been reviewed. Furthermore, details of how the interface between the analog and digital sections of mixed-signal models have been covered, enabling the designer to take a more intelligent approach to mixed-signal modeling in general.

We have also introduced numerous techniques for describing systems in fairly limited detail at this stage, but identified that there is both a desire and a need for more graphical approaches to modeling systems. The next chapter of this book will introduce such an environment and will introduce the key concepts involved for designers to be successful using this method.

References

[1] H.A. Mantooth, A.M. Francis, Y. Feng, W. Zheng, Modeling tools built upon the HDL foundation, IET Proc. Compu. Digit. Tech. 1(5) (2007) 519−527.

[2] A. Merdassi, L. Gerbaud, S. Bacha, Automatic generation of average models for power electronics systems in VHDL-AMS and modelica modelling languages, J. Model. Simul. Syst. 1(3) (2010) 176−186.

[3] J.R. Phillips, L. Daniel, L.M. Silveira, Guaranteed passive balancing transformations for model order reduction, Proc. Des. Autom. Conf. (2002) 52−57.

[4] J.R. Phillips, Model reduction of time-varying linear systems using approximate multipoint Krylove subspace projectors, IEEE/ACM Int. Conf. on Comput. Aided Des. (1998) 96−102.

[5] X. Lai, J. Roychowdhury, Macromodelling oscillators using Krylov-subspace methods, Proc. Asia and South Pac. Des. Autom. Conf. (2006) 527−532.

[6] J. Roychowdhury, Reduced-order modeling of time-varying systems, IEEE Trans. Circ. Syst. II: Analog and Digital Signal Processing. 46(10) (1999) 1273−1288.

[7]   N. Dong, J. Roychowdhury, General purpose nonlinear model-order reduction using piecewise-polynomial representations, IEEE Trans. Comput. Aided Des. Integr. Circuits Sys. 27(2) (2008) 249–264.

[8]   Arana Behavioral Modeling Platform, Orora Design Technologies, Inc., <http://orora.com/index.php?page=products-arana>.

[9]   Qfsm Graphical State Machine Editor, <http://qfsm.sourceforge.net/>.

[10]   Aldec State Machine Editor, <http://www.aldec.com/technologies/Technology.aspx?technologyid=90f4c713-4431-404b-b11c-167db6c8a565>.

[11]   StateForge Finite State Machine Diagram and Code Generator, <http://www.stateforge.com/>.

[12]   Ease Block and State Diagram HDL Entry, <http://www.syncad.com/hdlworks_ease_block_diagrams.htm?gclid=CJXFs-3hkawCFSwCQAod2hgCsw>.

[13]   IEEE Standard 1076, "VHSIC (Very High Speed Integrated Circuit) Hardware Description Language or VHDL", Latest Revision 1076–2008, IEEE.

[14]   IEEE Standard 1076.1, "VHSIC (Very High Speed Integrated Circuit) Hardware Description Language or VHDL – Extensions to Analog and Mixed Signal", Latest Revision 1076.1-2007, IEEE.

Further Reading

Electromagnetic transients program reference manual: (EMTP) theory book, W.D. Hermann, Bonneville Power Administration, Portland, OR, USA.

Modeling Approaches

This section of the book will describe and illustrate the variety of modeling approaches needed to fill the proverbial toolbox of engineers performing model based design. These approaches include both implementation methods as well as representation approaches. For example, the historical method of creating models has been through text editing of a modeling language of some type. However, there also exist graphical modeling tools that can, for those not yet experts in languages, make the engineer almost instantly productive in creating the various design representations they require to analyze and design their system.

The level of representation that a model will have depends on where you are in the design process and how much granularity you are seeking from computer analysis at a particular phase in the design. For example, when analyzing the system requirements and refining them for the purpose of determining the system's architecture, fairly high level models such as signal flow, Laplace or z-domain models, or block diagrams.

After reading this section of the book, the reader should be able to effectively create many types of models and be aware when one type of model is required versus another. Chapter 5 begins the section by describing graphical modeling as an implementation method. It introduces you to some tools that perform this task and illustrates the use of graphical modeling techniques. Chapter 6 is focused on the creation of system level models. Chapter 7 is a large chapter on analog modeling using continuous-time methods. It also covers multi-domain modeling of such systems as thermal, mechanical, and magnetic. Chapter 8 covers event-based modeling. The primary focus is on digital systems and sampled-data systems. Chapter 9 goes to the next step and covers what refer to as fast analog modeling methods including finite difference approximation, averaged modeling, and event-driven analog modeling. Chapter 10 covers optimization methods that, while not specifically a modeling approach, are nonetheless important in a model based design environment. This is because by including certain information in a model needed for optimization, robust design and design centering is enabled. The final chapter of Section 2 covers one of these important pieces of information – statistical modeling.

CHAPTER 5

Graphical Modeling

Einstein's Razor: "The right model is the simplest that will do the job, but no simpler." While it is often tempting to use complex and highly mathematical techniques to develop advanced models, more often than not this leads to poorly constructed, overly complex, and ultimately inaccurate models. Using a graphical approach is much more in keeping with an engineering philosophy, and not only has the benefit of displaying all the important information in a readily accessible way, but allows intrinsic validation of the model structures within the framework of the graphical interface and serves to document the model more clearly for others to understand.

5.1 Introduction

As we have seen thus far in this book, modeling is a complicated and sometimes difficult business to understand and get right! On top of the concepts involved, there is also the thorny issue of which simulator to use and, related to that, which modeling *language* to implement the models in. The plethora of choices available to the design engineer mean that while it is perhaps straightforward to identify a specific language or simulator to design or analyze a specific problem, it is less easy to define a single language that can be used for *all* the problems a typical project might require to be solved.

In addition, the variety of modeling approaches (e.g., control system, conserved energy system, and digital logic, to name but three) means that it is very difficult for a typical design engineer to become expert in more than one or two domains or languages. The obvious answer to this issue is to use a graphical technique that abstracts the modeling concepts to a level which can become essentially language independent. This chapter will develop this idea and demonstrate how this could be achieved practically by design engineers.

5.2 Modeling on Top of Languages

While we have said that we wish to separate modeling as a technique from the underlying representation in language form, this is not a simple process in practice. And even if we do manage this, the model will eventually be implemented in one language or another. Therefore, it is useful to think of modeling graphically as being a layer above the language models to a certain extent, therefore making the model at least "language aware". This does have significant implications for cases where different languages have limitations and constraints that must be considered by the graphical modeling approach to be used.

In the context of the engineering design of systems, we can limit our initial study to the general areas of systems design, circuits, and devices, where these can all be represented by lumped models that have a conserved energy implementation (e.g., electrical, mechanical, or hydraulic), a control system implementation (continuous time, but not conserved energy, and certainly without units), and discrete systems (logic, sampled data systems, digital signal processing). There are other types of systems we could consider, and some of these will be reviewed later in this book. However, this initial context does define the sensible starting point for many engineering systems. Whichever domain we use, and whatever level of model is required, the first step is that of model abstraction, where we simplify and distill the fundamental concepts behind the system into a form that we can conceptually turn into a model.

5.3 Model Abstraction

If we consider the concept of model abstraction in more detail, this requires a certain amount of engineering knowledge, and also basic modeling understanding. For example, if the device to be modeled is a resistor, then how does an engineer conceptualize this very simple component as a model? This depends on the perspective and requirements of the engineer. If the resistor model is to be used in a simple filter circuit, at a fixed temperature, where the goal is to calculate a nominal frequency response, then a very simple model based on Ohm's law will suffice. If a thermally-dependent model is essential, then temperature coefficients must be considered. Perhaps if the resistor is to be fabricated on a silicon integrated circuit, then a more detailed physical model is essential. In all these cases as described, the underlying representation or level of abstraction will be different. In addition, the implementation and infrastructure of the model must be considered. Is this model to be used in a circuit simulator? In which case, the model will need to have conserved energy connection points. In many respects this will return us to the concepts of a multi-faceted model with parameters, connections, variables, equations, and structure that was discussed earlier in this book. Taking that into account, being able to capture these myriad aspects graphically becomes very attractive. In the next section of this chapter we will introduce a framework that can handle the graphical modeling aspects of the model abstraction process and show how this abstraction can separate the intellectual analysis of the device from the implementation issues of coding a model in a specific language.

5.4 Getting started with ModLyng

There are very few tools available that engineers can use to create graphical models. Some simulators or circuit design software have limited point model generators or translators, but these tend to be platform or language specific. One new modeling tool available is ModLyng^{TM}, which is designed specifically as a graphical modeling tool, generating a variety of output language formats. The software is a standalone program that is essentially self-contained, and allows the definition of individual models or libraries of models. A screen shot of ModLyng is shown in Figure 5.1.

As with any software user interface there is a need to gain familiarity with the layout and structure. However, it can be seen that there is a schematic type interface (called the topology editor); spreadsheets for ports, parameters, and objects (such as variables); a library navigator; and a model navigator. This enables the engineer to be able to observe the structure of the model graphically, but also to develop hierarchical libraries, test benches, and use predefined libraries already in place.

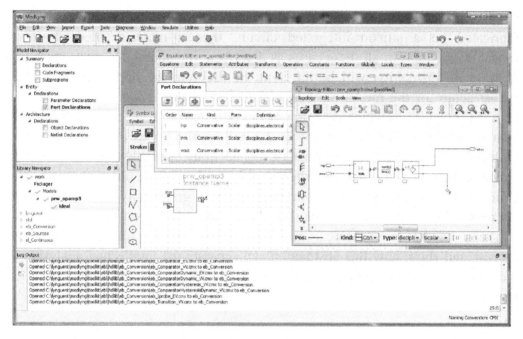

Figure 5.1:
ModLyng user interface showing some of the multi-paned windows to focus
the user's modeling activities on such things as symbols, ports, internal topology,
or equations

In order to illustrate the generic process of how to create models in ModLyng we will start from first principles and create some example models, and highlight the key functions involved in the important processes of creating a complete system model from scratch and using it in a simulator for design.

ModLyng is available on Unix systems, such as Linux, and also Windows. In Unix systems, after installation, the software can be run by using the *modlyng* command at the terminal command line. In Windows, the installation creates an icon on the desktop and a start menu item, either of which can invoke the software on a double click. The initial user interface is shown in Figure 5.2.

As can be seen from Figure 5.2, the user interface has a model navigator (empty), an initial library navigator, and a pane into which output is logged. To start the process of model creation is a classic choice of either starting from

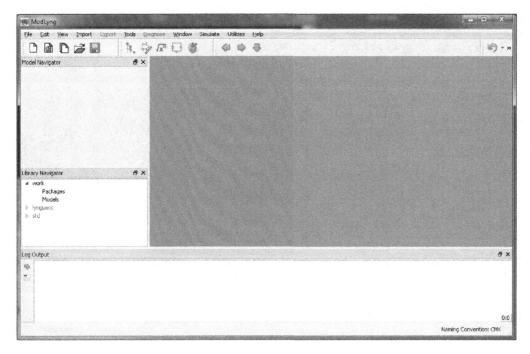

Figure 5.2:
Initial ModLyng user interface showing the starting point for a brand new model

scratch, or opening an existing model and saving with changes. In order to illustrate the process from the beginning, we will start by creating a model from scratch.

5.5 Creating a Simple Model

In the ModLyng interface, use the *File* command *New Model* to create a new empty model. This is shown in Figure 5.3. As can be seen from this figure, the model navigator now shows a tree with three main elements: summary, entity and architecture. The summary for the model is the area that defines top-level declarations and specific objects, such as subprograms or code fragments used later in the model. The model itself is divided into two main sections, the entity and architecture. The entity defines the name, interface, and parameters (i.e., the user interface to the model), and the architecture defines the core functionality of the model.

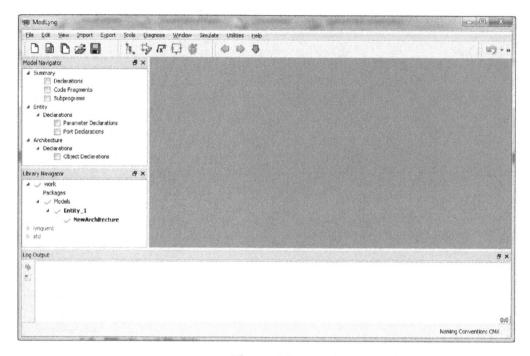

Figure 5.3:
New empty model in ModLyng

The library navigator shows a default library called `work`, which is used for all draft work, and a new model with the default name `Entity_1` is highlighted, with a default architecture called `NewArchitecture`. These are simply place-holders where the details of the model will be placed. In order to demonstrate the basic process in creating a new model, we will implement a simple ampli-fier, similar to an operational amplifier model, with a basic gain defined by a parameter `gain`, with the basic behavior of the single-ended output being the gain multiplied by the difference between the two input pins. The model is elec-trical in nature and is energy-conserved so we can use it in simple electrical cir-cuits. The first thing we will do is change the default name of the recently created entity from `Entity_1` to `op amp`, and the architecture name to `ideal`. We do these by right-clicking on the entity or architecture names and then selecting the *rename* option, bringing up the dialog window as shown in Figure 5.4.

This not only allows us to change the name of the entity or architecture, but we can also add comments and other information as required. These comments

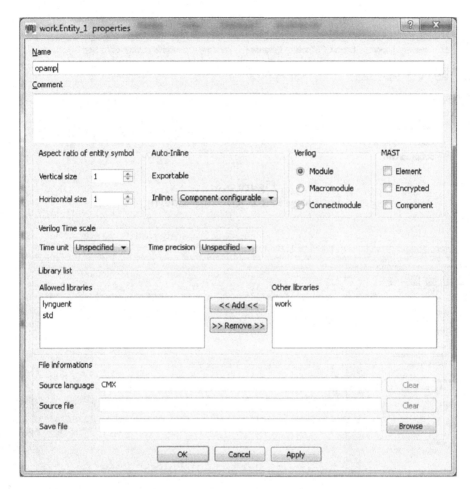

Figure 5.4:
Object properties form in ModLyng

will appear in the ultimately generated code for this model. In this example we can change the entity name to `op amp` and the architecture name to `ideal`.

At this point, there is a choice to be made as to how to enter the connection points for the model. We could either click on the *port declarations* in the model navigator to enter the port names as a list or we can use the schematic topology editor to add the ports on the model schematic. As this is the first time we've created a model, the schematic is probably the most familiar entry point, and so we will use that. Selecting the topology editor icon as shown in

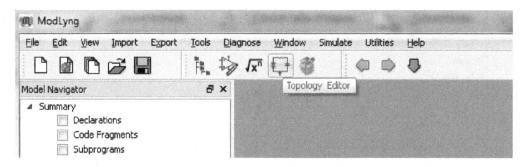

Figure 5.5:
Topology editor icon in ModLyng

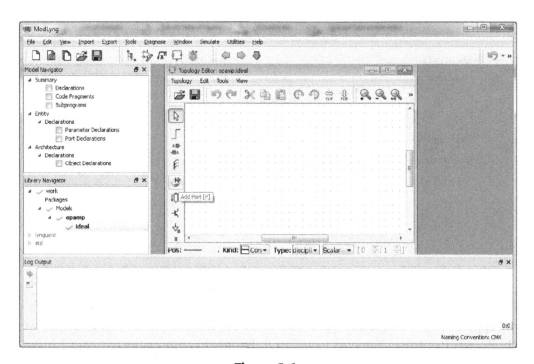

Figure 5.6:
Topology editor with Add Port icon selected

Figure 5.5 brings up a blank topology within the multi-pane workspace, and we can then select the *add port* icon as shown in Figure 5.6 to add ports `inp` (positive input), `inm` (negative input), and `vout` (output), respectively.

Figure 5.7:
Add Port entry form

When the *add port* icon is selected, the port entry form is opened, with the example of the positive input shown in Figure 5.7.

As can be seen from the form in Figure 5.7, the port name has been added (inp), and the type is defined as conservative and electrical. This is consistent then with any other electrical block and can therefore be connected in a circuit of electrical elements.

Using this approach the three pins inp, inm, and vout are added to the topology schematic as shown in Figure 5.8. Notice that although the output port was originally the same orientation as the inputs ports, we have used the "flip left-right" function in the topology editor to flip the symbol so it is oriented as an output on the schematic. It is interesting to now click on the *port declarations* check box in the model navigator, and the three new ports are now also listed in the port declarations form there. As can be seen from the table in Figure 5.9, new ports can be added here just as easily as in the topology editor. This also highlights the object-oriented feature of this integrated modeling environment (i.e., that a change in one place is reflected across the entire model definition).

Having defined the port connections, we can now add another necessity for the model which is the list of parameters. In this simple example, we only have

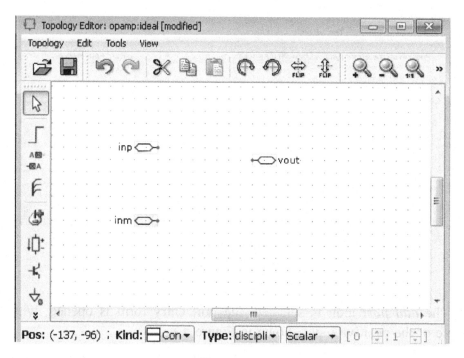

Figure 5.8:
Ports added to topology

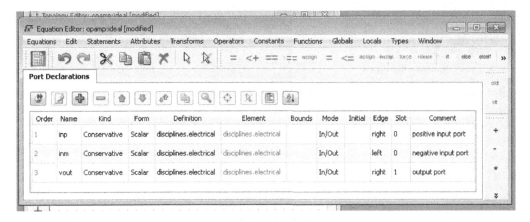

Figure 5.9:
Port declarations form

one parameter, which is the amplifier gain, defined by a parameter called gain. We can add this to the model by clicking on the parameter declarations tab in the model editor, and this brings up a similar form to the port declarations form, but for parameters to the model.

Clicking on the *add a new parameter* icon as shown in Figure 5.10, brings up the new parameter declaration entry in the form as shown in Figure 5.11.

To customize the new parameter is a simple matter of clicking on the name and modifying the initial property to reflect a default value, say 100 000, on the form. In this example, we can define this as shown in Figure 5.12.

At this point we have the complete entity definition in place (names, ports, and parameters), and we can concentrate on the model behavior. In the topology editor we can add components from libraries (a form of macromodeling − more on this later in the book) or define new equations. That is what we are going to implement in this case: define two equations, one for the input voltage (differential across the input pins) and one for the output voltage (with reference to ground).

In order to achieve this, we need to use a concept introduced previously in this book, that of the "branch". Recall that a branch is the relationship between two pins, a through and across variable (in electrical terms the current and voltage, respectively), defined in terms of an equation. In this example, we need to define two branches, one for the input voltage and one for the output voltage.

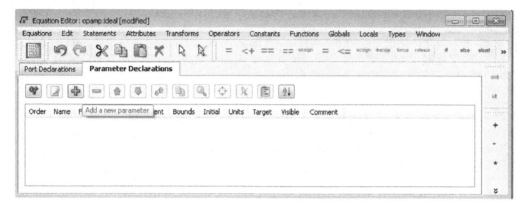

Figure 5.10:
Parameter declaration form

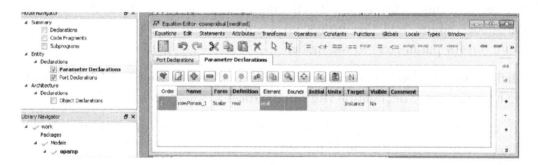

Figure 5.11:
New default parameter

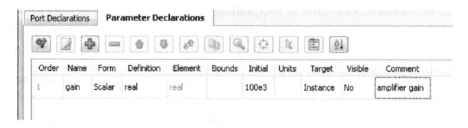

Figure 5.12:
Definition of the amplifier gain

In the topology editor, we need to select the *add branch* icon, as shown in Figure 5.13.

When the branch is added to the schematic, we need to connect it between two ports or signals in the topology, so, in turn, one branch must be added between ports `inp` and `inm`, and then between `vout` and a reference node (`ground`), also obtained from the topology editor. The default branch type is shown in Figure 5.14, and, as can be seen, the across variable is automatically the voltage *V()* and the through variable defined as a current *I()*. The alternative style is to define the names of the across and through variables manually, and this is accomplished by changing the branch style to "implied" from "explicit".

In this example, we will use the implied branch style and rename the through variable `i_in` and the across variable `v_in`. This now creates a "branch code fragment", which is the equation for this branch, and can be seen in the model navigator as shown in Figure 5.15.

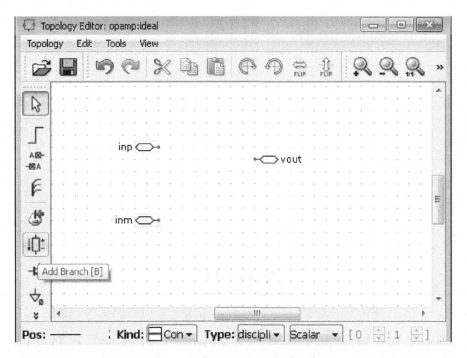

Figure 5.13:
Add Branch icon

Figure 5.14:
Default branch definition

The equation can be implemented in a variety of styles and, in fact, can be defined in the underlying HDL of choice for implementation, which can be very useful if that is what the engineer is familiar with. However, the generic

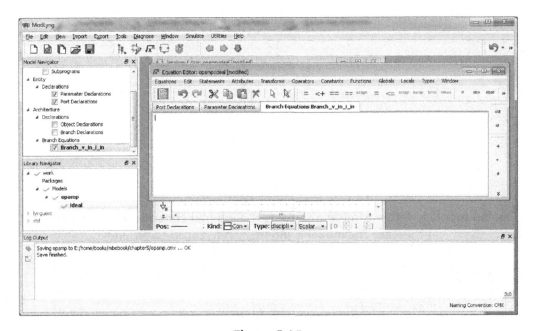

Figure 5.15:
Branch equation definition

form of equation is to use either a variable assignment (sequential ==) or a simultaneous equation (<+). This syntax was inspired by the operators used in VHDL-AMS. In this case, an ideal operational amplifier has zero differential input voltage, and so the equation would be of the form:

$$v\_in <+ 0$$

An output branch is also added, and then the basic equation of the amplifier can be implemented using the basic equation:

$$v\_out <+ gain^*v\_in$$

The resulting topology is shown in Figure 5.16.

The basic model is almost complete, and can be exported to a modeling language such as Verilog or VHDL. To complete the export use the *Export* menu and choose the export language of choice, for example Verilog-A for Spectre 5.1 (Figure 5.17).

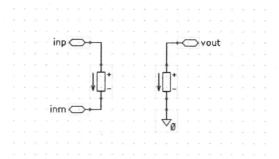

Figure 5.16:
Opamp topology within ModLyng showing the internal construction of the model; in this case the model is internally referenced to ground

Figure 5.17:
Choosing to export the model as Verilog-A

The resulting model is shown in the following listing.

```
/* pragma lynguent created by ModLyng-1.4.3-S443 */
/* pragma lynguent exported by Verilog-A Spectre 5.1 */
'include "disciplines.vams"
//an ideal operational amplifier
//****************************************************
//**
//* an ideal op amp for simple electrical circuit analysis *
```

```
//* *
//****************************************************
module op amp(inp, inm, vout);
//amplifier gain
parameter real gain = 100e3;
//positive input port
inout inp;
electrical inp;
//negative input port
inout inm;
electrical inm;
//output port
inout vout;
electrical vout;
electrical Ground_electrical;
ground Ground_electrical;
//
branch (inp, inm) Branch_v_in_i_in;
//
branch (vout, Ground_electrical) Branch_v_out_i_out;
analog begin
// pragma lynguent code fragment "Branch_v_in_i_in"
V(Branch_v_in_i_in) <+ 0;
// pragma lynguent code fragment "Branch_v_out_i_out"
V(Branch_v_out_i_out) <+ gain * V(Branch_v_in_i_in);
end
endmodule
```

The constructs in the language are fairly obvious based on the details we entered into the graphical tool. The port definitions, branch definitions and relations, and some comments are all there. And while this might be your first time reading Verilog-A, you can see how ModLyng can also assist in the teaching of this language, if appropriate. This model can now be imported into a Verilog-A simulator, such as Cadence Spectre, and used in a simulation. We have now created our first model in ModLyng.

5.6 Libraries and Models

Creating individual models is one thing, but to be re-useable we need to consider how to collect models into libraries that can be exchanged and used in more complex models. ModLyng supports the idea of a library, and we can demonstrate this by creating a new library and moving our recently created op amp model into it. In the *File* menu choose the option *new library* and this will create a new library placeholder in the library navigator (Figure 5.18).

If we right click on the library, we can rename the library to something specific, for example electrical_primitives, and then by right-clicking on the model op amp just created, the option *move to library* can be selected, which will allow the individual models to be moved into new libraries. The resulting library navigator view is as shown in Figure 5.19.

Now, while we have created a new library and put a model into the library, the most effective way of incorporating such a model into a new design is to instantiate a symbol. ModLyng provides the ability to create a symbol for a model by selecting the *symbol editor* icon (Figure 5.20).

This will invoke a simple symbol editor with the ports already roughly placed around the perimeter of the symbol area by default (Figure 5.21).

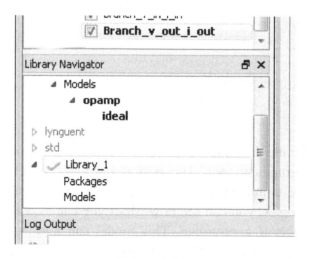

Figure 5.18:
Creating a new library

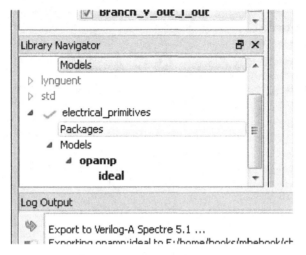

Figure 5.19:
New library with added model

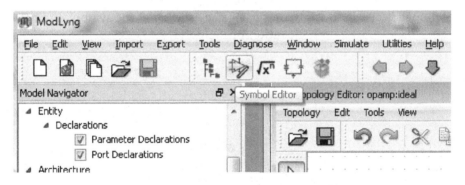

Figure 5.20:
Symbol editor icon

We can move the ports around and draw graphics to make the symbol more meaningful and useful for placing in hierarchical schematics, such as the familiar op amp symbol in Figure 5.22.

At this point we can now use this model in the schematic of a hierarchical model by selecting this symbol in the topology editor. For example, if we create a new model, and use the topology editor "add component" function, we can select the op amp model we have just created and instantiate it on a higher-level schematic (Figure 5.23).

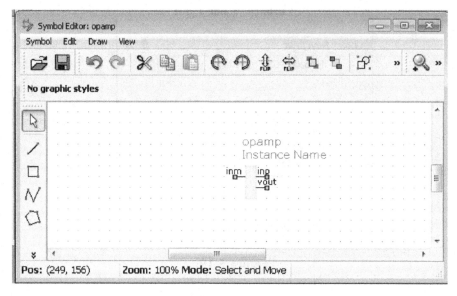

Figure 5.21:
Symbol editor window for composing the artwork of the symbol and the
location of the ports

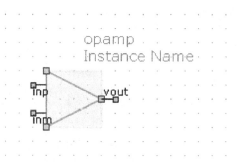

Figure 5.22:
Op amp symbol in ModLyng

From a modeling perspective this is incredibly powerful as we are now able to
build up a set of libraries which can be reused and validated extensively. This
is a massive time-saving approach, even if the ultimate model is equation-
based rather than graphical.

One of the interesting features of the ModLyng software is that the model can
be considered as a hierarchical model, which references previously defined

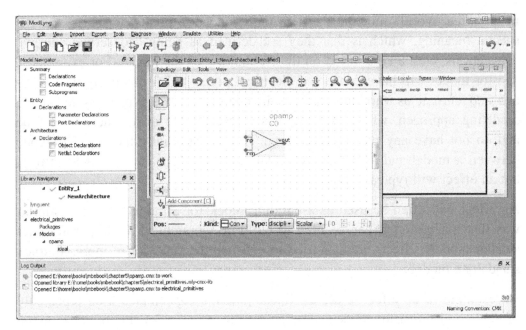

Figure 5.23:
Selecting the op amp symbol and placing it in the new model schematic

models, or it can convert a hierarchical model into what is called an *inlined* version. This is where the equations are "flattened" so that the complete hierarchy is implemented in a single model. This can be useful when the model needs to be delivered as a single piece of code without references to outside entities.

5.7 Effects and Models

One of the interesting discussions within modeling as a discipline is how to structure models and handle their different uses and implementations. In particular, it is important to be able to extend or modify models and to create generic behavior that is perhaps not a complete standalone model, but rather a specific "effect" that can be used in a wide variety of models.

To this end, within any graphical modeling environment it is essential to differentiate between standalone, self-contained "models" and these so-called "effects" which cannot, or may not, stand alone, but still need to be represented.

In any "real world" modeling system the connection points or "ports" have a special meaning, which is that they are defined as being "conserved energy" connections, with a type based on the technology of the model. For example, a resistor model will have conserved electrical ports, which obviously have voltage and current as variables. This is in contrast with a typical system-level modeling approach, where the variables or signals are of a generic real type and do not have any physical meaning. This is essentially the key difference between a model and an effect. A model can standalone in a physical setting, but an effect will typically not.

An effect will often be used as a building block for a model and, in a similar way that macromodels can be used to build up hierarchical models from other smaller models, effects are also used to enhance the behavior of more complex standalone models.

An example of an effect is the nonlinear junction capacitance used repeatedly in diode and transistor models. The governing equation of this effect is given in Eq. (5.1).

$$C_j = \frac{C_{jo}}{(1-(V_J/V_\phi))^m}$$

(5.1)

In this case, you can imagine simply making a capacitor model with this behavior rather than think of this as an effect. This is surely true and so this is an example of where the modeling engineer would have the choice to make an effect or a standalone model. However, the elegance of the graphical environment is that whatever choice is made the item — whether an effect or a model — is instantiated the same way and appears seamless to the user of the item.

Another example of an effect model is an assertion. An assertion is an exception handling mechanism in an HDL that can be used to issue warnings, errors or messages. The assertion can also be used to trigger alternative behavior in a model, such as a failure mode or an aging effect. This effect will not standalone, but can be created generically and reused in many settings involving fault modeling. One way in which such effects have been used is in modeling the effect of radiation on semiconductor devices and circuits. A radiation analysis is performed where different doses, dose rates, and single events are programmed into the test bench as a family of scenarios (i.e., such as different trajectories through space). The radiation effect reacts to this radiation and simulates the accumulation of damage and the appropriate behaviors are altered within a base model (say a transistor or circuit).

One of the advantages of a graphical modeling approach is that as designers we are not constrained by the implementation to either models or effects, but can, in fact, use both, using a hierarchical approach of model design in a schematic editor, as described in the next section.

5.8 Hierarchical Models — Using the Schematic Editor

As we have seen briefly already, it is possible to use a schematic editor to create models from branches, ports, and other basic behavior, but can we use the same approach to create hierarchical models using ModLyng? If we take the example of a first-order Butterworth filter, and implement it using the standard built-in SPICE primitives for resistor (R) and capacitor (C), plus the op amp used earlier in this chapter, we can create a hierarchical block of the Butterworth filter stage as shown in Figure 5.24.

Now when the model is exported, if inlining is not used, then the model will reference the lower-level building blocks (R, C, and op amp) from the libraries and will assume that coded models have also been exported. If the inlining

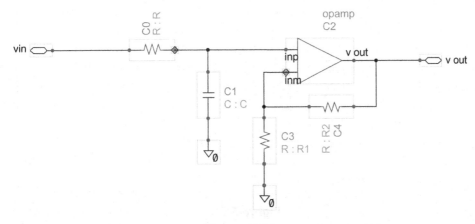

Figure 5.24:
Hierarchical model — first-order Butterworth stage

```
/* pragma lynguent created by ModLyng-1.4.3-S443 */
/* pragma lynguent exported by Verilog-A Spectre 5.1 */
`include "disciplines.vams"
module butterworth(vin, vout);
//
parameter real R = 1k;
//
parameter real C = 1u;
//
parameter real R1 = 1k;
//
parameter real R2 = 1k;
//
inout vin;
electrical vin;
//
inout vout;
electrical vout;
electrical Ground_electrical;
ground Ground_electrical;
//
electrical N4;
//
electrical N12;
parameter real inlined_IC_C1 = -1.0e38;
//
branch (vin, N4) inlined_Branch_V_I_C0;
//
branch (Ground_electrical, N4) inlined_Branch_V_I_C1;
//
branch (N12) inlined_Branch_V_I_C3;
//
branch (vout, N12) inlined_Branch_V_I_C4;
op amp C2 (N4, N12, vout);
analog begin
// pragma lynguent code fragment "(spl_R_resistor) resistance eqs"
// pragma lynguent code fragment "(spl_C_capacitor) eqs"
if (analysis("ic") && inlined_IC_C1 != -1.0e38)
V(inlined_Branch_V_I_C1) <+ inlined_IC_C1;
else
I(inlined_Branch_V_I_C1) <+ ddt((C * V(inlined_Branch_V_I_C1)));
// pragma lynguent code fragment "(spl_R_resistor) resistance eqs"
// pragma lynguent code fragment "(spl_R_resistor) resistance eqs"
// pragma lynguent code fragment "Branch_V_I"
I(inlined_Branch_V_I_C0) <+ V(inlined_Branch_V_I_C0) / R;
// pragma lynguent code fragment "Branch_V_I"
I(inlined_Branch_V_I_C3) <+ V(inlined_Branch_V_I_C3) / R1;
// pragma lynguent code fragment "Branch_V_I"
I(inlined_Branch_V_I_C4) <+ V(inlined_Branch_V_I_C4) / R2;
end
endmodule
```

Figure 5.25:
Butterworth filter inlined code example in Verilog-A

approach is specified, then the complete model will be exported as a single consolidated code model. The example model exported as inlined Verilog-A is shown in Figure 5.25.

As can be seen from Figure 5.24, we have defined the four key parameters for the filter as model parameters of the filter. This model can also have a symbol defined for use in a complex filter design (Figure 5.26).

In fact, we can use the symbol to create a test circuit (or test bench) to test the behavior in a simple simulation. We will add a sinusoidal voltage source to the input and measure the frequency response of the filter stage (Figure 5.27).

If we export this model we can carry out a frequency domain analysis (small signal AC), and this will provide the Bode plot of magnitude and phase versus

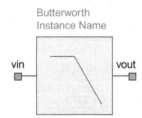

Figure 5.26:
Butterworth filter stage symbol

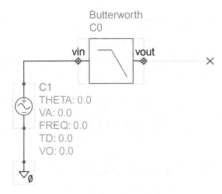

Figure 5.27:
Butterworth test circuit

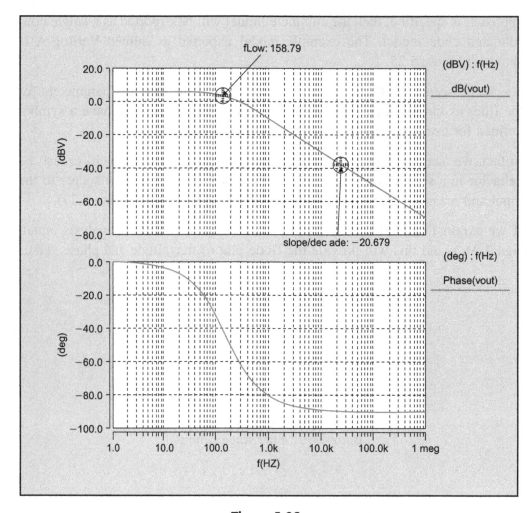

Figure 5.28:
AC analysis results of simulating the Butterworth filter stage

frequency. As can be seen from the results in Figure 5.28, the -3 dB point is measured at about 158 Hz, and the roll-off is approximately -20 dB/decade.

5.9 Test Benches and Model Validation

As we have seen in the previous section, a basic test bench can be used to carry out simulations to enable the designer to debug the model, and to understand its basic behavior. The approach of using "test benches" or "test circuits" is therefore an important concept to introduce at this stage, to ensure that the designer follows a logical sequence of tasks to validate the model.

For every model, there needs to be at least one test bench to carry out a test on the model behavior; however, often there is a need for a number of simulations to be carried out. In the Butterworth filter example at least the AC and time domain analyses should be carried out to ensure the correct operation of the model. This opens up the discussion as to what a test bench actually needs to contain. Obviously, the first and most important aspect is a circuit that has not only the circuit-under-test, but also has the relevant stimuli and loads to correctly exercise the model. The second aspect is the simulations that need to be performed. For example, in the Bode plot, we need to know the approximate range of frequencies to test the model. In this example, the -3 dB cut-off frequency is 158 Hz; therefore, it would be inappropriate to start the frequency sweep at 10 kHz. The related aspect of this is which analyses need to be undertaken. Is there a need for a linear, small-signal analysis, or is it necessary to run a nonlinear time domain simulation? Finally, the third aspect is how to measure the performance of the model. This is often done by the designer inspecting a graph or waveform, but, as we have seen, it is extremely useful to be able to automatically measure specific aspects of the simulation result (in this example that could be the -3 dB point). There is a technique used commonly in digital design of an "assertion", which is a very useful concept in the assessment of models. The basic idea is to carry out a test, and then have a logic output, which is effectively either "pass" or "fail". Assertions can be defined within the ModLyng graphical environment by placing specific assertion functions into the test circuit, and thereby provide pass/fail criteria for a variety of functions.

For example, we could add a crossing assertion block to the test bench. This allows a specific test point to be monitored automatically during a simulation. We can use the voltage crossing with a set threshold to check that the filter correctly allows (in this case) the signal to pass through enough to trigger the crossing of the 1.0 V threshold check (Figure 5.29). If we define a set simulation time (in this case 10 ms) and a set input frequency of 1 kHz, then in the simulation time there should be ten crossings corresponding to the positive parts of the output waveform. In the assertion, the test can be to count the

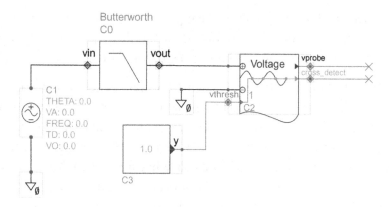

Figure 5.29:
Test bench with an assertion on the output voltage

number of crossings and then set the "cross_detect" detection only if that occurs. The placement and specification of these assertions defines the model validation criteria. This is precisely how executable specifications are captured for circuits. A set of assertions are instantiated and the pass/fail conditions specified that reflect the performance specs that the circuit, must meet.

Using the model for the filter the behavior including the test bench and assertions can be simulated. As can be seen from the waveforms in Figure 5.30, on each occasion that the threshold of 1.0 V is crossed, the internal variable in the assertion "count" is incremented until the final cycle of the test at which point the criteria have been satisfied and the detection logic signal is set to '1' and the test has been passed.

5.10 Examples

It is illustrative to look at some additional examples of model creation using the graphical modeling approach. The op amp is one circuit that all electrical engineers can easily relate to, but in the following examples we will introduce the instantiation and use of effects and model reuse as a more efficient starting point for model creation. This will prove to be the second valuable way in which model reuse is promoted with graphical tools. The first was the hierarchical instantiation described earlier. The focus of these next two examples is

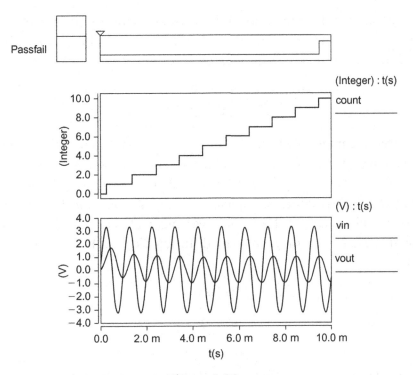

Figure 5.30:
Simulation results of filter with threshold crossing assertion test

on diodes in order to keep the topic simple and the focus on the modeling concepts, but we could have chosen any device, circuit, or system.

5.1.1 Example 5.1

In our description of design analysis in Chapter 3 we used a simple nonlinear diode to illustrate some concepts. Let us return to this simple nonlinear diode, model it, and add a nonlinear junction capacitance to it in the form of an effect. This nonlinear capacitive effect was introduced earlier in this chapter. The governing equations for these effects are repeated here as Eqs (5.2) and (5.3) with some slight variable naming modifications for convenience.

$$i_d = I_{SAT}\left(e^{\frac{v_d}{nV_T}} - 1\right) \tag{5.2}$$

Eq. (5.2) is the well-known exponential current relationship for a *p-n* junction, where I_{SAT} is the saturation current, V_d is the junction voltage, $V_T = kT/q$ is the thermal voltage, and n is the emission coefficient.

$$C_j = \frac{C_{jo}}{(1-(V_d/V_j))^m} \tag{5.3}$$

Eq. (5.3) represents the capacitance of the depletion region. (Note: our simple implementation of the diode is not charge-conserving here because we are implementing a capacitance rather than a charge-based model for simplicity. A charge-conserving, more complete implementation can be found in the Appendix to this chapter.) There are two more effects that the simple diode has, which are charge storage due to diffusion Eq. (5.4) and series resistance Eq. (5.5).

$$Q_s = \tau \cdot i_d \tag{5.4}$$

$$V_s = R_s \cdot i_d \tag{5.5}$$

where $V_s + V_d = V_{pn}$, the total drop across the diode. We will illustrate the creation of Q_s and C_j as effects and then place them into the diode model just as if we were instantiating actual models into a macromodel. This fine-grained, object-oriented composition approach is what adds tremendous efficiency over modeling directly in languages because pulling things from a library and dropping them down onto a schematic and having all instance-based book-keeping performed for you as a user makes this process more efficient. This keeps the designer's attention on the conceptual level of the modeling task. There is no need to switch contexts between programming and modeling.

Following the same flow as described for the op amp in this chapter, we being by defining the model name (diode), ports (dp, dm), and parameters (rseries, isat, n, m, cjo, vj, tau). We will begin by instantiating a branch for the drain current and a branch for the series resistance (these also could be reusable effects, but we simply implement them here to illustrate a non-reusable effect). These two branches are placed in series creating an internal node which we have named pi for dp internal. The actual diode junction voltage V_d is taken

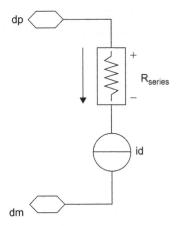

Figure 5.31:
The starting implementation of the diode model without capacitances

from `pi` to `dm`. The voltage V_s is taken across the resistance and the total voltage drop across the diode is the sum of these two as indicated earlier. We will now save this model away for the time being and return to it once we have completed the two capacitive effects. The work we have done on the diode model so far is summarized in Figure 5.31.

The first effect we will address is the junction capacitance C_j. We start a new model just like before and create the effect in the very same way we create a model. It has two ports (`p, n`), three parameters (`m, vj, cjo`), and is called `cjunction`. Further, it consists of a branch with the expression from Eq. (5.3) implemented, but as it is an electrical effect it must satisfy the rules of an electrical model. As such, the equation

$$i_{cj} = C_j \cdot \frac{dV_d}{dt} \tag{5.6}$$

must be implemented as the branch equation, while the capacitance equation of (5.3) is the nonlinear capacitance C_j. This work is summarized in Figure 5.32 where the equations and simple topology are shown. This model satisfies the rules of a standalone electrical model in that it has enough equations to describe the behavior of the branch between the two nodes and therefore it can be saved as a separate model in its own right. The basic implementation of

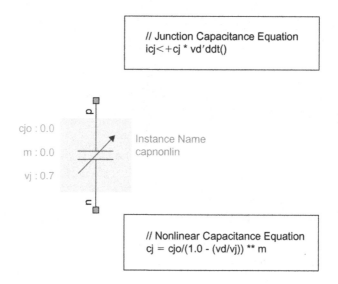

Figure 5.32:
The implementation of the nonlinear junction (or depletion) capacitive effect

the model consists of the branch equation Eq. (5.6) and sequential code for the nonlinear capacitance itself (Eq. (5.3)). The topology of the model is therefore shown in Figure 5.32.

Next, we create the diffusion, or charge storage, capacitive effect. This is actually most easily created using the charge expression of Eq. (5.4). The branch equation is:

$$i_s = \frac{dQ_s}{dt} \tag{5.7}$$

Figure 5.33 summarizes this simple effect based on two ports (p, m), one parameter (tau), and the name qstored.

We now return to our diode model and instantiate these effects the same way we instantiate any other library element as shown in Figure 5.34. Note that the symbol created for each of these capacitive effects is different. One indicates a nonlinear capacitor (the one with the arrow through the symbol), while the other indicates a charge-based capacitive effect (the capacitor with the box around it). These graphics provide information to the designer at a quick glance at all steps in the design in the same way that there are different

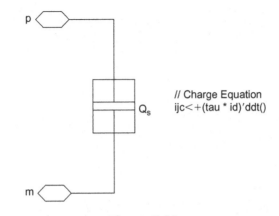

Figure 5.33:
The implementation of the charge storage effect

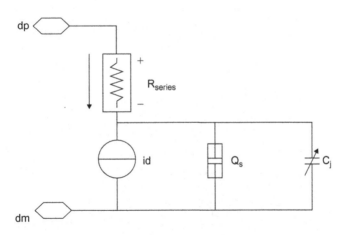

Figure 5.34:
The final implementation of the diode model with the capacitive effects

symbols for bipolar and MOS transistors. The generated code for this diode is shown in Figure 5.35. This code implementation illustrates that, even though we instantiated graphics in a macromodel-style fashion, our model turned out as a flattened, efficient model for simulation purposes because of the use of effects.

```
/* pragma lynguent created by ModLyng-1.4.3-S443 */
/* pragma lynguent exported by Verilog-A Spectre 5.1 */
`include "disciplines.vams"
module diode(dp, dm);
    //
    parameter real rseries = 0.0;
    //
    parameter real isat = 1.0e-15;
    //
    parameter real n = 1.0;
    //
    parameter real m = 1.0;
    //
    parameter real vj = 0.7;
    //
    parameter real tau = 0.0;
    //
    inout dp;
    electrical dp;
    //
    inout dm;
    electrical dm;

    //
    electrical N1;
    //
    branch (dp, N1) Branch_vr_ir;
    //
    branch (N1, dm) Branch_vd_id;
    //
    branch (N1, dm) Branch_vjc_ijc;

    capnonlin C1 (N1, dm);
    analog begin
      // pragma lynguent code fragment "Branch_vr_ir"
      I(Branch_vr_ir) <+ V(Branch_vr_ir) * rseries;
      // pragma lynguent code fragment "Branch_vd_id"
      // Id Drain Current Equation (5.2)
      I(Branch_vd_id) <+ isat * (exp(V(Branch_vd_id) / (n * $vt)) - 1.0);
      // pragma lynguent code fragment "Branch_vjc_ijc"
      // Charge Equation
      I(Branch_vjc_ijc) <+ ddt((tau * I(Branch_vd_id)));
    end
endmodule
```

Figure 5.35:
The Verilog-A realization of the diode model

5.12 Example 5.2

In this example we will illustrate the method by which we can avoid starting a model from scratch by starting from the result of Example 5.1, the nonlinear diode. We will modify this model to create a simple silicon power diode model and see how we can rapidly derive new models from existing ones in far less time. A more complete and realistic power diode model can be found in [1], so this model is for illustration purposes only.

We start by opening the existing diode model, adding the new effects to be modeled, and then saving it under a different name (power_diode). Knowing that we need to add more effects, we start by stating the governing equations of the effects to be added. For this illustrative exercise we will be adding an additional series resistance that varies with the moving depletion boundary in a power diode. This effect is important in the high-voltage blocking layer of a *p-i-n* diode structure. Again, details are best found in [1]. This resistance is a constant value in parallel with a voltage varying value. These two components are connected into the diode model on the n-side of the junction to decouple them from the former series resistance that was implemented on the p-side. The equations that govern these two effects are given in Eqs (5.8)–(5.10):

$$i_{\mathrm{mod}} = \frac{(\mu_n + \mu_p) \cdot V_m \cdot Q_s}{w^2} \tag{5.8}$$

where v_m is the voltage across the conductivity modulated region, Q_s is the charge in the intrinsic region from Eq. (5.4), μ_n and μ_p are the mobilities, and w is the active width (undepleted) of the region that is also changing with bias according to:

$$w = \sqrt{\frac{2 \cdot \varepsilon_{Si} \cdot (V_d - V_\phi)}{q \cdot N_b}} \tag{5.9}$$

where V_d is the voltage across the junction, V_ϕ is the built-in junction potential, and N_b is the background doping of the lightly doped region. The constant resistance portion is given by Ohm's law as:

$$V_m = R_{\text{mod}} \cdot i_{r\text{mod}} \tag{5.10}$$

In addition to this conductivity modulation resistance, two additional charge effects are added to the base diode to reflect the charge sweep-out effects present owing to the high fields in power diodes and a significant recombination that occurs in the lightly doped drift region. These two effects sum to equal the charge Q_s in Eq. (5.8) and are given by:

$$Q_R = (1 - ALPH0) \cdot Q_s \tag{5.11}$$

$$Q_{SW} = ALPH0 \cdot Q_s \tag{5.12}$$

where Q_s is given as before in Eq. (5.4), the charge storage effect that is proportional to the forward bias current i_d through the carrier lifetime. Eqs (5.11) and (5.12) partition this charge into a weighted sum for fitting purposes using the fitting parameter $ALPH0$. So, in our final model implementation the Q_s charge storage effect in the original diode is deleted and replaced with these two effects. The final implementation in ModLyng is shown in Figure 5.36. The generated Verilog-A code is given in Figure 5.37.

Conclusion

In this chapter we have introduced the key functions of graphical modeling and, in particular, the software ModLyng. Many of the key concepts in graphical modeling have been discussed briefly, and the remainder of this section of the book will investigate those in more detail.

Appendix

In this appendix we have included, just for reference, a more complete implementation of the small-signal diode model that is charge-conserving. It might

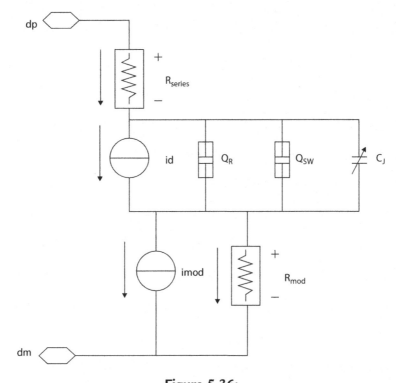

Figure 5.36:
The final implementation of the power diode model with the new effects

be a useful starting point for some readers. We have not shown all the graphical screenshots, just the equations and final code in this case.

Instead of the equations used in Example 5.1, the following set of equations will produce a much better low-voltage diode model in terms of convergence and predictability. The first three expressions are the same as before:

$$i_d = I_{SAT}\left(e^{\frac{V_d}{nV_T}} - 1\right) \tag{5.13}$$

$$Q_s = \tau \cdot i_d \tag{5.14}$$

$$V_s = R_s \cdot i_d \tag{5.15}$$

The junction capacitance formula needs to be integrated with respect to the junction voltage to obtain a charge expression. Furthermore, this expression

```
/* pragma lynguent created by ModLyng-1.4.3-S443 */
/* pragma lynguent exported by Verilog-A Spectre 5.1 */

`include "constants.h"
`include "disciplines.vams"
`include "ext_constants.va"

module power_diode(dp, dm);
//
parameter real rseries = 0.0;
parameter real isat = 1.0e-15;
parameter real n = 1.0;
parameter real m = 1.0;
parameter real vj = 0.7;
parameter real tau = 0.0;
parameter real rmod = 0.0;
parameter real alpha0 = 0.0;
parameter real vphi = 0.0;
parameter real nb = 1.0;
parameter real un = 1.0;
parameter real up = 1.0;

inout dp;
electrical dp;
inout dm;
electrical dm;

electrical N1;
electrical N14;
real w;
//
branch (dp, N1) Branch_vr_ir;
branch (N1, N14) Branch_vd_id;
branch (N1, N14) Branch_vjc_ijc;
branch (N14, dm) Branch_vmod_imod;
branch (N1, N14) Branch_vqsw_iqsw;
branch (N14, dm) Branch_vrmod_irmod;

capnonlin C1 (N1, N14);
analog begin
// pragma lynguent code fragment "calc_w"
// calculate w (5.8)
w = sqrt(2 * `P_E_ST * (V(Branch_vd_id) - vphi) / (`P_Q * nb));
// pragma lynguent code fragment "Branch_vr_ir"
I(Branch_vr_ir) <+ V(Branch_vr_ir) * rseries;
// pragma lynguent code fragment "Branch_vd_id"
// Id Drain Current Equation (5.2)
I(Branch_vd_id) <+ isat * (exp(V(Branch_vd_id) / (n * $vt)) - 1.0);
// pragma lynguent code fragment "Branch_vjc_ijc"
// Charge Equation
I(Branch_vjc_ijc) <+ alpha0 * ddt((tau * I(Branch_vd_id)));
// pragma lynguent code fragment "Branch_vmod_imod"
// calculate imod (5.7)
I(Branch_vmod_imod) <+ (un + up) * V(Branch_vmod_imod) * (tau * I(Branch_vd_id)) / (w * w);
// pragma lynguent code fragment "Branch_vqsw_iqsw"
// Charge Equation
I(Branch_vjc_ijc) <+ (1.0 - alpha0) * ddt((tau * I(Branch_vd_id)));
// pragma lynguent code fragment "Branch_vrmod_irmod"
V(Branch_vrmod_irmod) <+ I(Branch_vrmod_irmod) * rmod;
end
endmodule
```

Figure 5.37:
The Verilog-A realization of the power diode model

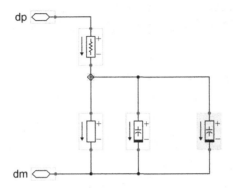

Figure 5.38:
The final implementation of the charge-conserving diode model

changes piecewise linearly as you transition about a crossover point related to the built-in junction potential V_ϕ. So, it is important to maintain continuity in the charge and to have the appropriate expressions in each region for accuracy and ultimately for fitting the model to experimental data. The resulting charge equations are given as follows.

For $V_d <$ fc V_ϕ,

$$Q_{depl} = V_\phi \cdot C_{jo} \cdot \frac{(1-(V_d/V_\phi))^{1-m}}{m-1} \tag{5.16}$$

And for $V_d \geq$ fc $\cdot V_\phi$,

$$Q_{depl} = V_\phi \cdot C_{jo} \cdot \frac{(1-fc)^{1-m}}{m-1} + \left[\frac{C_{jo}}{(1-fc)^{1+m}} \right]$$
$$\cdot \left[1 - fc \cdot (1+m)) \cdot (V_d - (fc \cdot V_\phi) + \frac{m \cdot (V_d^2 - (fc \cdot V_\phi)^2)}{2 \cdot V_\phi} \right] \tag{5.17}$$

where fc is a fitting parameter for the crossover point and all other parameters were defined previously. This charge Q_{depl} is another nonlinear charge effect placed in parallel with the Q_s to form the total diode charge and the C_j effect is removed from Figure 5.33. Now, the model is charge-conserving and continuous across the fc $\cdot V_\phi$ boundary. The junction capacitance that results is also accurate across a wider range of voltages than that in Example 5.1. The

```
/* pragma lynguent created by ModLyng-1.4.3-S443 */
/* pragma lynguent exported by Verilog-A Spectre 5.1 */

`include "disciplines.vams"

module diode_modified(dp, dm);
 parameter real rseries = 0.0;
 parameter real isat = 1.0e-15;
 parameter real n = 1.0;
 parameter real m = 1.0;
 parameter real vj = 0.7;
 parameter real tau = 0.0;
 parameter real cjo = 1.0e-15;
 parameter real vphi = 1.0;
 parameter real fc = 0.0;

 inout dp;
 electrical dp;
 inout dm;
 electrical dm;

 electrical N1;
 real Qdepl;
 branch (dp, N1) Branch_vr_ir;
 branch (N1, dm) Branch_vd_id;
 branch (N1, dm) Branch_vjc_ijc;
 branch (N1, dm) Branch_vdepl_Idepl;

 analog begin
 if (V(Branch_vd_id) < fc * vphi)
 Qdepl = vphi * cjo * (pow((1.0 - (V(Branch_vd_id) / vphi)), (1.0 - m))) / (m - 1.0);
 else
 Qdepl = vphi * cjo * ((pow((1.0 - fc), (1.0 - m))) / (m - 1.0))
 + (cjo / (pow((1.0 + fc), (1.0 + m)))) * (1.0 - (fc * (1.0 +
 m)) * (V(Branch_vd_id) - (fc * vphi)) + (m * (V(Branch_vd_id) * V(Branch_vd_id)
 - pow((fc * vphi), 2.0)) / (2 *
 vphi)));
 I(Branch_vr_ir) <+ V(Branch_vr_ir) * rseries;
 I(Branch_vd_id) <+ isat * (exp(V(Branch_vd_id) / (n * $vt)) - 1.0);
 I(Branch_vjc_ijc) <+ ddt((tau * I(Branch_vd_id)));
 I(Branch_vdepl_Idepl) <+ ddt(Qdepl);
 end
endmodule
```

Figure 5.39:
The Verilog-A realization of the charge-conserving diode model

final implementation of this low-voltage diode model as captured in ModLyng is shown in Figure 5.38 with the Verilog-A realization in Figure 5.39.

Reference

[1]   H.A. Mantooth, J.L. Duliere, A unified diode model for circuit simulation, IEEE Trans. On Power Electronics 12(5) (1997) 816–823.

Further Reading

H.A. Mantooth, M. Feigenbaum, Modeling with an Analog Hardware Description Language, Kluwer Academic Publishers, Boston, MA, 1995.

CHAPTER 6

Block Diagram Modeling and System Analysis

6.1 Introduction

The most common starting point in any system design is to establish fundamental or key concepts with a highly abstract model. This applies whether the design is low level or system level, but is especially relevant in systems design. The systems designer is most often concerned with overall structure, algorithms, signal flow, and establishing the structure of the individual elements that make up the system. As such, one of the main requirements for the systems designer is to establish that the overall specification is sensible and feasible, and to define the main techniques by which the ultimate design will achieve the overall goals of the project.

At this high level of abstraction it is often useful to ignore the detailed partitioning of the design into hardware/software and within the hardware area, between digital, analog, and other disciplines. Taking this approach allows the systems designer to take a more global view of the overall requirements and decide the best mix of functionality for the final system. Within this area, there are a number of techniques at the system designer's disposal, and these are described in this chapter of the book. The techniques vary from abstract system modeling of analog and digital systems to more mathematical aspects of modeling including Laplace and sampled data systems.

6.2 Signal Flow Modeling

From a systems perspective there is one very common starting point which is the so-called "signal flow" model. This is a system model that describes the overall system behavior, usually at a very high level of abstraction, but

Model-Based Engineering for Complex Electronic Systems.

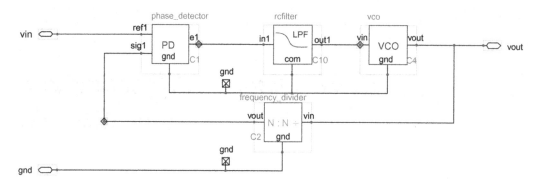

Figure 6.1:
System model of phase locked loop design

encapsulates the key concepts that the design requires. For example, consider the Phase Locked Loop (PLL) model shown below in Figure 6.1.

In this model each of the main blocks has been taken from a system level library and, as such, they allow very high-level analysis of the system performance, for example the loop response, without detailed timing behavior being required. Upon closer inspection, consider the loop filter (block named `rcfilter`) and what is required at different levels of analysis. For a systems analysis all we need to know initially is the order and cut-off frequency of the filter, so we do not care how it will ultimately be implemented at this stage to establish the response of the design as a whole. For example, in the first instance the key requirement of the system designer would be to establish the filter cut-off frequency, voltage controlled oscillator (VCO) gain, and feedback gain (divider) to obtain the correct output frequency for a known input frequency. Consider the case where the input frequency is 100 kHz and the output frequency is required to be 800 kHz; this enables us to establish that the divider ratio must be 8:1 so that the fed-back frequency is 1/8 of the output frequency to match the input 100 kHz. The choice of VCO gain and filter response becomes a matter for the system designer, but the important aspect of the process is that there is no need to implement any circuitry or to run detailed simulations – in this case the variable is effectively in units of frequency. The frequency response can be calculated directly and the result is shown in Figure 6.2.

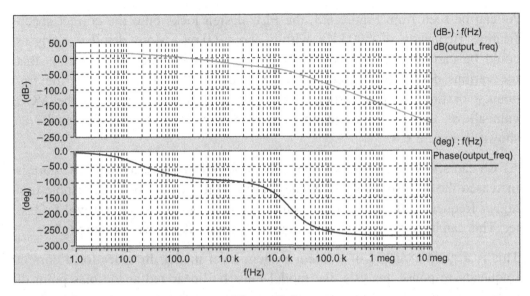

Figure 6.2:
System model frequency response (Bode plot)

One of the common misconceptions is that system-level modeling does not include units. This does not have to be the case. It is true to say that a system-level model is not necessarily a conserved energy model. It can range from an abstract conserved energy model, through models with units, to completely abstract models with only "pure" numeric values for variables. In this case, although the model is defined with a non-conserved, numeric system, the implication is that the units are, in fact, frequency (Hz) and they could be defined explicitly as such.

Where this type of model also has a significant advantage is that a time domain (or transient) analysis can also be carried out — again with no need to implement any detailed circuit models to observe the frequency behavior against time. As the model is not using any detailed switching circuits or running *at* the frequency in question, the analysis is extremely fast and gives a direct insight into the overall transient behavior of the system with the design coefficients chosen by the designer. For example, we can establish that the PLL will lock onto the demanded frequency within a certain time using a step input.

As can be seen from Figure 6.3, the PLL design locks onto the correct specification frequency of 800 kHz within about 60 ms. More detailed analyses could be carried out on this system design; however, the key aspect is that the various design parameters can be varied and the effect on both the frequency or time domain responses checked. For instance, varying the VCO gain allows the overall frequency response to be tested with the results as shown in Figure 6.4.

As we can see from Figure 6.4, the Bode plot indicates that as the VCO gain is increased the corresponding frequency response moves the first major pole to a higher frequency; therefore, in principle, the response of the circuit will be faster. This can be tested using a time domain analysis as shown in Figure 6.5.

This is a good example of a *linear* system-level model. In order for a correct frequency response analysis, the model must be linearized at the bias point of interest, so the static operating point is crucial for a correct result. In a nonlinear system, either the operating point must be correctly chosen, for a

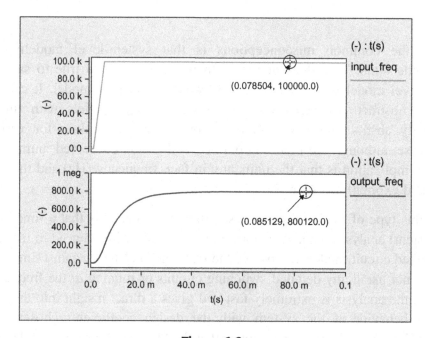

Figure 6.3:
System design transient response

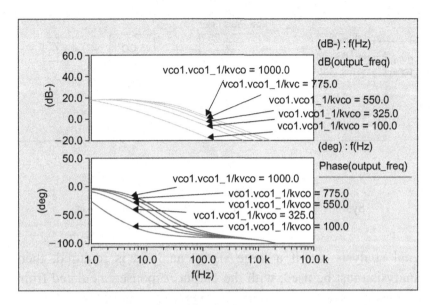

Figure 6.4:
System model PLL — varying VCO gain

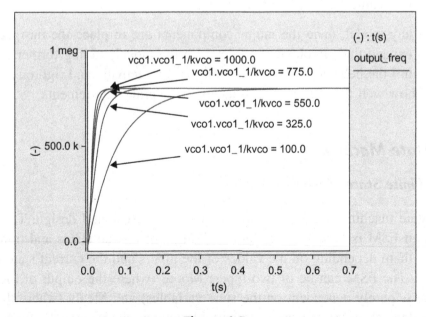

Figure 6.5:
System-level PLL — time domain response varying with VCO gain

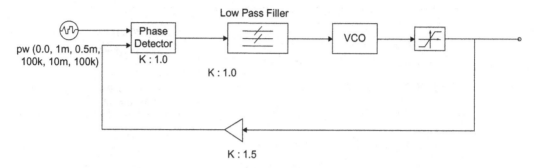

Figure 6.6:
System model of a PLL with VCO frequency limit

small signal analysis, or, if a large signal analysis is required, then a time domain analysis must be used, with the output response calculated from a number of these analyses across the range of operating conditions.

For example, the VCO is modeled as a linear block in this case. In practice there may be a limit to the range of operation of the device, so the model could be modified with a limit on the output (Figure 6.6).

As with any model, once the major components are in place the designer can begin to expand the complexity and detail in the model, adding further modifications until the behavior satisfactorily allows a systematic and rigorous assessment of how well the system design meets the overall requirements.

6.3 State Machines

6.3.1 Finite State Machines

Finite state machines (FSMs) are at the heart of most digital design. The basic idea of an FSM is to store a sequence of different unique states and transition between them depending on the values of the inputs and the current state of the machine. The FSM can be of two types: Moore (where the output of the state machine is purely dependent on the state variables) and Mealy (where the output can depend on the current state variable values *and* the input values). The general structure of an FSM is shown in Figure 6.7.

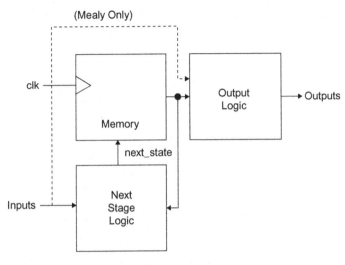

Figure 6.7:
Hardware state machine structure

6.3.2 State Transition Diagrams

The method of describing finite state machines from a design point of view is using a state transition diagram (bubble chart) which shows the states, outputs, and transition conditions. A simple state transition diagram is shown in Figure 6.8.

Interpreting this state transition diagram it is clear that there are four bubbles (states). The transitions are controlled by two signals ("rst" and "choice"), both of which could be represented by bit or std_logic types (or another similar logic type). There is an implicit clock signal, which we shall call "clk" and the single output "out1", which has an integer value.

6.3.3 Algorithmic State Machines

An alternative to the finite state machine approach is a technique called the algorithmic state machine (ASM), which is useful in that it can separate the control and data path structures in a relatively simple diagram. It looks a little like a flow chart, with decisions and actions, some of which are combinational and some of which are sequential (Figure 6.9).

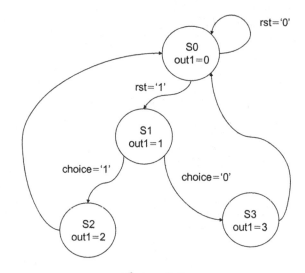

Figure 6.8:
State transition diagram

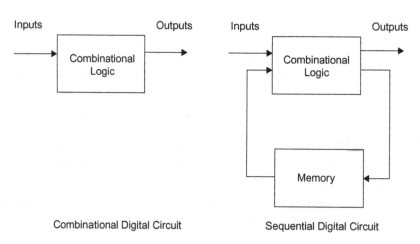

Figure 6.9:
Sequential digital circuit

The difference between an FSM chart and an ASM chart is that the ASM chart allows the explicit and clear definition of states, decisions and outputs, in a flowchart-like appearance, which makes the algorithm easy to see and understand — much easier than a state diagram. The three elements in an ASM chart are the STATE, DECISION, and OUTPUT blocks.

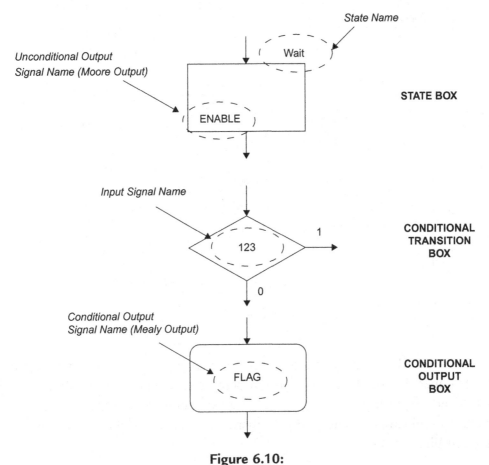

Figure 6.10:
Algorithmic state machine chart fundamentals

The STATE block holds the current state value, either as a name or as a numeric (hex, decimal, or binary) value representing the state variable. The state block can also hold the value of outputs depending on the state machine type (Mealy or Moore). The state block is represented by a rectangular box.

The DECISION block shows different conditional paths, depending on the value of input variables. For example, an input could be either 0 or 1, and, depending on this value, one or other paths out from the decision block can be taken. The decision block is represented as a diamond shape (Figure 6.10).

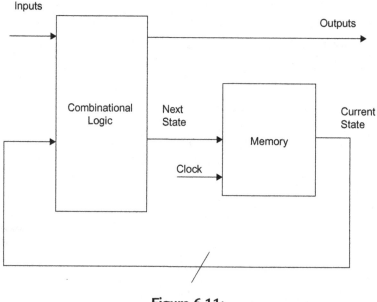

Figure 6.11:
Mealy machine

Using the ASM chart, the designer can divide the design into three separate phases depending on the type of design being undertaken. In a Mealy machine, there is no output logic, and so the designer is required to design the next state (input) combinatorial logic and establish the number of states. This is illustrated in Figure 6.11.

In the case of the Moore machine, there is also output logic and this is shown in Figure 6.12.

The beauty of the ASM approach is that in the first instance the algorithm is easy to establish and understand, crucial from a designer's perspective, and then it is relatively straightforward to convert this into the combinatorial logic and state logic using next state logic tables and Karnaugh maps.

If we take a simple example of a drinks dispenser machine to illustrate the design steps involved in an algorithmic state machine we can see how the designer translates the overall concept into a specific design for implementation in hardware. Consider the case where we have a simple coffee machine that dispenses a cup of coffee when sufficient coins have been placed in the dispenser.

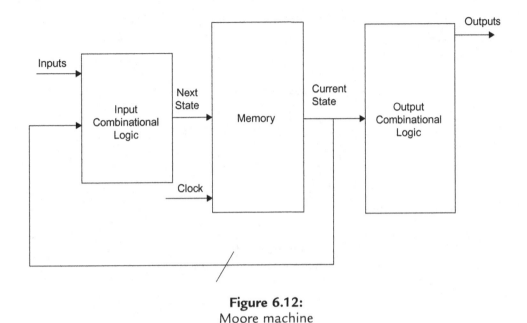

Figure 6.12:
Moore machine

The two coins recognized are 10 cents and 5 cents, respectively. A single cup of coffee costs 15 cents. If too much money is placed in the machine, then the machine will refund change, as well as dispensing the drink. The overall design concept implemented as a state machine is shown in the block diagram in Figure 6.13. As can be seen from the figure there is a clock, as well as the design inputs and outputs, ensuring synchronous operation.

Before designing the ASM chart it is worth thinking about the logic of the design. How many states will we need in principle? Clearly, in a complex design it is difficult sometimes to do this without some formal approach, but it is often the case that some preliminary analysis can make a big differ-ence in the efficiency of the final design. In this simple example, we have an initial state where the machine has no money, then if either 5 or 10 cents is put into the machine, there will be states where the machine contains either 5 or 10 cents (not enough to trigger the dispense). In the 5 cent state, if a 5-cent coin is put into the machine, then the state machine will move into the 10 cent state, and in the 10 c state, either 5 or 10 cents will be

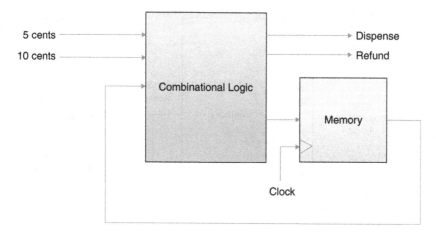

Figure 6.13:
Dispenser block diagram

enough to trigger a dispense. So we can say that we only need three states to accomplish this design.

Putting these elements into an ASM chart we can see the transitions between each state and also the outputs as a result of the transitions, as shown in Figure 6.14. We can establish from the fact that there are three states, that we need two state variables (for n state variables, we can have up to 2^n states).

We can define the state variables for this system (Q0 and Q1) and assign values for the states in the ASM chart.

State	Q0	Q1
Nothing	0	0
5 cents	0	1
10 cents	1	0
Unused	1	1

Using this assignment and the ASM chart we can define the next state logic table and also the output variables in the correct state.

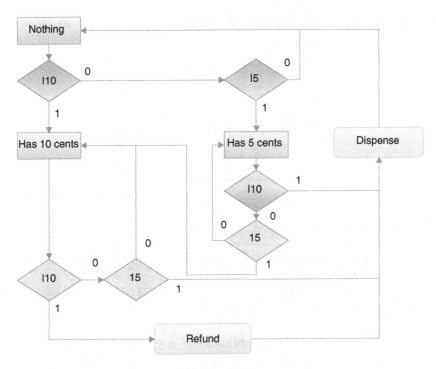

Figure 6.14:
Algorithmic state machine chart for the drinks dispense machine example

5 cents	10 cents	Q0	Q1	Q0'	Q1'	Dispense	Refund
0	0	0	0	0	0	0	0
X	1	0	0	1	0	0	0
1	0	0	0	0	1	0	0
0	0	0	1	0	1	0	0
X	1	0	1	0	0	1	0
1	0	0	1	1	0	0	0
0	0	1	0	1	0	0	0
X	1	1	0	0	0	1	1
1	0	1	0	0	0	1	0
X	X	1	1	0	1	0	0

This design can now be synthesized using Karnaugh maps to establish both the next state logic and also the output logic. For example, consider the case for

Q0'; this can be expressed using the Karnaugh map technique as shown in the following:

		Q0Q1			
		00	01	10	11
5 cents	00			1	
10 cents	01	1			
	11	1			
	10		1		

Which results in the logic equation

$$Q0' = \overline{Q0Q1}10c + \overline{Q0}Q15c\overline{10c} + Q0\overline{Q15c10c}$$

As a result, from the logic equation now synthesized, the appropriate logic can be instantiated as hardware to realize this design.

Using a similar approach, the combinational logic for Q1', DISPENSE, and REFUND can also be obtained, which, in this case, are as follows:

$$Q1' = 5c10cQ0Q1 + 5c10cQ0Q1 + 5c10cQ0Q1$$
$$DISPENSE = \overline{Q0Q1}110c + Q0\overline{Q1}110c + Q0\overline{Q1}5c\overline{10c}$$
$$REFUND = Q0\overline{Q1}110c$$

These can then be implemented in a hardware design, as shown in Figure 6.15.

This process is not in itself of huge interest to a modeler in that this is a standard digital design technique; however, the process is systematic and repeatable.

6.4 Algorithmic Models

6.4.1 Introduction

Algorithmic models can be considered as similar to the ASM models in many ways, except that there is no explicit clock or reset as such. They can be considered to essentially be programs, often described graphically using flowcharts, and therefore assumed to be running on a standard computer architecture.

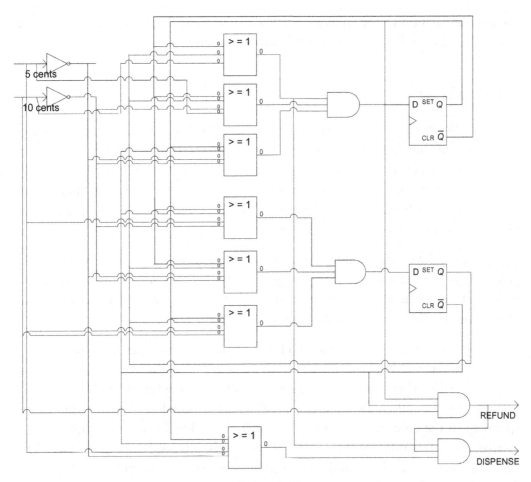

Figure 6.15:
Logic hardware for next state and output logic

These types of models are extremely useful for system-level modeling as there is no assumption about whether the model is describing hardware or software, and so the designer has much greater flexibility in many respects about the operation of the code, and decisions can be made about the partitioning of the algorithm into the relevant hardware and software parts.

Most algorithmic models will consist of code written in a standard programming language such as C, and, in fact, the system-level modeling language System-C is an example of this type of approach. In System-C, no assumptions are made about the implementation of the individual parts of the algorithm, and the code can be written in "standard" C++, or using hardware-oriented modules.

6.4.2 System-C

One advantage of System-C is that there is a built-in event-based simulator, which allows the implementation of synchronous systems, using clocks and resets in conjunction with higher-level algorithmic code. The code is essentially sequential, however, which means that there needs to be the appropriate infrastructure to handle events, an event queue, and transactions in the complete system-level model.

Using higher-level algorithmic code is particularly useful for designing a data path, where the various transforms required on data can be designed and tested in the correct sequence, and the control sequence can be established using standard programming constructs such as *if-then-else*.

For the drinks dispenser machine example, the code would take a slightly different form to the ASM approach already described. The outline code is shown first in Figure 6.16, where the overall loop (while) continues indefinitely, and, after carrying out the individual state checks, the user is asked to input a value corresponding to the coin input. Then the logical expression for the next state can be carried out. The code for the switch statement is provided in Figure 6.17. The case statement is used to handle each individual state of the state machine, and the user input effectively acts as a clock signal, stopping the algorithm so that the next state inputs can be obtained and then handled by the code.

Using the high level code model, the algorithm can be tested in a standalone compiled executable program or as part of a larger system in a code-based simulator, such as System-C. The key advantage of this type of modeling is that standard code development tools can be used to ensure the model is consistent and correct, and, ultimately, the algorithm could be implemented in either hardware or as software running in a processor.

This type of approach assumes a reasonably high level of coding knowledge and, for a more abstract approach, high level system models need to be

```c
#include <stdio.h>
#define nothing 0
#define has5c    1
#define has10c   2

main(){
  int i5c = 0;
  int i10c = 0;
  int state = nothing;
  int coin = 0;

  printf("Algorithmic Model - init\n");

  while(1){
    switch(state){
          case has5c:
                  // State has 5c code
          case has10c:
                  // State has 10c code
          default:
                  // State nothing code
      }

      printf("Enter a coin 1=5c, 2=10c : ");
      scanf("%d",&coin);
      printf("\nCoin = %d \n",coin);
      if(coin==1){
              i5c=1;
              i10c=0;
      } else {
              if(coin==2){
                      i10c=1;
                      i5c=0;
              } else {
                i5c=0;
                      i10c=0;
              }
      }
  }
}
```

Figure 6.16:
Outline code for drinks dispenser machine

```
switch(state){
    case has5c:
        if(i10c==1) {
          state = nothing;
          printf("DISPENSE\n");
        } else {
          if(i5c==1) {
            state = has10c;
          } else {
            state = has5c;
          }
        }
        break;
    case has10c:
        if(i10c==1) {
          state = nothing;
          printf("DISPENSE\n");
          printf("REFUND\n");
        } else {
          if(i5c==1) {
            state = nothing;
            printf("DISPENSE\n");
          } else {
            state = has10c;
          }
        }
        break;
    default:
        if(i10c==1) {
          state = has10c;
        } else {
          if(i5c==1) {
            state = has5c;
          }
        }
        break;
}
```

Figure 6.17:
Switch statement for algorithmic model in C

employed. An example of this is transfer function modeling, which is described in the next section of this chapter.

6.5 Transfer Function Modeling

6.5.1 Introduction

For many systems the definition of the highest level of behavior is using some form of transfer function model. This is similar in many ways to the signal flow approach described earlier in this chapter. Using transfer functions in a "pure" sense does allow the definition of a more rigorous mathematical treatment to systems when they are defined using S domain (continuous, linear, analog) or Z domain (sampled, digital, potentially nonlinear). It is possible to define systems in the S domain using nonlinear functions; however, the assumptions we make are contingent on the system being linear (e.g., Bode plots).

The most ubiquitous software for transfer function and system level modeling is Matlab®, which has become pervasive in university curricula worldwide over the last couple of decades. The first implementation of Matlab was clever in that a basic solving engine was supplied, with a series of analysis and visualization "toolboxes". This enabled rapid testing and evaluation of design concepts and equations from extremely abstract equations to quite detailed physical models of real systems. With limited versions available for students, it has become an essential feature of the education of many engineers and scientists.

The initial appearance of the Matlab command window is less promising than this introduction would, perhaps, suggest. Figure 6.18 shows the default user interface of Matlab with a fairly nondescript collection of windows in a standard graphical user interface layout. This initial appearance is a bit of a "wolf in sheep's clothing". Once the tool is used, the flexibility and capability become apparent. By typing in the command window it is straightforward to create variables and then begin to type equations, as shown in Figure 6.19 where we have created two variables $a = 10$ and $b = 5$, and then by typing "c = a*b" Matlab automatically creates the variable c and calculates the correct value. This is a very simple example and does not really go beyond a

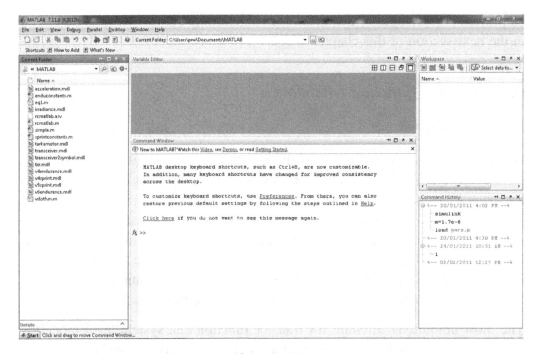

Figure 6.18:
Matlab user interface

programmable calculator; however, everything else that follows is based on this principle no matter how complex the equations become.

In this form of modeling the concept of a "block" is not defined by a physical drawing of a block on the screen, but rather the model description as a discrete entity. This could be a module of code, for example. This is, in essence, the approach in system-level modeling tools where analysis functions, modeling functions, and post-processing functionality are captured in this modular form. As has been described previously in this book, when we talk about a "Matlab model" we actually mean a series of declarations of variables, and a collection of function calls to custom-made or pre-existing modules of code.

If we revisit the code model used to solve a simple RC network that was introduced in Chapter 4 we can see how this actually works:

```
1:   r = 1.0;
2:   c = 1.0;
```

```
3:   vinit = 1;
4:   tmax = 10;
5:   n = 100;
6:   deltat = tmax/n;
7:   tspan = [0:1:n];
8:   options = odeset('RelTol',1e-2);
9:   [t1,v] = ode23(@eq1,tspan*deltat,vinit,options,r,c);
10:  v2 = v.*v;
11:  ecc = 0.5*c*v2;
12:  i = v/r;
13:  pr = i.*v;
14:  er = cumtrapz(t1,pr);
15:  plot(t1,er);
16:  grid
```

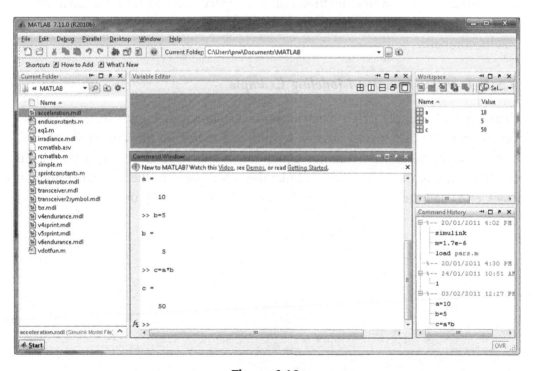

Figure 6.19:
Creating variables and simple equations in Matlab

Looking at this code, we can break the model into the key parts. Lines 1–5 are the initial parameter declarations. Line 6 is an equation, which we can consider in a couple of ways. It could be seen as an algorithmic or procedural piece of code, much like the C code example already discussed in this chapter, or we can see this as a system-level gain block, where the output is called **deltat**, the input is **tmax**, and the gain is **1/n**. Lines 7 and 8 define more parameters, and lines 9–14 are the functional heart of the model in that the key equations are solved in this section. Lines 15 and 16 are also interesting in that they are functions for post-processing the output data. Although the distinction between model functionality and analysis is somewhat blurred in Matlab, it does have the advantage that the "model" contains not only the variables and equations, but also the mechanism to analyze the model built in to the code.

With this type of modeling, we do have a choice of methods for describing systems, either using transfer functions or state space. To illustrate the differences and similarities a simple example will be described using both methods and the results compared.

6.5.2 Transfer Function Modeling Example

Consider the example of an electric motor as shown in Figure 6.20. A typical simple example of a mixed domain system is a motor, in this case a simple DC motor. Taking the standard motor equations as shown in Eqs (6.1) and (6.2), it can be seen that the parameter K_e links the rotor speed to the electrical domain (back emf) and the parameter K_t links the current to the torque.

$$V = L\frac{di}{dt} + iR + K_e\omega \tag{6.1}$$

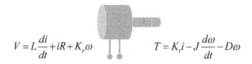

Figure 6.20:
Simple motor model and equations

$$T = K_t i - J \frac{d\omega}{dt} - D\omega \qquad (6.2)$$

The motor model has a number of specific parameters which define its behavior when driven by a voltage source. These are defined as follows:

- L: Inductance of the motor
- R: DC resistance of the motor
- K_e: Motor constant (for back EMF)
- K_t: Motor torque constant
- J: Motor inertia
- D: Motor damping.

Now, we have two equations (6.1) and (6.2), and the first step is to represent them using Laplace equivalents, where, essentially, derivatives are replaced by the Laplace operator *s*. For example, if we have an equation for the derivative of the current (di/dt) this would be represented with the Laplace equivalent sI. It is also worth noting that in this standard form, the speed is defined as the variable (ω); however, in practice this is the derivative of the position (θ), and so we can rewrite these equations as shown in Eqs (6.3) and (6.4):

$$V = L \frac{di}{dt} + iR + K_e \dot{\theta} \qquad (6.3)$$

$$K_t i = J \ddot{\theta} + D \dot{\theta} \qquad (6.4)$$

We can rewrite these equations in terms of the Laplace operator to get the equivalent "s domain" equations as shown in Eqs (6.5) and (6.6):

$$[Ls + R]I = V - K_e s\theta \qquad (6.5)$$

$$[Js^2 + D_s]\theta = K_t I \qquad (6.6)$$

These two equations can be rearranged by eliminating the term for the current *I*, and combining them into a relationship between the voltage *V* and the position θ, as shown in Eq. (6.7):

$$\frac{\theta}{V} = \frac{K_t}{(Js^2 + D_s)(Ls + R) + K_e K_t} \qquad (6.7)$$

This is now a transfer function for the model of the motor and electrical circuit which can be simulated to either obtain the small-signal analysis (frequency response and Bode plot) or time-domain response. Using this model, with the parameters defined in the following list, the step response can be calculated as shown in Figure 6.21.

L: 1 mH
R: 1 Ω
Ke: 0.1
Kt: 0.1
J: 0.01
D: 0.02.

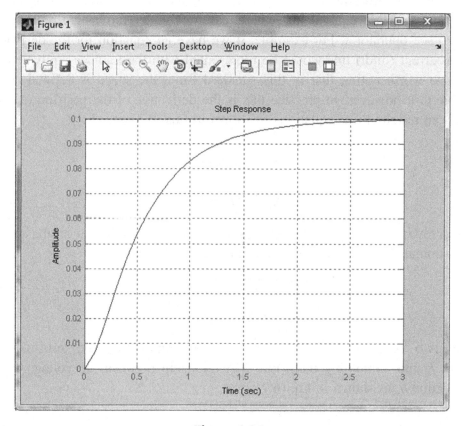

Figure 6.21:
Step response of the transfer function model

This approach is good for theoretical models, where it is relatively easy to establish the overall system equations. It is clearly going to be more challenging in circumstances where there are more complex connectivity and physical interdependencies, such as circuit analysis. It is also useful to be able to see both the time-domain and frequency-domain characteristics; however, it must be stressed that the premise for a Laplace model interpretation is that the equations are *linear*. It is therefore assumed that the model equations have been linearized at some stage in this process, which is obviously difficult to do in many cases.

A similar, related approach to the single transfer function modeling approach is a technique called *state space*, where instead of a single transfer function, the model is classified using a number of discrete differential equations in terms of a number of so-called *state variables*. This will be described in the next section of this chapter.

6.5.3 State Space Modeling

As we have seen in the previous section, the goal of the transfer function model was to establish the time-domain and frequency-domain relationship between the input (voltage) and output (speed) of the motor. The basic idea with a state space approach is to define the inputs, define the outputs, and a *set* of equations to represent the behavior of the system. It is essentially identical to the transfer function in behavior (it has the same equations after all), but a slightly different representation, with sometimes the advantage that it is not necessary to complete a full derivation. If the current I and the speed ω are taken as the two state variables, then the general form of the state space representation is as follows:

$$[Derivatives] = [parameters][State Variables] + [parameters][outputs] \quad (6.8)$$

$$[outputs] = [parameters][State Variables] \quad (6.9)$$

So, in this case, the state space description will be as follows:

$$d/dt \begin{bmatrix} \omega \\ i \end{bmatrix} = \begin{bmatrix} -D/J & K_t/J \\ -K_e/L & -R/L \end{bmatrix} \begin{bmatrix} \omega \\ i \end{bmatrix} + \begin{bmatrix} 0 \\ 1/L \end{bmatrix} v \quad (6.10)$$

$$v = \begin{bmatrix} 1 & 0 \end{bmatrix} \begin{bmatrix} \omega \\ i \end{bmatrix} \qquad (6.11)$$

The step response and Bode plot can be recalculated and are shown in Figures 6.22–6.24.

Conclusion

This chapter has reviewed some of the commonly used techniques for block- and system-level design, ranging from typical block diagram design using standard library elements, to fundamental equation approaches, such as transfer function and state space modeling. While in many cases the approach is

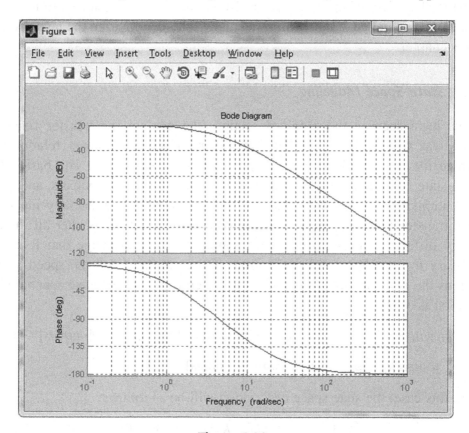

Figure 6.22:
Bode plot for the transfer function model

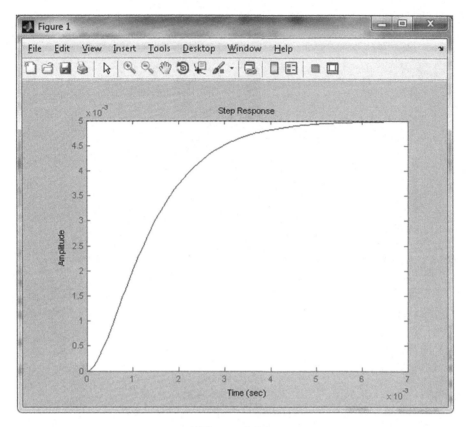

Figure 6.23:
State space step response

relatively simple, the examples also highlight that there is, in fact, a design *process* involved from the initial capture of the behavior in a graphical or equation form for obtaining usable results. For example, the transfer function modeling approach requires a significant level of experience and expertise to be able to interpret a medium-to-high complexity system and derive the fundamental equations from it.

While it may seem enough in the first instance to simply make a comment along the lines of "well, we know how to model systems now, don't we?", it is not as simple as that. What we have described is a series of tools and methods to assist in modeling and defining systems at a high level; however, we have not integrated this within a design process yet. From a *practical* perspective,

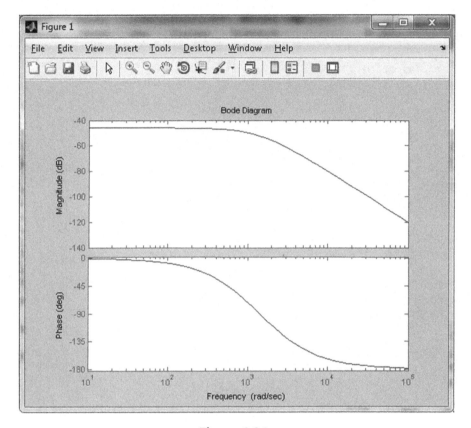

Figure 6.24:
Bode plot state space description

we have now gained useful skills enabling some of this to take place, but it is not enough to simply create a high-level model, demonstrate that it works, and then stop. Real design does not work like that and, in fact, we need to think about the integration of this within a design process that starts from abstract requirements and concepts, and then develops the design in more detail to the point where we can actually implement hardware. As the subsequent chapters of this book will describe, we need to understand some fundamental concepts of modeling and describing hardware systems in detail before we can truly embark on a complex system design.

CHAPTER 7

Multiple Domain Modeling

Sir Isaac Newton — "If I have seen farther than others, it is because I was standing on the shoulders of giants."

7.1 Continuous-Time, Conserved Modeling

7.1.1 Introduction

Being able to represent multiple disciplines and technological interfaces in a single modeling framework is incredibly powerful; however, some pitfalls present themselves to the unwary. The first potential problem is defining a common, standard interface such that models created within different simulators will have consistent definitions of natures, quantities, and terminals to allow them to be connected together. If this is not the case, then problems can obviously occur when models and libraries undergo deployment. Another related problem is in the underlying assumptions made in models with regard to constants. If the same name is used, but different value (or accuracy, for example) is used, then erroneous results can occur in simulations of ostensibly the same circuit. An example is the constant π. The U.S. National Institute for Standards and Technology defines the value of π out to many digits of precision and to simply approximate it as 3.14 in digital computer calculations can lead to errors. These errors become most relevant if all models do not use the same value of π, leading to inconsistencies and needless inaccuracies. Taking all of these issues into account allows a clear requirement for a set of multiple domain techniques that define basic physical constants and interface conventions.

As systems continue to increase in complexity, there is a greater need for multi-domain models within a single simulation environment to validate and verify designs against specification. Even a simple behavioral model of mechanical loads, motors, sensors, etc. can provide more realistic test benches (i.e., either loads or stimuli) for the design. This chapter focuses specifically on the need for, and implementation of, multiple domain models.

7.1.2 Fundamentals

7.1.2.1 Conservation of Energy

The fundamental concept of energy conservation is central to our understanding of the physical world around us and is a universal law of physics. Richard Feynman in his excellent series of lectures on the fundamentals of physics "6 easy pieces" uses a series of analogies to express the concept of energy conservation, and we will use the same basic ideas in this chapter [1]. The fundamental of energy conservation is that no matter what happens, the amount of energy in the universe never changes. If we apply this to our models of systems, we must consider that in any model we create, energy must be conserved at all times. We can convert stored (potential) energy into other forms, such as kinetic energy, and we can convert energy from one form to another (e.g., electrical into heat), but the overall energy equation must remain balanced.

Feynman uses the concept of "blocks" of energy, and in a simple example notes that if we start with a fixed number of blocks which represent energy, then no matter what we do, at the end we will always have the same number of blocks at the end. Applying this to a system, if it is energy conserving, then if we generate an amount of energy in a source, then this energy must be converted into other forms (such as potential or kinetic energy), so that the overall amount of energy is the same. For example, if we apply a voltage across a resistor and power is dissipated then what happens to the energy "lost"? In a true conserved energy system, the temperature will rise in the air around the resistor, which is simply the energy being converted from electrical to thermal. If the voltage comes from a charged capacitor, then as the energy is moved into the thermal domain through power dissipation in the resistor, the voltage will drop across the capacitor as the energy stored in the capacitor is reduced.

Given this premise of energy conservation, the models of this type need to be more sophisticated than those already used at the system level. In a conserved energy system, we need a new concept − the "branch" − which is described in the next section.

7.1.2.2 Domain Connections

In the conventional electrical world of circuit simulation, the default "type" of every terminal is *electrical*. Even when the system being represented consists of nonelectrical elements, for example, a motor, when the model is simulated, the results are expressed in terms of volts and amps. In reality, of course, the physical world is actually represented by numerous domains, each with their own definitions for variables, for example, thermal systems have a *temperature* difference rather than a *potential* difference across connection points in the system.

The method by which we can define these domains is usually implemented by defining a specific domain for every node in the system. This is important as each node must have one and only one technology definition, and cannot be connected to nodes of another technology type. Therefore, only electrical nodes can connect to other electrical nodes via a *branch*. The concept of a branch will be discussed in more detail in the next section; however; for now; we can make the assumption that this is the model element that describes the relationship between the variables relating to the two nodes connected to a branch.

We can define a set for the majority of physical systems in common use for engineering applications. This is shown in Table 7.1. With each technology, there is a combination of the through and across variable that will define the instantaneous power, and from this power the energy can be calculated.

For example, in the electrical domain, the power can be calculated from the product of the through and across variable, $V \cdot I = power$, and has the units of Watts. When the instantaneous power is integrated with respect to time, the energy is obtained.

7.1.2.3 Branches

In the definition of a conserved energy model, there are several new concepts to introduce, the most important being that of a physical "branch". This also

Table 7.1: Definitions for Through and Across Variables for Various Technologies

Technology	Through Variable	Through Symbol	Through Unit	Across Variable	Across Symbol	Across Unit
Electrical	Current	I	A	Voltage	V	V
Rotational	Torque	T	N·m	Angle	rad	radian
Rotational velocity	Torque	T	N·m	Angular velocity	ω	rad/s
Translational	Force	F	N	Displacement	P	m
Translational velocity	Force	F	N	Velocity	m/s	m/s
Thermal	Heat flow	Q	W	Temperature	T	K
Magnetic	Flux	F	Weber	mmf	mmf	A
Pneumatic	Flow	F	kg/m^3	Pressure	P	Pa
Radiance	Luminous flux	lm	Lumen	Luminous intensity	Candela	cd

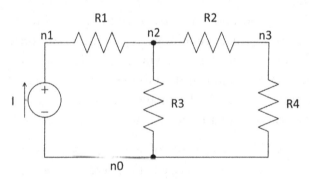

Figure 7.1:
Simple electrical circuit

requires the definition of terminals that can handle the physical types for each domain. The easiest way to introduce these concepts is using an example. In a particular circuit or system, such as the circuit shown in Figure 7.1, there are four resistors (*R1, R2, R3,* and *R4*), four nodes (*n1, n2, n3,* and *n0*) and a single current source (*I*). We can now look at this circuit and redraw it in terms of the nodes and the branches, where a branch defines the behavior between two nodes. We redraw the circuit diagram in Figure 7.1 in branch diagram

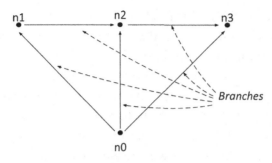

Figure 7.2:
Branch diagram

form as shown in Figure 7.2, where each individual component is replaced by a branch.

So, what do we mean by a branch? Well, in a conserved system, we are defining the relationship between the potential difference between two nodes at either end of the branch (in electrical circuits this is the voltage) and the flow from one node to the other (in electrical circuits the current). Therefore, each node has the definition of both through (current) and across (voltage) variables. As we discussed in Chapter 3, there is the fundamental concept of *dependent* and *independent* variables to consider, and how this relates to the terminology also used for *through* and *across* variables.

As we saw with a simple example of the definition of a *stamp* for a resistor in Chapter 3, there is a relationship between the *dependent* variable i and the *independent* variables vp and vm as defined in Eq. (7.1) for a resistor:

$$i = \frac{v_p - v_m}{R} \tag{7.1}$$

In this case, the dependent variable i is also classified as the *through* variable, and the difference between the two independent variables vp and vm is the *across* variable.

7.1.2.4 Interdomain Energy Transfer

With the definition of specific technology connections and branches between the nodes of that specific technology, there is still the need to transfer energy from one domain to another if true energy conservation is to be modeled. In modeling a multiple domain system, there has to be such a mechanism in place for this to happen, and usually this is implemented using the equations containing variables from both domains. For example, if a resistor generates power which is to be dissipated as heat into a volume of air, then as well as the equation for dissipated power, there also has to be a calculation of the heat flow to ensure that the energy generated by the electrical element is transferred correctly into the thermal domain. However, electrical conservation laws, such as Kirchhoff's current and voltage laws, can be represented properly without having to account for generated heat, etc. if energy conservation is not required.

7.1.3 Procedure for Model Creation

While there are no hard and fast rules for the sequence of events required to create new models, there is clearly a set of tasks that need to be undertaken to do so. This is particularly the case for conserved energy models which are dependent on the technology domain of the model. With this in mind, in this chapter a sequence of tasks will be defined which are the bare minimum required to create a new model in any technology. If these steps are followed, then, at the least, the model will have the necessary basic elements and then it will be up to the designer to ensure that the behavior is correct. The steps are described as follows.

7.1.3.1 Step 1: Define a New Model

This sounds obvious to say the least; however, there is more to consider than just the mechanical task of creating a new empty model. The name of the model should be chosen to be meaningful, but brief. Verbose naming conventions can become cumbersome and unwieldy. The model itself will have potential dependencies on global parameters, or perhaps libraries of functions or models.

7.1.3.2 Step 2: Define the Model Connections

The second task is to define the connections to the model. Again, this sounds obvious, but in a domain-specific context, the type of connection pins needs to be defined explicitly, for example, to be "electrical" or "thermal". The names should again be meaningful but brief.

7.1.3.3 Step 3: Define the Model Parameters

The model will likely have a number of parameters to characterize the behavior of the model, and these should be defined early in the process. Whether a graphical or language-based approach is required, it is difficult to debug the model equations or structure if parameters have not been defined first. Also, whichever model capture method is used, parameters should be given some brief documentation with at the least a comment as to its function, default value, and perhaps range where appropriate. Also, in many cases, the type needs to be explicitly defined (e.g., real or integer).

7.1.3.4 Step 4: Define the Model Topology

With the model's external interface in place (name, connections, parameters), it is now time to start defining the internal behavior of the model. The model itself could be equation based or structural (i.e., based on pre-existing models) or some combination thereof, but in either case there is an intrinsic structure to the model that must be defined. Generally, in hardware description language (HDL)-based models, it is a very *ad hoc* process to define the model topology, and fraught with potential for errors as there is no explicit structure in place; however, the basic premise is to establish the branch structure between the nodes. Therefore, there are two aspects to be defined in the model, the first being any internal nodes (which like the connection ports must be of the correct technology type) and then the network of branches to provide the framework for equations to be defined.

7.1.3.5 Step 5: Create the Model Equations (Sequential and Simultaneous)

If we have taken a "branch"-based approach rather than a structural/macro-model one, even though the branches are in place at this point, there is no behavior defined yet. For example, a branch in our simple network shown in Figure 7.1 may have a voltage, current, and two connection points or nodes; however, there is still the necessity to define the behavior intrinsic to the branch. Models may use sequential and simultaneous type equations to define behavior, but it is important to remember that each branch must have one and only one simultaneous equation to define it. There can be as many sequential, procedural, and function call equations as is necessary to define the details of the behavior, but only one simultaneous equation.

The most important point to make regarding the model equations is that they must be continuous. Lack of continuity will certainly lead to disastrous results in simulation (i.e., convergence problems). In fact, it is best if the equations continuous in all derivatives. However, sometimes this is difficult to achieve. The typical requirement most often quoted is that the equations be continuous and the first derivatives also continuous owing to the nature of the Newton—Raphson algorithm. If the designer simply keeps this rule in mind when implementing piecewise defined behaviors, then it is unlikely that problems will be encountered in terms of convergence.

At this point, for a HDL-based approach, this would be a complete model; however, in a graphical or schematic-based approach, we need to define a symbol.

7.1.3.6 Step 6: Create a Symbol

The model symbol represents the model connectivity and parameters in a hierarchical context, and is required for a graphical or schematic-based design approach.

At this point, we have a complete model ready for use in analysis and design of more complex systems.

7.1.4 Electrical Domain

7.1.4.1 Introduction

Electrical modeling is very common for circuit analysis, clearly, with excellent support for circuit simulation using either purely electrical simulation (such as SPICE) or potentially mixed domain (Saber/VHDL-AMS/Verilog-A). The key aspect of any domain modeling, and electrical is no exception, is to know the basic definitions of terminals, branch variables (through and across), power, and energy.

Electrical connections or nets have the units of voltage (V) for potential difference and current (A) for through current flow. KCL is used by circuit simulators to solve for these variables at nodes in schematics using standard matrix techniques described previously. All the connection points that are considered electrical will have an associated voltage and current. HDL-based behavioral modeling languages require the explicit definition of connection points as electrical. Thus, using a resistor as an example, its connection points p and n need to be explicitly defined as electrical, implicitly defining the voltage and current across and through variables for the pins in question.

If we take the example of our resistor again, with the two pins p and n of type electrical, then in order to construct a complete model it is essential to ensure that the through and across variables are defined either explicitly or implicitly, but most important is the through variable. Some simulators allow the definition to be loosely defined and automatically insert implicit equations (such as for a voltage source). If we take the resistor equation $V = I \cdot R$, then this is a compact definition of the device equation with both the through variable (I) and across variable (V) defined, and so is a complete equation that can describe the model. We can illustrate how these work in a graphical modeling context by creating a simple example model in ModLyng — in this case an electrical diode.

7.1.4.2 Step 1 : Creating a New Diode Model

The first step in the creation of a new model is to select the "File -> New Model" option from the menu or choose the "New Model" icon on the icon bar. This will create a new empty model with a default name for the model and its behavioral architecture. ModLyng uses an Entity-Architecture structure for the model definition, where the Entity defines the interface and parameters, and the Architectures (there can be more than one) define the model behavior and structure.

When the model has been created, if the entity name is right-clicked, the details of the model are provided, as shown in Figure 7.3.

We can change the entity name to the desired name for the model and add a comment to describe the basic behavior in the model comments window.

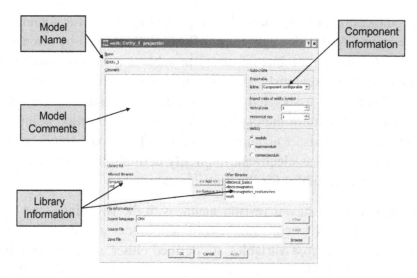

Figure 7.3:
Defining the information for a simple diode model

In this case, we can call the model "diode" and add a suitable comment to describe its behavior. We can do the same for the individual architectures if more than one level of model is required.

7.1.4.3 Step 2: Diode Model Connections

This model has two conservative, electrical pins called p and m. These can be added in a number of ways, such as adding ports to a topology (structure) of the model, or explicitly in the model interface itself using the port declarations section, as shown in Figure 7.4.

Ports are added using this approach by clicking on the "+" symbol to add extra ports to the model and double-clicking the individual fields such as the name or comment. These ports will appear automatically in the model topology as we will see in step 4. They can also be edited and/or created in the model topology.

7.1.4.4 Step 3: Model Parameters

Most models will have some parameters, and these can be added using the parameter declarations form where the individual parameters can be added in a

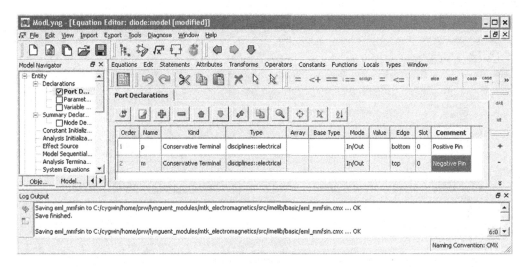

Figure 7.4:
Electrical diode model port declarations

similar manner to the port additions in the previous section. In this simple example, we have three parameters I_{sat} (saturation current), n (emission coefficient), and the current temperature, T, in degrees Kelvin.

Note that in addition to the parameter names, we can define the types (important in some simulators) and a default value. It is also possible to define a valid range for parameters, which has a different meaning in different hardware description languages; however, this removes the need for the designer to understand these individual language nuances as the graphical model will resolve this automatically, as shown in Figure 7.5.

7.1.4.5 Step 4: Define the Model Topology

With the ports defined, the designer can work in two ways. One is to use equations to explicitly define the behavior of the model and another is to create a topological representation of the model to which equations can be attached. In this simple example, the basic diode equation is a single branch between the positive and negative pins of the diode, which can be described with a single

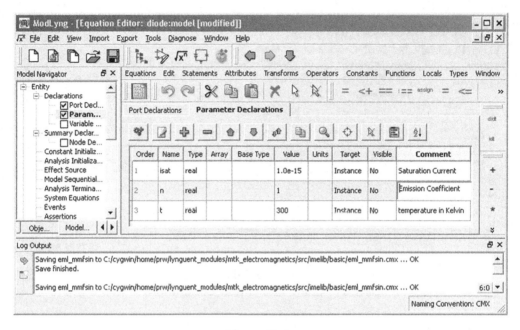

Figure 7.5:
Electrical diode model parameter definition

equation. If we open the topology editor, we can connect a branch across the ports and then edit the definition of the branch when it has been placed as in Figure 7.6.

There are two types of branch that can be placed: explicit, where the appropriate across and through variables are obtained directly from the branch; and implicit, where they are named by the designer specifically. Explicit branches are simple and less prone to error than implicit naming and this is what has been used in this example. As can be seen from Figure 7.6 we have named the branch *diodeeqn*. This is, effectively, a placeholder for the equations to be implemented.

7.1.4.6 Step 5: Create the Model Equations (Sequential and Simultaneous)

There are two types of equations that can be defined at this point: sequential and simultaneous. For simple models, particularly those like this with a single

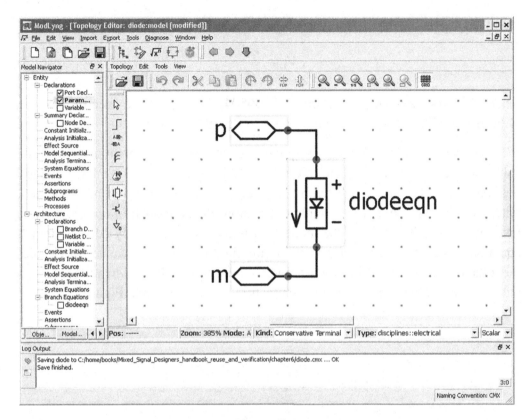

Figure 7.6:
Topology editor — simple diode example

equation, the choice of equation type is largely a matter of personal preference, as the choice will generally not make much difference to the final output model. One advantage, however, of implementing sections of the equation sequentially is the ability to plot the equations directly, without the need to export a model for simulation. This can make the process of model debug much simpler and more direct.

In this case, for example, the model equation is defined as follows:

$$I_{\mathrm{d}} = I_{\mathrm{sat}}(e^{(Vd/nV_T)} - 1) \tag{7.2}$$

where V_T is defined as

$$V_T = \frac{kT}{q} \tag{7.3}$$

where k is Boltzmann's constant ($1.3806504e^{-23}$), q is the electron charge ($1.602176e^{-19}$), and T is the temperature in degrees Kelvin.

We can implement Eq. (7.2) as a sequential expression and then assign the value calculated for the current to the through variable of the branch we have already defined. Similarly, the voltage can be assigned to the across variable of the branch. Alternatively, we can assign a single expression for the branch in one step. If we take this approach first, then the model editor allows us to simply edit the branch equation and enter the correct definition directly. In both cases, we need to implement the constant calculation for V_T, and we accomplish this by creating an internal variable of type constant in the architecture. This is shown in Figure 7.7.

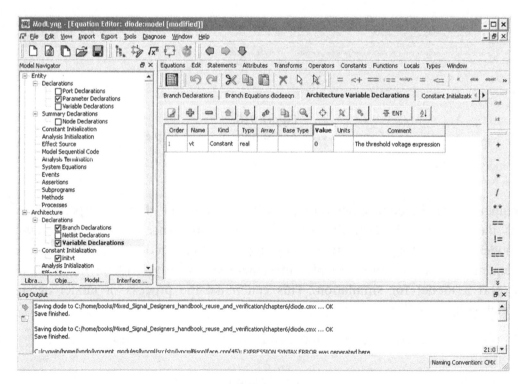

Figure 7.7:
Defining a variable declaration for V_T

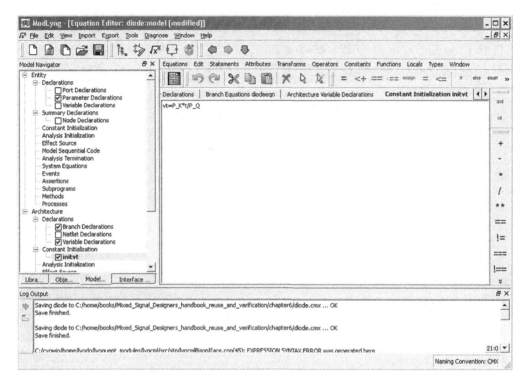

Figure 7.8:
Defining an equation for V_T

With the new internal variable (of type constant) defined, we can create a simple constant initialization equation to set the correct value according to Eq. (7.3). We can see this in Figure 7.8, and it is worth noting that ModLyng has the standard math and physical constants already available, so these can be included without the designer needing to reference them.

With the branch defined, and the equation for V_T in place, the designer can now implement the equation for the diode itself in the equations section of the code manager. This is shown in Figure 7.9 and note the generic use of the " $<+$ " operator to denote a simultaneous equation. It is possible in ModLyng to choose a "native" style (say Verilog or VHDL) if the designer is more

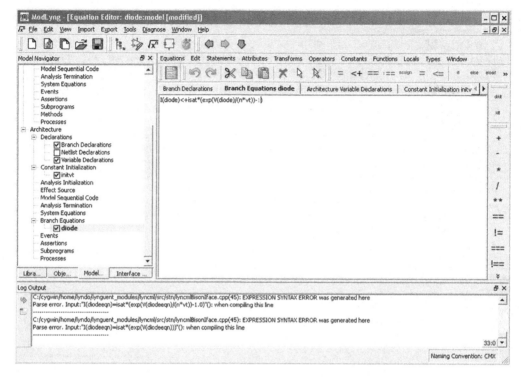

Figure 7.9:
Simultaneous equation for the diode example

comfortable with a standard language format or use this more generic format, which is language independent.

Another option for creating the equations is to define two new sequential variables, `id` and `vd`, and then have two sequential code fragments for each. `vd` is simply the voltage across the diode branch and `id` is the diode equation in Eq. (7.2). The branch current is then set to this value in the simultaneous equations. Mathematically, the two methods are identical; however, using this approach allows the sequential version of the diode equation to be graphed and verified without the need for a simulator. For example, the resulting $I-V$ curve can be plotted from the sequential version of the same diode model (Figure 7.10).

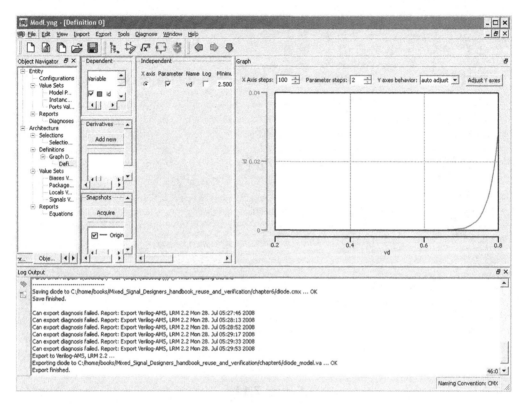

Figure 7.10:
Equation graph for the diode example

7.1.4.7 Step 6: Create a Symbol

If we wish to use this model in hierarchical models it is necessary to create a symbol. We can do this using the standard symbol editor within the modeling tool. With this tool, the user can create a simple graphical representation of the model which means it can be used in other hierarchical models as a building block (Figure 7.11).

7.1.4.8 Step 7: Export the Model

The primary reason for this technique of creating models is to export them for simulation. We can therefore export the model into a variety of formats,

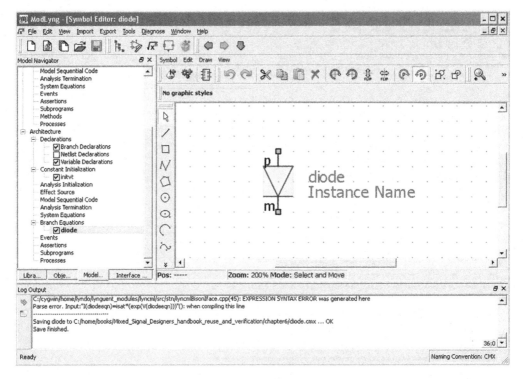

Figure 7.11:
Symbol for the diode model

including VHDL-AMS and Verilog-AMS. If we export this model to Verilog-AMS, we get an exported standard model as follows:

```
/* pragma lynguent created by ModLyng-1.3.0-S297 */
`include "disciplines.vams"
`include "constants.vams"
//A simple diode model
module diode(p, m);
    //Positive Pin
    inout p;
    electrical p;
    //Negative Pin
```

```
inout m;
electrical m;
//Saturation Current
parameter real isat = 1.0e-15;
//Emission Coefficient
parameter real n = 1;
//temperature in Kelvin
parameter real t = 300;
//The threshold voltage expression
localparam real vt = 0;
//
branch (p, m) diodeeqn;
analog
    // pragma lynguent code fragment "diodeeqn"
      I(diodeeqn) <+ isat * (exp(V(diodeeqn) / (n * (('P_K * t) / 'P_Q))) -
1);
Endmodule
```

Note that all the comments inserted in the model editor, ModLyng, have been reproduced in the exported model.

If the VHDL-AMS export option is selected, then in one step a new model is created for the VHDL-AMS simulator with the exported code as follows:

```
-%PRAGMA LYNGUENT CREATED BY ModLyng-1.3.0-S297
-A simple diode model
library ieee;
use ieee.FUNDAMENTAL_CONSTANTS.all;
use ieee.MATH_REAL.all;
use ieee.ELECTRICAL_SYSTEMS.all;
entity diode is
    generic (
-Saturation Current
        constant isat : in real := 1.0e-17;
-Emission Coefficient
        constant n : in real := 1.0;
```

```
-temperature in Kelvin
      constant t : in real : = 300.0
    );
    port (
-Positive Pin
      terminal p : ELECTRICAL;
-Negative Pin
      terminal m : ELECTRICAL
    );
end entity diode;
architecture model of diode is
-The threshold voltage expression
    constant vt : real : = 0.0;
-

    quantity Across_V_diodeeqn across Through_I_diodeeqn through p to m;
begin
    Through_I_diodeeqn = = isat * (EXP(Across_V_diodeeqn / (n * (PHYS_K *
t / PHYS_Q))) - 1);
end architecture model;
```

7.1.4.9 Summary

This section has demonstrated how to create electrical models graphically using ModLyng and, as such, the resulting models are standardized in format, checked and validated without the designer being required to carry out simulations to check syntax and basic equation behavior checks already performed.

7.1.5 Thermal System Modeling

7.1.5.1 Introduction

Thermal modeling uses the same type of construction as electrical with only definitional differences. The through variable is defined as heat flow (W) and the across variable is temperature (°C or K). This makes sense from a physical point of view, as the heat flows through the physical elements, and this causes a temperature difference between points.

Modeling thermal networks in simulators is then simply a case of assigning thermal types to the system connections, where the across variable represents the temperature and the through variable represents the heat flow.

Consider the case of a thermal resistance. The definition of a thermal resistance (R_{TH}) is that the relationship between the temperature difference (T_D) and heat flow (Q) is linear, with the equation defined in Eq. (7.4):

$$T_D = Q \cdot R_{TH} \tag{7.4}$$

As we have discussed already, the temperature only makes sense when it is considered with reference to another node, just as in an electrical system the voltage is a potential *difference* between two nodes, not an absolute value. This is obviously similar to the electrical domain, although, of course, in the thermal domain there are physical constraints, such as absolute zero (0 K). The *through* variable is defined as heat flow, which is analogous to current in the electrical domain, with the units of W (watts). A useful way to think about thermal models is that power applied to the thermal resistor will cause a temperature change. As the relationship between heat flow and temperature in a thermal resistor is instantaneous, it is also necessary to consider the case of thermal capacity, where there is a lag between the applied power and the resulting temperature change, and also the entity will hold the temperature when the source of heat is removed and will gradually decay over time (thermal energy storage). The equation for thermal capacitance is given in Eq. (7.5):

$$Q = T_C \frac{dT}{dt} \tag{7.5}$$

where Q is the heat flow through thermal capacitance, T_C is the thermal capacitance, and T is the temperature. We can create models for these elements, such as the thermal capacitance using the same standard procedure as we illustrated in the electrical domain. In this section, we will create models for a thermal resistance and capacitance, and then show how they can be connected in a thermal network to model the behavior of a simple thermal system to change in temperature.

7.1.5.2 Step 1: Thermal Resistor Model

As we did for the electrical diode, we can create the model for a thermal resistor and define the top level details of the model. In this case, the model name is defined as r_thermal.

7.1.5.3 Step 2: Thermal Resistor Connection Ports

The thermal resistance is a two-port model with the names defined as p and m. When these are created in a graphical modeling software tool, such as ModLyng, the types are defined from the existing library of conservative (thermal) types, as shown in Figure 7.12.

7.1.5.4 Step 3: Define the Model Parameters for the Thermal Resistor

A basic thermal resistor only has a single parameter − the thermal resistance. This is defined as having the name rth, is a real number type, and we can

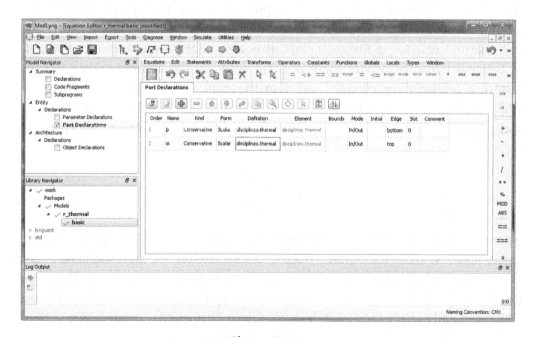

Figure 7.12:
Port definitions for thermal resistor

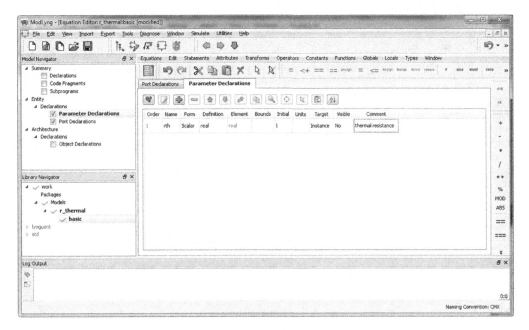

Figure 7.13:
Parameter declaration for thermal resistance

define a default value of 1°C/W, so for every watt of heat flow (power) applied, the temperature will increase by a single degree (Figure 7.13).

Now the basic elements for the model framework are in place, we can define the model internals, starting with the structure.

7.1.5.5 Step 4: Define the Model Topology for the Thermal Resistor

As we have seen already, the model for the thermal resistor is defined by a single fundamental equation relating the temperature difference and heat flow through and across the model, respectively. We can therefore state that we need a single branch to be defined between the two ports p and m. This is easy to define in a graphical form, as shown in Figure 7.14.

There are two options for defining the behavior in this topology. If we use an "implied" branch approach, we can name the through variable (q) and the across variable (td), and the branch will then take the name branch_td_q,

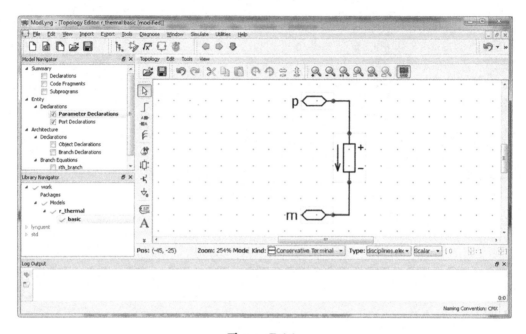

Figure 7.14:
Topology for thermal resistor

constructed from those names. We can also use an "explicit" branch definition approach, where the branch is named (say `rth_branch`), and the across variable becomes `HF(rth_branch)` and the across variable becomes `K(rth_branch)`.

Now with the branch in place, we can define the governing equation for the model.

7.1.5.6 Step 5: Define the Fundamental Equations for the Thermal Resistor

Having established the branch topology for the model, we can implement the equation for the thermal resistance, using the basic equation we introduced at the beginning of this section in Eq. (7.4) and the branch we have just created in the topology editor, directly in the graphical modeling software. This is shown in Figure 7.15.

At this point we have a complete model for a thermal resistance and we could use this model as a primitive element to create a more complex thermal network.

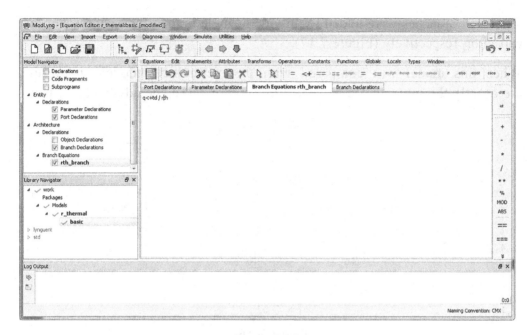

Figure 7.15:
Equation added for the thermal resistance

7.1.5.7 Thermal Network Modeling

As we have a basis for creating thermal primitive models, we could use a schematic-based approach using these primitives to create more complex thermal networks including a variety of elements. This is a straightforward macro-modeling technique used commonly in electric circuits with electric primitives. With the graphical modeling software tool, however, we have another option, which is to create a more complex topology and include all the relevant equations directly in a single model.

Consider the thermal circuit shown in Figure 7.16. It has a single input and output, and the network is with reference to a thermal reference. In this instance, the topology will reflect the global nature of the reference; however, it could in principle be a relative reference and connected to a port on the network model.

Modifying the simple thermal resistance model, the port connections were first changed from p and m to heat_in and heat_out, respectively, an extra parameter was added cth for the thermal capacitances, and the biggest change was the topology. Branches were added to provide a single branch for each resistor and capacitor, respectively (Figure 7.17).

When this model was compared with the hierarchical model, the results are exactly the same as can be seen from Figure 7.18. The advantage of using

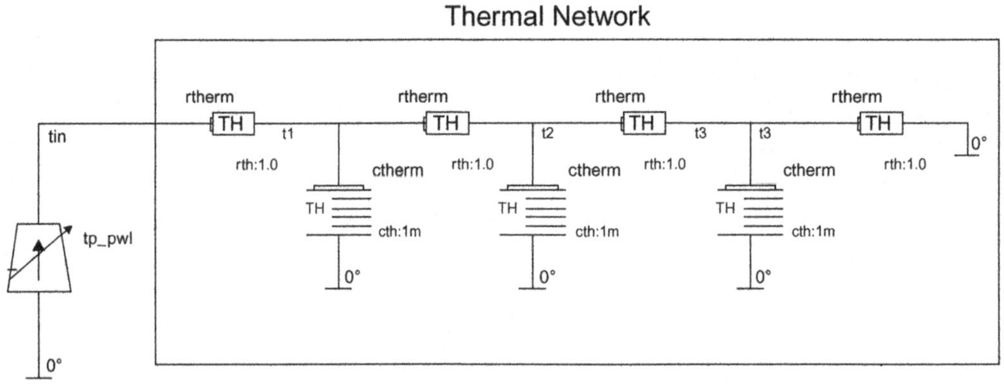

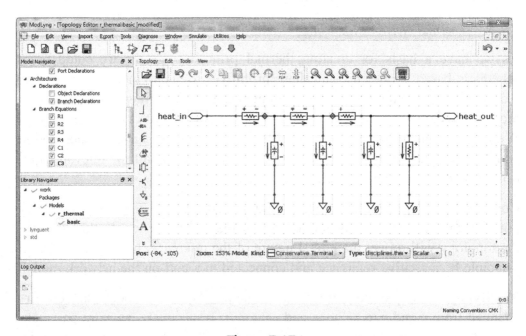

Figure 7.16:
Thermal network using thermal resistance and capacitance primitives in Saber

Figure 7.17:
Topology of thermal network

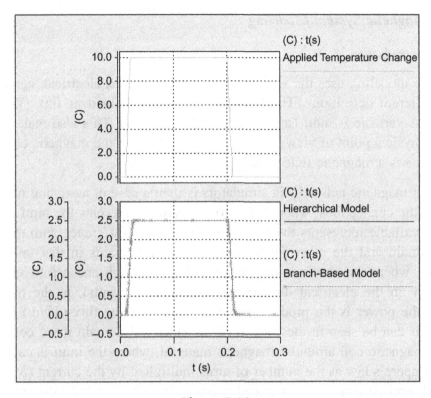

Figure 7.18:
Thermal model — hierarchical and branch-based simulation results

the branch-based approach is that the model is self-contained, which can be very useful, and also there is no overhead in equations being used to manage interconnections between individual models, but rather the branch equations are as efficient as it is possible to be. As we saw in Chapter 3 on design analysis, every time an equation is added to the matrix to be solved, the simulation time will generally increase, and so keeping unnecessary additional equation variables to a minimum is desirable. The test case applies a step temperature change of 10°C to the input of the thermal network, and the output of the network is shown in Figure 7.18, where the thermal network spreads the temperature over the thermal resistors and the thermal capacitance provides a time constant to the behavior of the temperature change.

7.1.6 Magnetic System Modeling

7.1.6.1 Introduction

Magnetic modeling uses the same type of construction as electrical, again with some different definitions. The through variable is defined as flux (Wb) and the across variable is mmf (magneto motive force) (A). This also makes sense from a physical point of view, as the flux flows through the magnetic elements, which causes a magnetic field difference between points.

Modeling magnetic networks in simulators is then a case of assigning magnetic types to the connections, where the across variable represents the mmf and the through variable represents the flux. There is a slight difference with the magnetic domain and the other fundamental physical domains in that unlike the majority, where the power is the product of the through and across variables (e.g., $I \cdot V$ in the electrical domain, giving power in watts), in the magnetic domain the power is the product of the *derivative* of the flux ($d\Phi/dt$) and the *mmf*. This can be seen in the context of the electrical domain if we consider a simple magnetic coil around a magnetic material, where the mmf is calculated using Ampere's law as the number of turns multiplied by the current ($N \cdot I$), but the back emf generated is calculated using Faraday's law $N(d\Phi/dt)$. So, if we calculate the *electrical* power and then substitute the *magnetic* equivalents for voltage and current, we end up with the following relations:

$$\text{Power} = I \times V = \frac{mmf}{N} \times N\frac{d\Phi}{dt} = mmf \times \frac{d\Phi}{dt} \qquad (7.6)$$

So, as can be seen from this, the power is obtained from the rate of change of the flux, not the flux itself. This is an issue when we need to create a primitive element that *dissipates* power, analogous to an electrical resistor, in the *magnetic* domain. In this instance, we need a primitive element that is a linear relationship between the across variable (mmf) and the through variable *derivative* (rate of change of flux), which we can see is very similar to an *electrical inductor* equation:

$$mmf = R_m \times \frac{d\Phi}{dt} \qquad (7.7)$$

where R_m is the "magnetic resistance" and is, in fact, despite the derivative, a power loss. In a similar manner, we can establish what the element for storage of the energy in the magnetic domain is. In this instance we can use the analogous form of the capacitor in the electrical domain, especially if the charge is used as the fundamental unit.

Consider the case where we relate charge Q to current i:

$$i = \frac{dQ}{dt} \tag{7.8}$$

And the equation for a capacitor is

$$i = C \cdot \frac{dv}{dt} \tag{7.9}$$

We can substitute the equation for the current and charge into the capacitance equation to give the following:

$$\frac{dQ}{dt} = C \cdot \frac{dv}{dt} \tag{7.10}$$

Which we can rewrite, integrating both sides as

$$Q = C \cdot v \tag{7.11}$$

This situation is directly analogous to the magnetic domain, where we can replace charge with flux, v with mmf, and so a "flux storage" primitive will be an energy equivalent to an electrical capacitor. However, it requires a linear relationship between the flux (charge) and the mmf (voltage):

$$\Phi = C_m \times mmf \tag{7.12}$$

where we have defined the capacitance C_m as the "magnetic capacitance" of the element.

This is a counter-intuitive model primitive in the same sense as the "magnetic resistor," as a *linear* expression of the flux and mmf will result in a storage (active) element, rather than a dissipative primitive (owing to the derivative of the flux being the variable in the power calculation). It is worth noting, however, that it is important in such cases to ensure that the initial conditions are satisfied correctly, as the integration assumes that the initial conditions have been established correctly.

Calculating the energy in the system from the power is carried out in the same manner as the other domains, so the energy is then the integration of the power. In the magnetic domain, this is the integration of the mmf and the derivative of the flux:

$$\text{Energy} = \int mmf \times \frac{d\Phi}{dt} dt \tag{7.13}$$

7.1.6.2 Step 1: Linear Magnetic Core Model

Now that we have established the fundamental behavior of the magnetic domain in general, we can define some basic primitive elements, based on the flux and mmf, and the simplest element is a magnetic material with a linear permeability. The relationship between the flux and the mmf in the linear core model is defined as follows:

$$mmf = \Phi \times \Re \tag{7.14}$$

where the reluctance ($\Re$) is related to the size of the core and the permeability. In most practical systems the designer is concerned with the relationship between the flux density (B) and the field strength (H), which are related to the flux and mmf, respectively, with the area (A) and length (L) of the core material:

$$B = \Phi/A \tag{7.15}$$

$$H = mmf/L \tag{7.16}$$

In fact, the core model is often expressed in terms of B and H, and the reluctance is then composed of the permeability, A and L:

$$B = \frac{\mu_0 \mu_r L}{A} \times H \tag{7.17}$$

where μ_r is the relative permeability of the core material.

It is common, in fact, to compose magnetic material models by calculating variables for B and H, and then expressing the relationship using the permeability (which is often nonlinear in practice).

7.1.6.3 Step 2: Linear Magnetic Core Connection Ports

The linear magnetic core is a two-port model, with the names defined as p and m. When these are created in a graphical modeling software tool, such as ModLyng, the types are defined from the existing library of conservative (magnetic) types as shown in Figure 7.19.

7.1.6.4 Step 3: Define the Model Parameters for the Linear Magnetic Core Model

As we have discussed already, the linear core model can be defined using reluctance; however, in practice, most designers will use the relative permeability, which means that the cross-sectional area and effective magnetic path length will also need to be defined. These definitions are shown graphically in Figure 7.20.

Now that the basic elements for the model framework are in place, we can define the model internals, starting with the structure.

7.1.6.5 Step 4: Define the Model Topology for the Linear Magnetic Core Model

As we have seen already, the model for the thermal resistor is defined by a single fundamental equation relating the flux density (B) and the magnetic

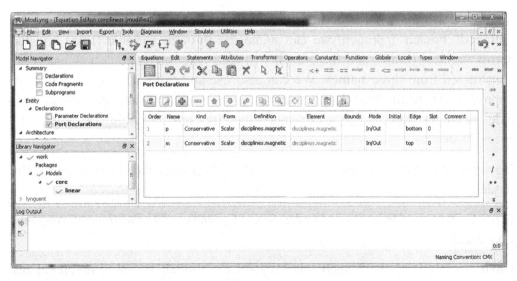

Figure 7.19:
Port definitions for linear magnetic core

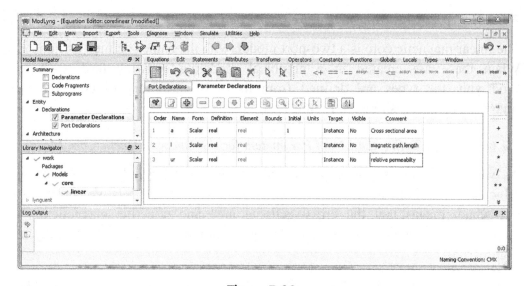

Figure 7.20:
Parameter declaration for linear magnetic core model

field strength (*H*) through and across the model, respectively. We can therefore state that we need a single branch to be defined between the two ports p and m. This is easy to define in a graphical form as shown in Figure 7.21.

There are two options for defining the behavior in this topology. If we use an "implied" branch approach, we can name the through variable (flux) and the across variable (mmf), and the branch will then take the name branch_mmf flux constructed from those names. We can also use an "explicit" branch definition approach, where the branch is named (e.g., core_branch), and the across variable becomes MMF(core_branch) and the through variable becomes PHI (core_branch).

Now with the branch in place, we can define the fundamental equation for the model.

7.1.6.6 Step 5: Define the Fundamental Equations for the Linear Magnetic Core Model

Having established the branch topology for the model, we can implement the equations for the linear magnetic core using the basic equations we introduced

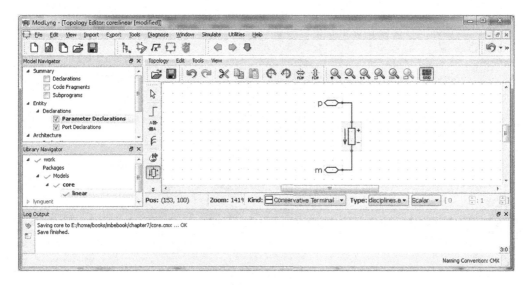

Figure 7.21:
Topology for linear magnetic core model

at the beginning of this section, and referencing the branch we have just created in the topology editor we can create the required equation directly in the graphical modeling software. In this instance, however, we have the added issue of the related equations for flux density (*B*) and magnetic field strength (*H*), which we are using to make the governing equation more accessible to designers. We therefore need *three* equations in total to define the behavior completely. Only one of the equations is, in fact, a *branch* equation in terms of *B* and *H*. Prior to the creation of this expression, we need to create the two expressions for *B* and *H* using the *quantities* definition within the graphical modeling software (Figure 7.22).

Now we can define expressions for *B* and *H* within the model and then the overall expression for the flux and mmf (Figure 7.23).

At this point we have a complete model for a linear core and can use this to define magnetic circuit elements. Exporting a model allows some simple simulation to take place and, in this case, a linear core model can be exercised using a simple mmf or flux source, as shown in Figure 7.24.

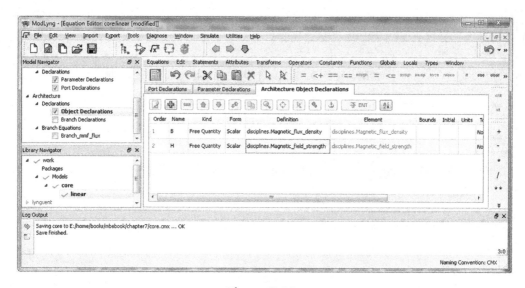

Figure 7.22:
Definition of *B* and *H* magnetic model free quantities

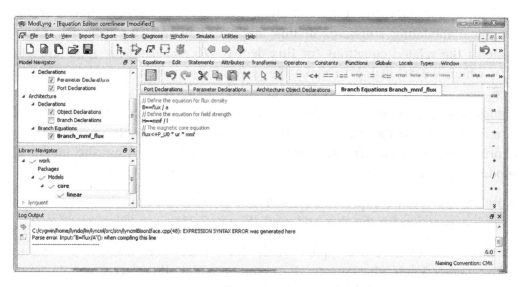

Figure 7.23:
Equations for the linear core model

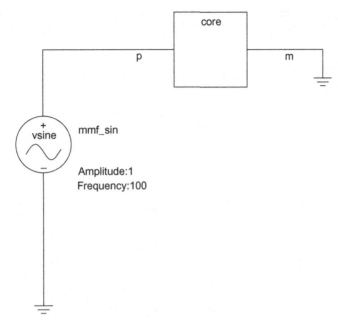

Figure 7.24:
Linear magnetic core test circuit

The resulting time domain waveforms for magnetic field strength (H) and flux density (B) can be seen in Figure 7.25.

These curves show the fact that the flux density is linear and in phase with the magnetic field strength as one would expect for a relatively low value of permeability (10). The resulting flux density is also very low. Note that in this case, the area and length are normalized to 1, whereas in most practical instances, the ratio of length to area would give a scale factor of something on the order of 1000. We can gain more insight into the behavior by plotting the flux density B with respect to the field strength H to produce a $B-H$ curve and, as we can see from Figure 7.26, it shows a straightforward linear relationship.

7.1.6.7 Nonlinear Magnetic Models

In practice of course, there is usually a highly nonlinear relationship between B and H in most magnetic materials (such as ferrites or metals), and these need to be modeled in a real system, to establish saturation or power loss.

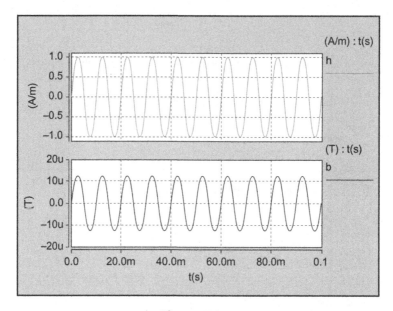

Figure 7.25:
B and *H* in the linear magnetic core model

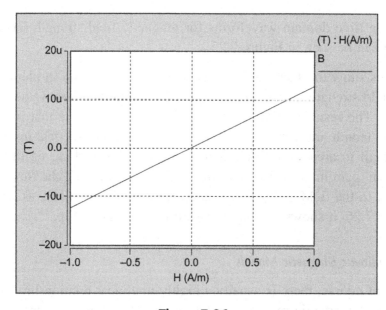

Figure 7.26:
B–H characteristic of linear magnetic material model

Jiles and Atherton developed a theory of ferromagnetic hysteresis (1983–1986) which separates the hysteresis function into the reversible, or anhysteretic, and the irreversible, or loss, magnetizations [2–4]. The magnetization is the lumped change in magnetic state when an external magnetic field is applied to the magnetic material and gives rise to an equivalent magnetic flux. Jiles and Atherton explain how the behavior of individual magnetic particles and domains can be treated as a bulk material and an effective lumped expression derived for the magnetization. The total lumped magnetization as derived by Jiles and Atherton is given in Eq. (7.18) and illustrated in Figure 7.27:

$$M_{TOTAL} = M_{IRREVERSIBLE} + M_{REVERSIBLE} \tag{7.18}$$

The normalized anhysteretic function, M_{an}, is approximated by the Langevin function as given in Eq. (7.19), where H_e is the effective applied magnetic field and A is the parameter modifying the curvature of the function. The function must be scaled by the saturation magnetization M_s to obtain the actual reversible magnetization:

$$M_{an} = \frac{1}{\tanh(H_e/A)} - \frac{A}{H_e} \tag{7.19}$$

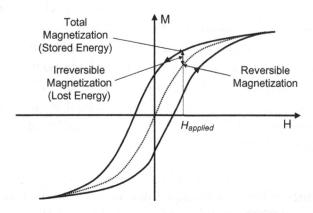

Figure 7.27:
Jiles and Atherton reversible and irreversible magnetization

The rate of change of the irreversible magnetization, M_{irr}, is obtained using Eq. (7.20), which is a lumped model of the losses caused by domain wall movement and distortion. M is the total magnetization, δ is the direction of the applied field strength H ($+1$ for positive, -1 for negative slope), μ is the permeability, and k is the model parameter that defines the hysteresis of the loop. The parameter A defines interdomain coupling and is effectively a proportion of the magnetization:

$$\frac{M_{irr}}{dH} = \frac{(M_{an} - M)}{(\delta k/\mu) - \alpha(M_{an} - M)} \tag{7.20}$$

The total magnetization rate of change is calculated using Eq. (7.21), where c is the parameter dictating the relative proportion of reversible and irreversible magnetizations:

$$\frac{dM}{dH} = \frac{1}{1+c} \frac{M_{an} - M}{(\delta k/\mu) - (M_{an} - M)} + \frac{c}{1+c} \frac{dM_{an}}{dH} \tag{7.21}$$

This model has been implemented in a variety of commercial simulators and is the *de facto* standard model for most nonlinear core modeling using circuit simulators.

Implementing the nonlinear equations for the Jiles and Atherton model highlights a number of issues for practically modeling, such a nonlinear system common to many such situations. Taking the anhysteretic magnetization defined in Eq. (7.19), for example, it can be seen that for small values of magnetic field strength, there will be an issue when H approaches zero, leading to an infinite value of magnetization, and, in fact, most simulators use a small linear region close to the origin, such that the potential divide by zero error is avoided. If the equation in Eq. (7.19) is plotted, the effect of the saturation can be seen easily (Figure 7.28).

If the ratio of H/a (where a is a model parameter) is taken as the basic variable in the equation, then it can be calculated that for a value of 0.001, the resulting magnetization is 0.000333, and therefore for values of H/a less than 0.001, a simple linear expression can be used instead.

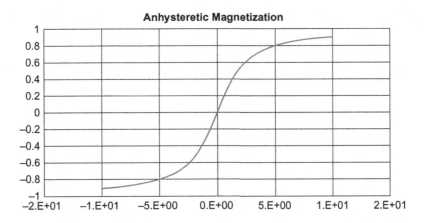

Figure 7.28:
Langevin anhysteretic magnetization

The resulting model equations can be implemented as an alternative architecture to the linear core model, with the name "Langevin" after the function in this case and the equations implemented as follows:

```
// Define the equation for flux density
B = = flux / a
// Define the equation for field strength
H = = mmf / 1
// calculate the ratio of H over the JA parameter a
H_a = = H / ja_a
// calculate the magnetization
if ( abs(H_a) < 1.0e-3 ) then {
mag = = 0.333 * H_a
} else {
mag = = (1 / tanh(H_a)) - (1 / H_a)
}
```

The resulting model behavior can be seen in Figure 7.29.

This clearly correlates with the ideal function, and the case where the function crosses through the origin is also taken care of by the addition of the linear stage where *H/a* is less than 0.001. The result of having this ability to implement a nonlinear function is that the saturation of the core can now be taken

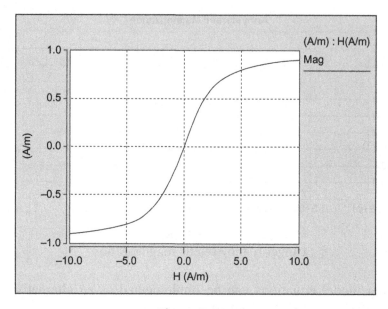

Figure 7.29:
Langevin function modeled and tested

into account. This model will be discussed and extended in much more detail later in this book.

7.1.7 Electromagnetic System Modeling

7.1.7.1 Introduction

The power of multiple domain modeling becomes apparent when different technologies are required to be simulated in the same circuit. A good example of this is simulating a magnetic component in the context of an electrical circuit. If we consider a winding, it has electrical connections to the circuit with voltages and currents, but there are also the magnetic variables, flux, and magnetic field in the magnetic domain. As has been stated previously, if a purely electrical simulator is used, then all the variables will be electrical anyway, but analogies are used to infer the correct type. If HDLs are applied, then the correct units can be stated explicitly in the models. The example of a winding has four connections, as shown in Figure 7.30.

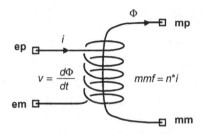

Figure 7.30:
Electromagnetic winding model

The connections ep and em are electrical (voltage and current) with the connections mp and mm magnetic (flux and mmf). The nature of the relationship between the domains is defined in Figure 7.30. Note that the transfer from electrical to magnetic is straightforward, with the mmf proportional to the current in the winding, but that the magnetic to electrical transfer has the voltage proportional to the *derivative* of the flux. This has important implications in regard to operating points and offsets in the domains, and a magnetic offset will not be recognized directly in the electrical domain. One possible approach therefore is to use the derivative of the flux as the through variable, but this precludes the setting of DC flux conditions in the model. This multiple domain model can still be modeled using a graphical modeling approach, and, as we have seen already with the electrical and magnetic domains, defined using appropriate connections and parameters. In this case, we have two sets of equations for the translation of energy from the electrical to magnetic domains, defined as we have seen in the section on magnetic modeling by Ampere's law and Faraday's law (in Eqs (7.22) and (7.23)).

$$mmf = N \cdot I \tag{7.22}$$

$$v = -N \cdot \frac{d\Phi}{dt} \tag{7.23}$$

As can be seen from these equations, in addition to the four connections shown in Figure 7.30, there are two branch equations for each side of the winding, an electrical equation that relates the flux to the voltage and a magnetic equation that relates the current to the mmf.

7.1.7.2 Definition of Parameters

As we have done in single domain models, after the basic model has been created (in this instance called winding), the next step is to define the parameters. In this model, there is one parameter for the two equations, which is the number of turns (N). This is defined as shown in Figure 7.31.

7.1.7.3 Definition of Multiple Domain Connection Points

The model is interesting as it contains not only magnetic pins, but also electrical pins, and these are defined using the same graphical approach as was used previously for the single domain models. Note that pins p and m are defined as magnetic, and ep and em are defined as type electrical (Figure 7.32).

7.1.7.4 Multiple Domain Equations

The final stage in the model is to create the two governing equations that define the behavior of the model linking the electrical and magnetic domains as defined in Eqs (7.22) and (7.23), respectively. Firstly, the two branches were created in each of the domains, although they can be in the same model topology as long as the connections are made correctly (Figure 7.33).

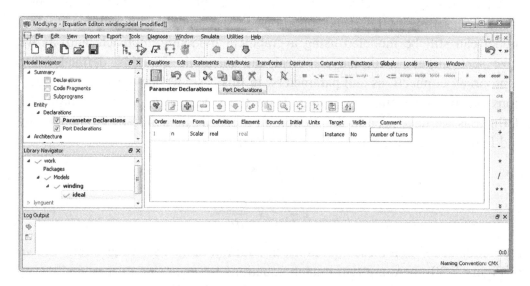

Figure 7.31:
Definition of N, number of turns, in winding model

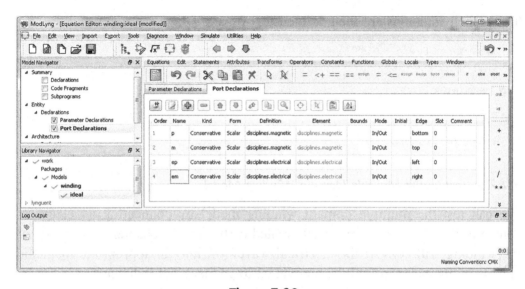

Figure 7.32:
Multiple domain connection points in winding model

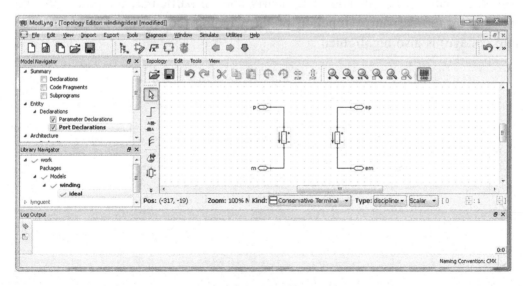

Figure 7.33:
Multiple domain topology diagram for winding model

The branch equations are then defined for Ampere's law:

```
// Amperes Law
mmf< + n * i
```

and Faraday's law:

```
//Faraday's law
v< + -n * flux'ddt()
```

7.1.7.5 Testing the Multiple Domain Winding Model

We can test the multiple domain winding model by connecting an electrical source to the magnetic core model created earlier in this chapter. The model of the winding will convert the electrical waveform into an appropriate magnetic signal and the behavior will be equivalent to an electrical inductance.

The test circuit is as shown in Figure 7.34.

The simulation results are shown in Figure 7.35, and these clearly demonstrate the multiple domain model in action with mmf and flux in the magnetic domain, and voltage and current in the electrical domain. The transformation of one domain to the other is seen clearly and, in particular, the phase change induced by the derivative function of the winding (converting flux derivative into voltage) is also highlighted.

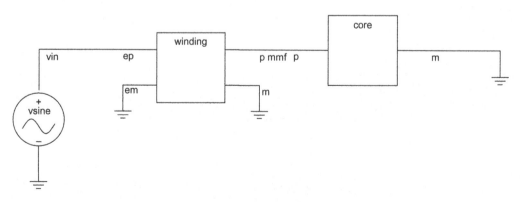

Figure 7.34:
Test winding circuit

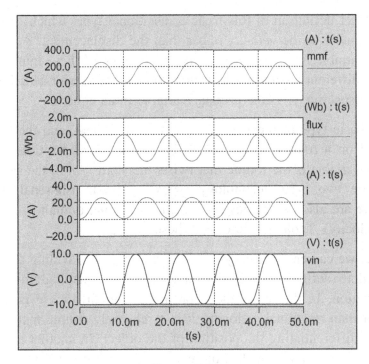

Figure 7.35:
Multiple domain simulation results of winding and core models

This model is an excellent demonstration of the ability of the model-based engineering approach to give insight into the details of the physical system and greatly simplify the complexity of the design.

7.1.8 Mechanical System Modeling

7.1.8.1 Introduction

Mechanical modeling is slightly different to the domains we have covered thus far as there is more than one way to describe systems. There are two common types of mechanical systems: translational and rotational. Translational systems exist where the movement is limited to one linear dimension, and the through variable is defined as the force on an object and the across variable is the resulting displacement.

There is another definition of this basic translational type, which is to consider the across variable as the *velocity* rather than the displacement. Again, though, the through variable is still defined as the force. Both of these implementations are useful in different contexts, where it is convenient to operate using distance or velocity, respectively, as the primary variable in an equation.

The second basic type of mechanical system is rotational, and the type definition is based on a rotating shaft, which has a rotational position or velocity. The through variable is the torque that twists the shaft. So, in the same manner that there are two basic technology types in the translational mechanical domain, there are also two basic types in the rotational domain (rotational and rotational velocity). Table 7.1 described these types.

For example, we can envision a system where a mass is traveling along a track, and we need to define the behavior in terms of the applied forces and the resulting position. In this case, the appropriate domain to use is the translational, rather than rotational, velocity. It is a relatively simple matter to obtain the speed by differentiating the position of the object. In another example, this time of a rotating motor, it makes more sense to describe the model using the rotational velocity domain, as the primary equations consider the velocity and torque as shown in Eqs (7.24) and (7.25):

$$V = L\frac{di}{dt} + iR + K_e\omega \tag{7.24}$$

$$T = K_t i - J\frac{d\omega}{dt} - D\omega \tag{7.25}$$

We will show how the two types of mechanical system can be defined using either a translational or rotational approach, and this can be used to create simple and effective mechanical models.

7.1.8.2 Mechanical Translational Systems

Mechanical systems in a translational context may be defined in multiple dimensions; however, the basic principle is the same regardless of configuration. The basic idea is that a force applied to an object will result in a change of position along the direction of the applied force as shown in Figure 7.36.

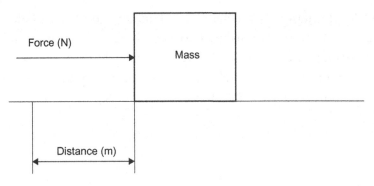

Figure 7.36:
Translational mechanical system

In a modeling context how do we turn these into the appropriate variables for modeling? In the same way as we have defined a physical *branch* that encapsulates the ability of a system to move energy around, this mechanical system is no different. In this context we have force and distance as the two variables of interest, and we can define force as the through variable and distance as the across variable. In an analogous way to applying a current to a resistor, we achieve a certain voltage in the mechanical system, and we can apply a force to a mass and the mass will move a certain distance. Also, analogous to the electrical system where the power is the product of voltage and current, in the translational mechanical system, power is the product of force and distance.

So, how will a mechanical system such as a moving mass be modeled in practice? As in any conserved energy system the key is in the basic equations for the entity itself. For example, a mass is governed by the forces applied to it, according to Newton's second law, that a body of a certain mass m, which has a force F applied to it, will accelerate in the same direction as the force applied to an acceleration a, as defined in Eq. (7.26):

$$\text{Force} = \text{mass} \cdot \text{acceleration} = m \cdot a \qquad (7.26)$$

Thus, the model of a mass will have a position relative to the translational reference and a governing equation:

$$F = m \cdot a \qquad (7.27)$$

Other typical translational mechanical elements might include a spring, where there is a linear relationship between the force applied and the position of the spring: the higher the force, the shorter the distance. This is defined in Hooke's law, which states that the displacement of a spring is directly proportional to the force applied to it and is defined in Eq. (7.28):

$$F = -k \cdot x \qquad (7.28)$$

where k is the spring constant, x is the displacement from the equilibrium, and F is the force applied. As can be seen from the equation, the force is inversely proportional to the displacement, so as the displacement increases the force increases to resist the displacement change.

We can envisage a system where there is a sprung mass, which will oscillate if released perpetually; however, in practice there is always either a mechanical damper force or some kind of dissipation of energy (through heat perhaps) to gradually slow a system down, and the key aspect of a mechanical damper is that the force is proportional to the velocity of the object, as defined in Eq. (7.29):

$$F = -k \cdot v \qquad (7.29)$$

We could, therefore, demonstrate the behavior of a simple mechanical system by creating a model of a mass, connected to a spring and a damper, and then applying a force to the mass and observing its behavior. For example, the mass could be suspended from a fixed beam and then allowed to be released with the effect of the force of gravity; then, we can establish what the equilibrium position would be (Figure 7.37).

With this system there are three elements that need to be modeled: the mass, spring, and damper. We can combine the equations for each of them to develop the resulting overall model of the system. At any particular time there is an overall force equation that must be made equal to the force applied by gravity; the equation is as follows:

$$F = ma - k_d v - k_s x \qquad (7.30)$$

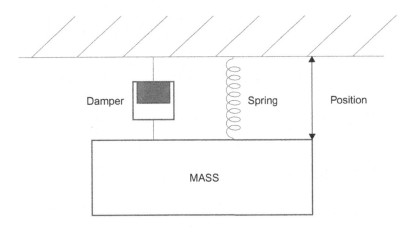

Figure 7.37:
Spring mass with damper

We can also express the velocity (v) and acceleration (a) in terms of the position (x) as follows:

$$F = m\frac{dx^2}{dt^2} - k_d\frac{dx}{dt} - k_sx \qquad (7.31)$$

which is a second-order differential equation system and can therefore be modeled either as a collection of elements or as a consolidated equation. Using a schematic approach is consistent with the model-based engineering philosophy of reusing models, and the resulting design is shown in Figure 7.38.

The results indicate that when the mass is released, the mass overshoots the equilibrium position; however, the spring ultimately provides a force that balances the gravitational pull and the damper ensures that the oscillation ceases after a few seconds (Figure 7.39).

Another approach to modeling mechanical translational systems is to use force and velocity; however, as we have seen, this is not appropriate for specific elements, such as springs, which depend on the absolute position, as initial conditions become difficult to manage with a complete system as a result (with the implicit integration required to calculate the instantaneous position in such models). Despite this, we may wish to use velocity rather than displacement if we are considering momentum as our primary interest so that the models can be defined in terms of mass and velocity.

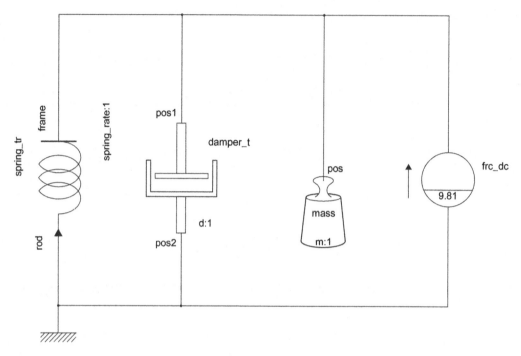

Figure 7.38:
Translational position mass, spring, and damper system test

7.1.8.3 Mechanical Rotational Systems

As with the mechanical translational system models, where there is the concept of a force applied to an object generating a change in position, which is analogous to a current applied to an electrical element generating a change in voltage (or emf or potential difference), there is a similar approach to modeling mechanical *rotational* systems.

The general background to a mechanical rotational system is that in the system the elements are rotating, for example on a shaft, with a certain rotational velocity or angle. In order to generate the rotation, a torque must be applied to the shaft and so the two variables which define the through and across variables in a mechanical rotational system are *torque* and *angular velocity*, respectively. It is also possible to use rotational angle as the variable; however, the main reason for using the angular velocity is that the power is simply the

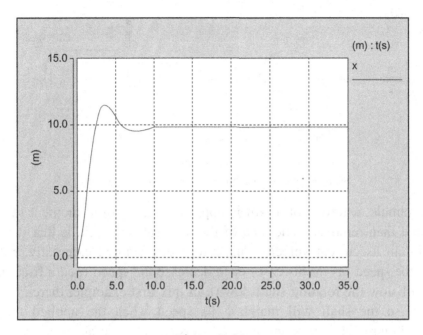

Figure 7.39:
Position response to gravity force application to mass of 1 kg

product of the torque and angular velocity, so the system is implemented in a similar manner to the other physical domains:

$$\text{Power} = \text{Torque} \cdot \text{Angular velocity} \qquad (7.32)$$

As we have seen previously in this chapter, once we have established the basic physical behavior of an element using equations, it is a relatively straightforward matter to convert these into suitable models. For example, consider a simple moment of inertia model. The basic equation for a moment of inertia is given in Eq. (7.33), where the torque is proportional to the acceleration, multiplied by the inertia (j). Note the negative sign, which indicates that the inertia *resists* the applied torque:

$$T = -j \cdot \frac{d\omega}{dt} \qquad (7.33)$$

We can create a simple model for a moment of inertia and demonstrate its behavior using the simple circuit shown in Figure 7.40.

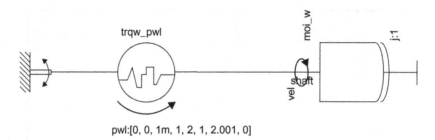

pwl:[0, 0, 1m, 1, 2, 1, 2.001, 0]

Figure 7.40:
Simple test of mechanical moment of inertia

In this example, a torque of 1 Nm is applied for 2 s to the moment of inertia ($j = 1$) and then removed. The idea of the moment of inertia is that it will try and maintain its current velocity unless torque is applied externally, either to increase the speed (as in this case) or to decrease the speed (i.e., a friction loss, which will slow the rotating shaft down). In this first example, there is no friction loss, so the shaft will maintain its speed when the applied torque is removed (Figure 7.41). In many respects, this can be seen as an energy storage mechanism, where the energy is stored in the rotating moment of inertia, until it is released into some other element. The energy stored in a rotating inertia (perhaps a flywheel) is defined in Eq. (7.34):

$$E = \frac{1}{2}j\omega^2 \tag{7.34}$$

which can be seen to be very similar to the energy storage equation for a capacitor (except the capacitance is replaced by the moment of inertia j and the voltage is replaced by the angular velocity ω) (Figure 7.41).

In a practical system, there will be some form of energy loss. Just as in an electrical system, the resistor dissipates energy through i^2R loss, in a similar manner the rotational mechanical system needs a dissipation mechanism, and this is through friction. There are actually a number of different types of friction in a practical system, and these can be roughly divided into viscous friction (where the friction torque is proportional to the rotational speed), the Coulomb torque (which is constant no matter what speed is applied), and a breakaway torque, which is the torque at which the system needs to overcome to begin moving (usually higher than the Coulomb torque). If we consider a very simple

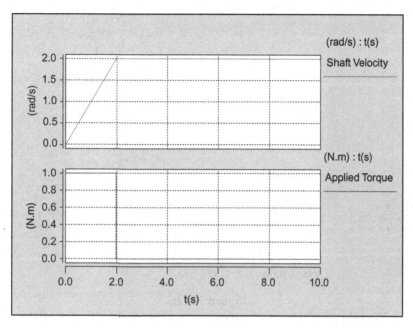

Figure 7.41:
Effect of applied torque on shaft speed for a moment of inertia

case, where we have a model of a viscous friction, then we can model the effect of this in a simple equation relating the torque to the speed:

$$T = F_v\omega \tag{7.35}$$

where F_v is the viscous friction. If a model for viscous friction is added to our previous moment of inertia model, then the effect will be observed as a dampening of the velocity (Figure 7.42).

The first thing to notice is that with the friction in place, the speed reached is much lower than the ideal case for the same torque. And then, when the torque is removed, the speed reduces to zero. This is clearly shown in the same simulation test, with the same applied torque as for the ideal moment of inertia, with the results given in Figure 7.43.

As for the translational mechanical domain, it is possible to use rotational *angle* rather than velocity. However, it is necessary to calculate the velocity in this domain to obtain the power and hence energy stored in these models.

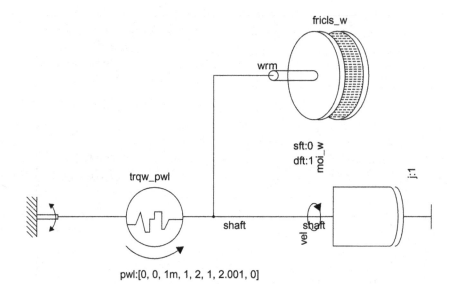

Figure 7.42:
Mechanical moment of inertia with an added friction model

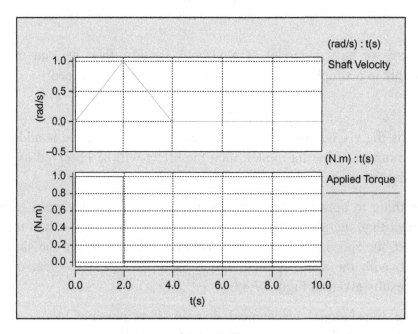

Figure 7.43:
Mechanical moment of inertia and added friction loss

7.1.9 Fluidic Systems

7.1.9.1 Introduction

In many applications, there is an interface to either microfluidics (such as in micro-electromechanical systems, or MEMS) or in larger systems to hydraulic pumps or actuators. We can describe such systems in two ways in the fluidic domain: compressible and noncompressible. If the fluid is effectively uncompressed, this is a simple situation to manage as the volume does not change and we can consider the through variable as the volume flow rate. The resulting pressure change is the across variable in such systems, and for laminar flow the models are relatively easy to characterize. If the flow becomes turbulent, we can still use the same basic definitions of types; however, the equations become much more complex. Fortunately, in many hydraulic systems, the fluid is viscous and dense, and therefore obeys these simpler laminar rules.

If the material is compressible (e.g., air), then the situation is different and we need to consider the *mass* of the material in the system; therefore, the through variable becomes the mass flow rate, while the across variable remains the pressure.

7.1.9.2 Simple Hydraulic Systems

If we consider a simple pneumatic system, where the fluid is effectively not compressible, then obviously this is a classical hydraulic system to be modeled, where we use pumps to move fluid around, chambers to store the fluid, and a series of pipes to connect the system together.

Consider a simple system where there is a cylinder forcing fluid into a pipe, and this, in turn, is pushing another cylinder out at the other end, as shown in Figure 7.44.

Figure 7.44:
Simple incompressible hydraulic system

In this system, the pressure applied is calculated in Eq. (7.36):

$$Pressure = \frac{Force}{Area}$$

$$\therefore P_{in} = \frac{F_{in}}{A_{in}} \tag{7.36}$$

In this system, as the pressure is constant (as the fluid is incompressible), the result is that the force on the output will therefore be calculated as follows:

$$P_{in} = \frac{F_{in}}{A_{in}}$$

$$P_{out} = P_{in} \tag{7.37}$$

$$\therefore F_{out} = P_{out} \cdot A_{out} = \frac{F_{in} \cdot A_{out}}{A_{in}}$$

In effect, therefore, a cylinder is an interface model which converts a translational force (and, by implication, position) into a hydraulic pressure. If the system has an equal pressure, then that pressure can then be converted back into a force by another cylinder. The "pipe" model connecting the two cylinders does not change the volume of fluid, and so there is no pressure change. In order to achieve a pressure change, there must be a flow of fluid, expressed for incompressible fluids as a volume flow rate (m^3/s).

7.1.9.3 Hydraulic System Elements

If a pipe has a constriction, then the effect will be to restrict the flow of fluid in the system, and this will cause work to be done, thus dissipating energy. This is analogous to an electrical resistor and can be modeled in a similar manner, where the pressure difference required to achieve the required flow rate depends on the amount of constriction in the pipe (Figure 7.45).

The basic equation for this model will be to calculate the relationship between the pressure difference at either end of the pipe, and this will be proportional to the flow rate, where the scale factor is a function of the pipe constriction. The greater the constriction, the greater the pressure difference for the same flow rate.

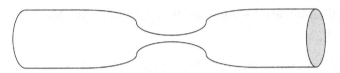

Figure 7.45:
Hydraulic pipe with constriction

Another useful element is that of a "valve", which is analogous to the electrical diode. A valve will allow fluid to flow in only one direction.

7.1.10 Optical Systems

Optical systems are interesting as they can be considered in a variety of ways. Large-scale optical systems, such as solar panels or lamps, can be modeled using lumped elements, and, as such, the common definitions of optical (also known as radiant) systems are based on luminance and light intensity. The luminance is defined in terms of its luminous intensity, which is a measure of brightness, usually with the unit of candela (basically how many candles it takes to get this brightness!) and a through variable in units of lumens.

Another related term is "irradiance", which is used in solar panel designs to describe the equivalent power per unit area incident on a panel, so that the output electrical power can be calculated. These terms are useful in lumped or large-scale models, and not appropriate for quantum-level modeling, where physicists are interested in photon-level models. In that context, the focus is more on energy transfer at a fundamental level, using very basic low-level units and definitions.

There is limited application of optical system modeling in most engineering systems, except perhaps in the calculation of irradiance incident onto photovoltaic (PV) cells for energy generation, and for this case the overall energy equation can be calculated.

The radiation from the sun reaches the earth in the wavelength range of a few hundred nanometers to several micrometers. Depending on the location of the PV array (space, mountain top, or sea level), the characteristics of the radiation will be different. The two main spectra used are either Air Mass Zero (AM0, for extra-terrestrial applications such as in space) or Air Mass 1.5 (AM1.5, for earth-based analysis).

For earth-based systems, the "standard total" (also referred to as "global tilt") spectrum results in an irradiance of 1000 W/m² and is commonly referred to as AM1.5G. The spectra along with supporting information can be found on the NREL web site [5]. The actual data can be downloaded and implemented as a data source in a model, as shown in Figure 7.46.

The AM 0 (air mass zero) response can also be obtained online [6], and this is used primarily for space-based applications.

The actual irradiance is the integration over the wavelength (shown in nanometer) of all the individual spectral responses. For a PV installation, the overall irradiance is the important figure; therefore, this will be the source data to the model. In addition to the spectral response of the irradiance, there is a profile over the daily cycle of the irradiance called the irradiance profile. This is

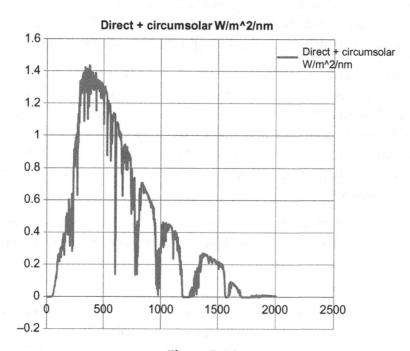

Figure 7.46:
Irradiance data implemented in a system model

obviously location- and calendar-specific. The profile can be modeled using a simple equation, as shown in Eq. (7.38), using a simple Gaussian distribution curve as a basis:

$$Irr = ae^{\frac{-(x-b)^2}{2c^2}} \qquad (7.38)$$

where b is the offset from zero (say, 12 noon), c is the width of the curve, and a is the maximum amplitude.

Obviously, each day will have a different profile, so each of the parameters a, b, and c can be characterized from the observed weather data to provide a complete model of the annual irradiance for system modeling.

This can be implemented using a building block approach (Figure 7.47).

With the irradiance calculation in place, the next step is to model a PV cell. The basic principle of PV cell modeling is to have a current source, with an output proportional to the applied irradiance, an ideal diode, and a shunt resistance, all in parallel, with a series source resistance as shown in Figure 7.48.

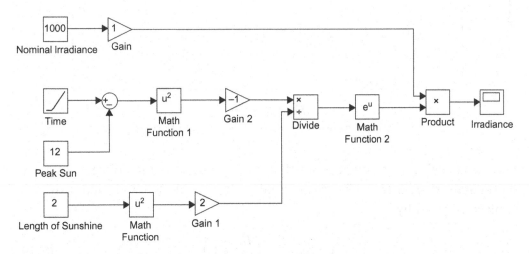

Figure 7.47:
System model using building blocks of irradiance

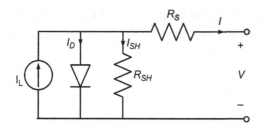

Figure 7.48:
Simple photovoltaic cell equivalent circuit model

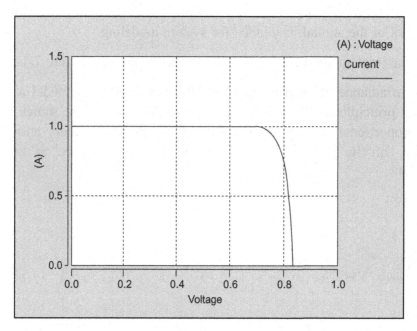

Figure 7.49:
Photovoltaic cell model *I-V* characteristic

The basic cell can be characterized using its open circuit voltage (V_{oc}) and short circuit current (I_{sc}). The open circuit voltage with the diode and resistance in place is given by

$$V_{oc} \approx \frac{kT}{q} \ln\left(\frac{I_L}{I_0} - 1\right) \qquad (7.39)$$

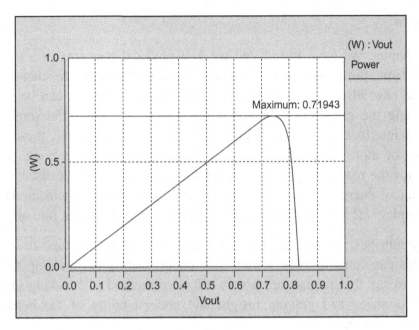

Figure 7.50:
Maximum power point from a photovoltaic cell

The short circuit current with the diode and shunt resistance shorted out is

$$I_{sc} \approx I_L \tag{7.40}$$

Implementing this simple model with $R_{sh} = 1$ kΩ and $R_s = 1$ mΩ, with a source current of 1 A yields the *I-V* characteristic as given in Figure 7.49.

This simple model can now be used to model larger, more complex PV systems. There is also a direct relationship between the temperature and the performance of the solar cell, and this can be included in the model using Eq. (7.41):

$$I(T) = I^*(1 + TK1^*[T - T_{mean}]) \tag{7.41}$$

In addition, the device (diode) also includes temperature dependence in circuit simulators, such as SPICE or Saber. With the *I-V* characteristic shown in the basic model, there is a simple calculation of power $(I \cdot V)$, which shows the maximum power achievable from this PV cell (Figure 7.50).

Conclusion

In this chapter, we have introduced the key physical concepts in a variety of domains from fluids to electronics and mechanical systems to magnetic systems. We have also introduced how such continuous systems can be modeled easily using the graphical techniques introduced in Chapter 5. Some of the important interfaces have also been described briefly; however, these will be the focus of more detailed analyses in examples later in this book. The key concept for the reader to grasp is that whatever system is under discussion, the movement of energy is the fundamental driver and for a system model to be truly complete *all* aspects of energy conservation must be taken into account.

Each domain has its own nuances and, in some cases, language to interpret. While this can sometimes be a barrier to understanding engineering solutions, it is true to say that models are a very effective method of breaking into these different domains and gaining insight and understanding of the behavior of devices and systems.

References

[1] R.P. Feymann, Six Easy Pieces: Fundamentals of Physics Explained, Penguin Press, 1998.

[2] D.C. Jiles, D.L. Atherton, Theory of ferromagnetic hysteresis (invited), J. Appl. Phys. 55 (6) (1984) 2115−2120.

[3] D.C. Jiles, D.L. Atherton, Theory of ferromagnetic hysteresis, J. Magn. Magn. Mater. 61 (1986) 48−60.

[4] D.C. Jiles, D.L. Atherton, Ferromagnetic hysteresis, IEEE Trans. Magn. 19(5) (1983) 2183−2185.

[5] Solar Spectral Irradiance for Air Mass 1.5, Renewable Resource Data Center, National Renewable Energy Laboratory, Golden, CO, USA. Available from: http://rredc.nrel.gov/solar/spectra/am1.5/#about (accessed 2012).

[6] Solar Spectral Irradiance for Air Mass Zero, Renewable Resource Data Center, National Renewable Energy Laboratory, Golden, CO, USA. Available from: http://rredc.nrel.gov/solar/spectra/am0/ASTM2000.html (accessed 2012).

CHAPTER 8

Event-Based Modeling

"Mach 2 travel feels no different," a passenger commented on an early Concorde flight. "Yes," Sir George replied. "That was the difficult bit." — Sir George Edwards, co-director of Concorde development — quote taken from "Concorde, New Shape in the Sky." The ability of design engineers to predict massively complex systems often relies on event-based and primarily digital simulation. There is a trade-off between speed and accuracy inherent in this approach and understanding that trade-off is essential for mixed-signal designers. Most digital designers take this efficiency completely for granted and have no understanding of this — much like the passengers on the supersonic Concorde aircraft. The opposite could be said of the analog and mixed-signal designers — much like the high altitude military aviators who observed Concorde passengers flying at 60,000 ft sipping Champagne while they wore helmets and full pressure suits!

8.1 Event-Based Modeling

8.1.1 Introduction

As we have seen so far in this book, modeling is shaped by multiple drivers, whether it is the specification or the platform or perhaps the fundamental technology to be used. In modern day engineering, however, there is usually one very "small" "elephant in the room", which is, generally, the computer that will control the overall engineering system. The key aspect of any computer since the 1970s has been that they are universally *digital* in nature. As we have already seen in the earlier chapters on simulation fundamentals, the basic premise of a digital system is that we are required to model variables using *discrete* levels and their behavior using a *discrete* time axis. Whichever digital representation is ultimately used by the engineer, these two fundamentals dictate how we handle *models* which can map into this digital world.

We actually have several rationales for digital modeling, however, which are not necessarily as obvious as it might appear. The first (and reasonably self-evident) reason for modeling systems digitally is that we intend to implement the design digitally using logic or computational hardware. In this instance, there is a well-understood design flow that design engineers can adopt to start from the standpoint of a reasonably abstract behavioral model and ultimately use software tools to *synthesize* the hardware to be realized. This process is called digital synthesis and we will discuss it briefly in this chapter. The second, and perhaps less obvious, rationale is to take advantage of the relative speed of digital simulation to implement a model of a system (which may not be digital in itself), which can, nevertheless, be simulated using a digital simulator. This encompasses a range of techniques, which we could collect under the general banner of "event-based" or "discrete" modeling and simulation. This set of techniques will be one primary focus of this chapter. The third rationale is more abstract, and perhaps beyond the scope of this book is that, ultimately, we will be running a software model on a digital computer, so *no matter which* modeling technique we choose, there will, ultimately, be a digital process involved somewhere along the line. While this may be one or two steps removed from the model itself, it is essential to bear this mind, even if it is simply the numeric resolution of the data to be considered.

As we have already introduced the key concepts in Chapter 2 on digital simulation, we can build on these concepts in this chapter by developing the techniques for logic modeling, illustrating how the systematic nature of digital systems leads almost directly to digital synthesis. Moving on from this, we will introduce the more general techniques of event-based models (sometimes referred to as sampled data systems or Z-domain).

Before embarking on a wide ranging review of event-based modeling approaches, however, there are some fundamental concepts that must be introduced first. As in any sampling system, the problem of resolution needs to be addressed, and therefore the basic nomenclature, definitions, and meaning of many of the event types, and how they relate to the analog world are first reviewed in this chapter.

8.1.2 Practical Issues

8.1.2.1 Type Definitions

Most simulators and modeling languages are strongly typed. This can be a particular issue when handling variables between the analog and digital domains. There are also sometimes restrictions in simulators about parameter types, as we shall see. This can lead to interfacing being the biggest single problem for the mixed-signal designer to consider *in the design,* and this, of course, feeds into the model as a result.

8.1.2.2 Type Mismatch

A common issue with modeling in any hardware description language is ensuring that types are consistent and correct in basic parameters. One of the issues for the designer is to make sure that a model that works in one simulator can potentially work in *all* simulators. This might seem a little overboard for single models, but tools migrate, cooperate, and change languages, as do companies, and so the largest common denominator in modeling is often a smart choice.

Consider the "humble" number 1.0. This is used in a huge number of equations, especially when inverting an expression, such as $1/x$. This simple usage can cause a number of issues, some subtle, when implementing in an analog or mixed-signal hardware description language (HDL) context. As we have seen already, the fundamental type in most analog simulators is the basic *real* type (usually based on a floating point number) and, depending on the simulator, this is checked rigorously. For example, take a simple expression of the form:

$$Y = \frac{1}{X} \tag{8.1}$$

In VHDL-AMS, if Y and X are defined as variables for analog simulation (the specific definition is QUANTITY — more on this in the next chapter), then this equation, as it stands, will fail a syntax check. The reason for this is that the number "1" is assumed to be of type INTEGER and is therefore incompatible with the other variables in the equation. However the MAST language (used in the Saber simulator) uses a generic number type (called "number"), which encompasses integers and real numbers. Thus, it is useful to note at this point that some HDLs and simulators are more strongly typed than others. This is an important issue for modeling and must not be overlooked. It is particularly an issue when translating models from pure equations into any implementation and certainly when translating from one language to another. In many ways, it is another strong argument for a graphical approach, as shown in Chapter 5, that abstracts the model away from any particular language definition.

8.1.2.3 Type Translation and Resolution

A typical issue for the mixed-signal designer is translation from the analog-to-digital domains and vice versa. This is an interesting problem, as the analog signals are usually defined as real number types, whereas the digital signals are usually defined with a certain number of digital logic connections in a bus configuration. How that digital value is described can vary enormously from a full floating point implementation consistent with the representation in the computer to a simple integer-based value using a relatively small number of bits. Whichever technique is used, however, requires some translation of types and then some additional functions to accurately specify the translation.

As a first example, consider the route from analog to digital. In this first example, we will define the digital representation as 8 bits wide, where the range of the analog input is restricted to 0.0–5.0 (a typical voltage range). The output digital word will therefore be between 0 and 255, where 0 corresponds to 0.0 and 255 corresponds to 5.0. In this approach, we first scale the input real number to the correct range, convert to an integer, turn this into a bit equivalent integer (unsigned), and finally translate this into a logic vector (Figure 8.1).

This sequence can also be written in a *pseudo-code* sequence; from this it is clear that whichever HDL is used, the translation functions are required to be in place to carry out these transforms:

```
Scaled_value : = input*255.0/5.0

Integer_value : = to_integer(scaled_value)

Unsigned_value : = unsigned(integer_value)

Output : = std_logic_vector(unsigned_value)
```

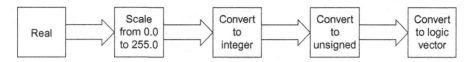

Figure 8.1:
Analog to digital conversion

It is also important to note that the assumption is made that the ranges of the scaled `integer_value`, unsigned, and logic vector variables are consistent. If this is not done carefully, then rounding or overflow errors may occur in the results. Clearly, this is not a simple process and has to be managed carefully to ensure the correct behavior of the final interface model. In this example, the issue of resolution is also apparent, where the smallest voltage resolution is dictated by the number of bits (8 in this case) and the input voltage range (5.0−0.0 V). The resolution is simple to calculate based on the power of 2 to the number of bits divided by the input voltage range. In this example, therefore, the smallest input voltage resolution is defined by:

$$\text{resolution} = \frac{\text{Vin}_{max}\text{-Vin}_{min}}{2^{\text{number of bits}}} = \frac{5.0 - 0.0}{2^8} = 5.0/256 = 19.53 \text{ mV} \qquad (8.2)$$

This can be improved by increasing the number of bits, but, clearly, this quickly becomes cumbersome and specific to each case. In more sophisticated models, either fixed or floating point resolution numbers will be used (e.g., in digital signal processing models).

In mixed-signal modeling including a string digital content, we have access to a range of types from bits and Boolean variables (which consist of two states "0" and "1" (or true and false), which are effectively enumerated types, through integer numbers (including positive and natural subtypes) and eventually we can use real numbers (floating point). Unfortunately, the big drawback is not necessarily what we can use, but rather what we can simulate in particular simulators or, perhaps, what is possible to synthesize in hardware. Despite recent research efforts and standardization efforts, there is still a limited availability of packages and libraries that support both fixed and floating point arithmetic. Where there is a need for some digital signal processing (DSP)-type application, generally a form of fixed point arithmetic will be adequate in most of these cases.

128	64	32	16	8	4	2	1
0	0	0	1	0	1	1	1

16 + 4 + 2 + 1 =23

Figure 8.2:
Basic binary notation

−128	64	32	16	8	4	2	1
1	1	1	0	1	0	0	1

−128 + 64 + 32 + 8 + 1 = −23

Figure 8.3:
Negative number binary notation

In integer arithmetic, unsigned, signed, or a logic vector, the basis of the number is a bitwise representation of an integer with no decimal point. For example, to represent the number 23, using 8 bits, we simply set a bit for each binary element required to construct the integer value of 23. This is shown in Figure 8.2.

If we require a negative number, then we use the "signed" approach, where the most significant bit (MSB) is simply the sign bit as shown in Figure 8.3. In fact, the standard "twos complement" notation can be obtained by inverting the bits and adding one to the least significant bit (LSB).

With this basic idea of handling numbers, we can extend the notation to a "fixed point" scheme by defining where the decimal point will go. For example, in the same number scheme shown, we have 8 bits. We can therefore define this in terms of 5 bits above the decimal point and 3 below it. This will give some limited fractional usage for the numbers. The way that this is implemented is by using fractions of 1 for each "negative" bit to the right of the decimal point. As an example, take the same number in terms of bits used in Figure 8.4 and use the new fixed point numbering system for the bits. In this case we get a value of -2.875.

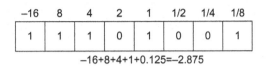

Figure 8.4:
Fixed point notation

The nice thing about this notation is that the bitwise functions defined for the integer-based ALU developed previously can also be applied to this new fixed point notation with almost no modification. The only difference is that we need to translate from the new fixed point type to a logic vector type and also consider how to handle overflow conditions.

For example, if two numbers are added together and the result is too large, how is this handled by the fixed point algorithms? Do we simply flag an overflow and output the result? Or do we set the maximum value and output this? Similarly, for numbers that may be too small and for which we can potentially lose precision, do we simply round up or flag another loss of precision condition? These are questions that the designer needs to answer for their application or may be defined by the particular package or library being used.

If we consider the route from digital to analog, a similar procedure is required to convert from logic vector, to unsigned (say), then to integer, and, finally, to a real number type. This is much easier to manage from a designer's perspective, as the ranges of numbers can propagate from the first definition of the logic vector through to the final number.

8.1.3 Digital Logic Modeling

8.1.3.1 Fundamentals

If we recall the basic definition of a logic signal, with values of either "0" or "1," from Chapter 3, and observe how it relates to a "real world" voltage, we can see how this looks in Figure 8.5. Note how the "real" voltage levels correspond directly to an absolute logic level. As was discussed in Chapter 3, there is also the definition of transition times, "strong" and "weak" logic, and also the "unknown" state X and high impedance Z.

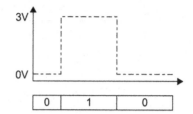

Figure 8.5:
Digital logic and voltage equivalence

The key aspect of the logic model, however, is the use of resolution tables to establish the output of logic conflicts in addition to logical equations to establish behavior.

8.1.3.2 Logic Operators

All digital simulators have a set of basic logic operators — "VHDL:Boolean operators" — built in, which are generally self-explanatory. The list of these operators are AND (&&), OR (||), and NOT (!). These operators can be applied to basic logic types such as:

```
out1 = A && B;

out2 = A || B;

out3 = !in3;
```

In addition to these basic logical operations, digital languages also have bitwise operators that operate bit by bit on the operands. These operators include NEGATION (~), AND (&), OR (|), XOR (^), and XNOR (~^, or ^~). These operators work on the operands using truth tables to calculate the correct output based on the four state logic described previously in this book. Taking each operator in turn, the truth table shows the correct output for the range of possible inputs.

For example, if we have two binary numbers, $A = 1010$ and $B = 1101$, then applying the bitwise operators to them we get the following code and results, which can be checked using the truth tables as shown in Figure 8.6.

	0	1	X		0	1	X		0	1	X		0	1	X		0	1
0	0	0	0	0	0	1	0	0	0	1	X	0	1	0	X	0	1
1	0	1	X	1	1	1	1	1	1	0	X	1	0	1	X	1	0
X	0	X	X	X	0	1	X	X	X	X	X	X	X	X	X	X	X

Bitwise AND   Bitwise OR   Bitwise XOR   Bitwise XNOR   Bitwise NOT

Figure 8.6:
Logical bitwise operator truth tables

```
A & B;    // A AND B:    Result = 1010 & 1101 which is 1000

A | B;    // A OR B:     Result = 1010 | 1101 which is 1111

A ^ B;    // A XOR B:    Result = 1010 ^ 1101 which is 0111

A ~^ B;   // A XNOR B:   Result = 1010 ^~ 1101 which is 1000
```

Using the combination of logic-type definitions, resolution tables and logical operators, it is possible to build up any logic expression. The crucial advantage of any digital system built on these logical foundations is that no matter what the language used, or the implementation chosen, ultimately the logic is always going to behave as specified. This is different to an analog circuit, for example, which may change behavior if the load value changes. This approach is used explicitly in combinational logic circuits, which are discussed in the next section.

8.1.3.3 Combinational Logic

Combinational logic circuits can be classified as digital logic circuits that do not have any storage elements. We can consider them in many respects to be purely a logic circuit, with an effectively instantaneous output change that reflects the change in inputs. From a modeling perspective, this could be implemented using logic equations, truth tables, lookup tables, or primitive logic gates, but the ultimate effect is the same.

For example, consider a simple logic expression:

$$Q = (A + B) \bullet (C + D) \tag{8.3}$$

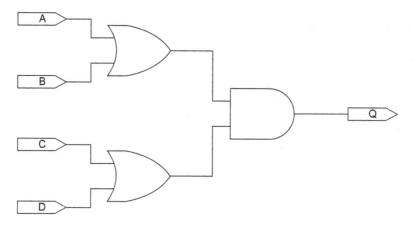

Figure 8.7:
Simple logic function implemented using logic primitives

This can be written in an English form as "(A OR B) AND (C OR D)" and perhaps represented using a set if logic primitives in a simple schematic as shown in Figure 8.7.

Another method might be to define the logic behavior using a lookup table (Figure 8.8).

Whichever approach is taken to implement these logic functions there is a direct and systematic match between the logic expressions and the implementation.

8.1.3.4 Behavioral Digital Design

There is a specific distinction drawn in the world of digital design between two different approaches to digital models, which stems from the ability (or otherwise) to automatically *synthesize* the design to hardware. Register transfer logic (RTL) is the generally accepted form of digital modeling (in whichever language is used) that can be synthesized, and behavioral high level models which are not guaranteed to be able to be synthesized. It is not always possible to tell the two forms apart; however, there are some important aspects which can highlight the differences.

The first important aspect is that RTL designs map data onto registers, with combinational logic between each register. The implication of such a system is

A	B	A or B	C	D	C or D	(A or B) and (C or D)
0	0	0	0	0	0	0
0	0	0	0	1	1	0
0	0	0	1	0	1	0
0	0	0	1	1	1	0
0	1	1	0	0	0	0
0	1	1	0	1	1	1
0	1	1	1	0	1	1
0	1	1	1	1	1	1
1	0	1	0	0	0	0
1	0	1	0	1	1	1
1	0	1	1	0	1	1
1	0	1	1	1	1	1
1	1	1	0	0	0	0
1	1	1	0	1	1	1
1	1	1	1	0	1	1
1	1	1	1	1	1	1

Figure 8.8:
Lookup table for logic function $(A + B)(C + D)$

that there is at least one clock (there may be more than one) to enable the transfer of data through the registers. The second aspect is that there will be a reset signal to allow a known start state of the system. Both of these are important; however, in many ways, the clock is the key to understanding the difference with high-level behavioral digital models.

Consider the case where there is a register or variable that will take on a new value at a certain time. How is this managed in an RTL model? In this case, the variable will be assigned *when the clock changes*. This could be a rising, falling, or both-edge trigger, or perhaps even a level shift, but, in any case, the effect is the same — the variable will only change or be assigned on the *clock event*. In a true behavioral model, there is more flexibility to the model, and we could, for example, use an absolute time definition to specify exactly when the variable is assigned. This is *not* synthesizable and is therefore true behavioral digital modeling.

The main criteria, in addition to the clock and reset, are essentially whether a data path and a control path can be established from the model definition, and, if so, then the digital model can be synthesized. As we have seen in previous chapters, digital *systems* can be represented using state machines and, lookup tables, and earlier in this chapter we showed how these different representations can be used to implement a model.

In general, we can say that a high-level behavioral model will provide information on the functionality of the system, not necessarily the architecture or implementation of the system.

8.1.3.5 Synchronous Logic

Synchronous logic systems are the easiest digital system to describe systematically as they have such a regular structure. The most obvious aspect is that there is a system clock and a global reset that governs the overall behavior of the system. As we have seen previously, the precise architecture of the system may follow a Moore or a Mealy approach. As a reminder, the Moore outputs are purely a function of the current state variables, and the Mealy outputs are also a function of the current inputs. The principle is that there is a set of logic (using flip flops) that defines the current state in a register, and a block of combinational logic which defines the next state.

From a modeling perspective, we therefore need to be able to define six elements to be able to implement such a system:

1. Clock
2. Reset
3. State variables or registers
4. Next state logic
5. Inputs
6. Outputs (with associated logic for a Mealy machine).

Once we have these in place, then the model can be implemented easily. Consider the case of a simple 4-bit counter with a clock and a reset. We can make some immediate decisions based on this limited information almost immediately, using the six elements defined earlier.

Criteria	Design Decision
Clock	We need a clock, and so this can be defined almost immediately. The only decision to make is the name of the clock signal. CLK is a universally recognized name for a generic clock. For a digital signal this will be of the standard Boolean logic type in the platform or language we have chosen, but will at least have a minimum of a zero state "0" and a one state "1".
Reset	A reset is required to initialize the system. In this example there is no preset input, and so the assumption may be made that we wish to initialize the output with the zero output value "0000".
State variables or registers	In the case of a 4-bit counter, we would need four flip-flops to implement the state at any instant of time — one for each bit. One option for a *model* of the counter would be to implement the counter using an integer that represents the digital state; however, if this is done, then care must be taken to ensure that the integer type is chosen to have enough bits to adequately represent the bits in the system. Another option is to define the variable using a bitwise form.
Next stage logic	Even though the system is a counter, it is still a state machine, and, as such, does have next state logic. The next state logic will ultimately be implemented in hardware as Boolean equations using primitive logic functions or a lookup table, but in the model, may be higher level logical expressions.
Inputs	In this simple example, there are no inputs; however, there could be the case of a preset counter, which initialized the counter to the state values on the input word, and this would require that logic to be implemented in the model. In this case, however, no such logic would be required.
Outputs	In this case, the output of the state variables maps directly into the outputs. This may not be the case, e.g., in a decoder. In such cases, output logic expressions would be required, and perhaps registers to ensure no glitches on the output as a result.

We now have several options to create the synchronous digital model using graphical methods or language-based ones. We can also implement a behavioral model or an RTL implementation. For example, we could create a HDL model of a counter that is based on the following *pseudo*-code:

```
Process
Begin
  Count = 0
  While(count<16) do
Begin
  Count = count + 1
End
End
```

This code will initialize the process with a count value = 0 and then loop until the counter reaches 15, at which point it will then exit the process and the process can begin again. This has no explicit clock or reset, and does not therefore define anything of the architecture or potential implementation. If we were to synthesize this we would need to carry out a preliminary step to add in the required clock and reset prior to a final synthesis of the design.

We could, however, redesign this model with a clock and a reset, using a pseudo-code something like this:

```
Process(clk,rst)
Begin
  If(rst = '0') then
    Count = 0
  Else
 if(edge(clk) && clk = '1') then
    if (count<16) then
    Count = count + 1
  Else
```

```
    Count = 0
  End if
End if
End if
End
```

Now we can see that the process has what is called a *sensitivity list*, which is the list of signals that will invoke activity in the process if they change value. The concept of a "process" is simply nomenclature for a separate parallel digital system with its own self-contained behavior. Note that the sensitivity list has both `clk` and `rst` in the list. This implies that the reset (`rst`) is *asynchronous* in the sense that whenever the `rst` signal changes, the process will invoke and check the logic to see if the counter would need to be reset. The alternative approach is to use a *synchronous reset*, which is shown in the following pseudo-code.

We could, however, redesign this model with a clock and a synchronous reset, where the process will only activate on the clock edge, and then test the state of the reset, unlike the previous example where the reset can asynchronously activate the process. The resulting code would look something like the following:

```
Process(clk)
Begin
 if(clk = '1') then
    If(rst = '0') then
      Count = 0
    Else if (count<16) then
      Count = count + 1
    Else
      Count = 0
```

```
    End if

   End if

  End if

 End
```

It can be seen that the sensitivity list no longer includes rst; the process will only be invoked on a clk event, and the model first checks for a rst value of "0," and only if that is not the case will the counter logic be checked.

This section has looked at some basic synchronous logic modeling elements and this relies, above all, on a governing system clock. Now we will look at asynchronous design.

8.1.3.6 Asynchronous Logic

The use of the term "asynchronous" when apportioned to logic design means different things to different designers. Each method uses different assumptions in its approach to the design of digital circuits. The fundamental definition of asynchronous logic in general is that there is no system clock to govern the overall system timing, and so the intrinsic delays in the logic elements dictate the overall system behavior.

The basic concept in a typical realistic asynchronous system is not in fact that the system is reduced to purely combinatorial logic *per se*, but rather that the flow of data through registers is managed using a handshake protocol.

For example, if we consider a typical synchronous register system as shown in Figure 8.9, the data is transmitted from one register to another on the clock signal edge (usually rising or falling edge).

In contrast, an asynchronous logic design does not have any explicit clock, but rather uses a handshake protocol to "request" and "acknowledge" when data are active. For example, the same series of registers would be represented using the following form as shown in Figure 8.10.

The important aspect to this design is that the "request" and "acknowledge" signals are transmitted down the control path and transparent for combinational

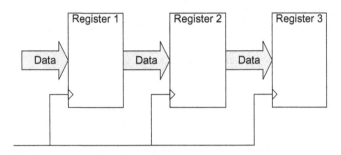

Figure 8.9:
Synchronous design utilizing the clock to transfer data

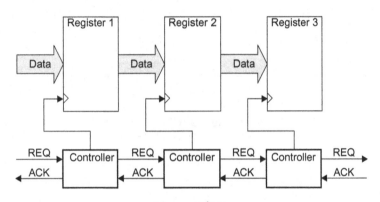

Figure 8.10:
Asynchronous logic design using handshaking to transfer data

logic. When the "request" signal is high, the data is valid and can be acted upon (Figure 8.11).

The possible advantages of an asynchronous approach to logic design is that the data are clocked through when it is necessary to do so, and therefore there is the possibility that an asynchronous circuit may be lower power than a synchronous equivalent. Clearly, however, there is an area overhead intrinsic to an asynchronous approach, where each register requires a "handler" or "controller" to manage the data flow through the system. Another potential advantage to this type of system is that there is the potential for very rapid data transfer, as the rate of data transfer is dictated by the ability of the register to change to the new value and acknowledge the receipt of the data, rather than a fixed global clock frequency, which is dictated by the combinatorial logic of the critical path.

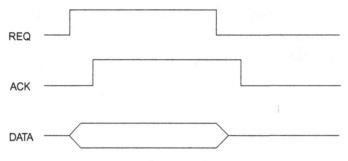

Figure 8.11:
Asynchronous data valid

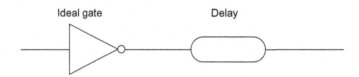

Figure 8.12:
Transport or ideal delay

8.1.4 Harsh Realities

8.1.4.1 Timing Behavior

Clearly, the obvious limitation of any logic system from a functional perspective is the timing delays through gates. Each gate will have a finite delay, which is generally modeled using a global gate delay parameter. For example, an inverter may have a delay defined as 1 ns, and any logic change on the input would be delayed by this amount. This type of delay is sometimes called "transport" delay and can be modeled as an ideal gate, plus a fixed delay on the output (Figure 8.12).

Of course, in practice, the delay may be more pronounced on a rising or falling edge, leading to an asymmetric value for the delay, which would be encapsulated in the model of the delay, giving two values (Low->High and

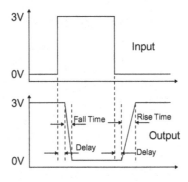

Figure 8.13:
Transport delay

Input	0	1	0

Ideal output	1	0	1

Output	1	⊠	0	⊠	1

Figure 8.14:
Ideal or transport delay in a model

High->Low, often defined as two parameters *tplh* and *tphl* — transport delay low to high and transport delay high to low, respectively) (Figures 8.13 and 8.14).

The asymmetric nature of the transition can be seen when we exaggerate the difference (Figure 8.15).

A more sophisticated delay model may depend on the capacitive loading on a node, and this is implemented in a digital model with a delay proportional to the capacitive load (Figure 8.16).

The final form of delay implemented in digital logic models is called *inertial* delay, and this defines a finite time where any changes occurring in a smaller time will be ignored by the simulator. This is very useful in ignoring "spikes" or unrealistically small transition times, and reflects the real scenario when, in practice, a spike would be damped out by the inherent capacitance on a gate (Figure 8.17).

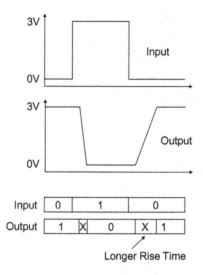

Figure 8.15:
Different rise and fall times

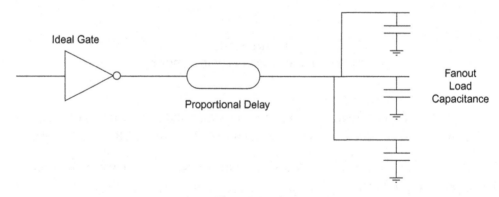

Figure 8.16:
Fanout proportional delay

8.1.4.2 Resource Management

One example of how a model can result in differing hardware implementations is when multiple resources could be shared. Consider the example of two parallel equations:

Y < = A + B

X < = C + D

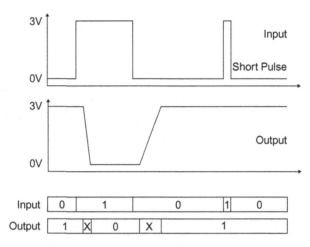

Figure 8.17:
Inertial delay cancellation

In this case, the two variables, X and Y, depend on independent input variables; however, they both use the same operator − in this case an addition function. In real hardware, this would result in either a very fast implementation where there would be two adders; however, this would potentially be quite large, or an alternative implementation where one adder is shared between the two equations (Figure 8.18).

8.1.5 Sampled Data Systems (Z-domain)

8.1.5.1 Introduction

Having looked at the discrete logic domain, where both the time and signal axes are quantized, and considering the truly analog domain where the time and signal axes are continuous, there is a third way to consider modeling signals. This is often given the generic term "sampled data systems" and occurs when we do not necessarily quantize the signal axes; however, the time axis IS quantized or sampled. This is also often referred to as the Z-domain.

This is an extremely useful and commonly used form of modeling as it maps directly onto most signal processing applications and inherently requires a

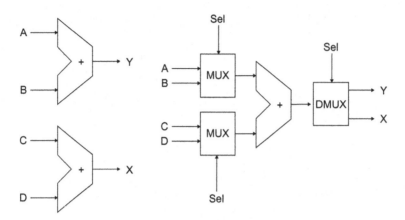

Figure 8.18:
Parallel or serial implementations of two adders

sampling function to take place somewhere in the system. This is obviously the case in a system that contains an analog-to-digital converter (ADC), and Z-domain modeling can be considered in largely the same manner.

One massive advantage in such systems from a modeling and simulation perspective is that once the signal has been sampled, it is essentially a *discrete* process and as such significantly faster to simulate than an analog equivalent. Models in this form can also be simulated in a digital simulator.

8.1.5.2 Sampling

The sampling process is an important one mathematically from a model perspective and can be considered as the generation of an ordered sequence of samples of a continuous signal $x(t)$, at regular, specific instants of time (Figure 8.19).

Mathematically, this can be represented by the multiplication of the analog signal by an impulse signal (Figure 8.20).

We have an obvious limitation in practice, which applies to both the "real world" and also to models which is that defined by Shannon's theorem, that the maximum frequency we can consider is exactly half the sampling frequency. We can see how this operates if we assume that $x(t)$ is band limited,

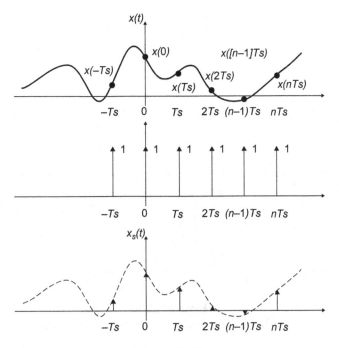

Figure 8.19:
The sampling process

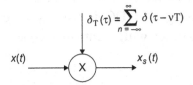

$$\delta_T(\tau) = \sum_{n=-\infty}^{\infty} \delta(\tau - vT)$$

Figure 8.20:
Mathematical representation of sampling

i.e., $X(\omega) = 0$ for $\omega > \omega_m$, the highest frequency in $X(\omega)$ is smaller than half the sampling frequency (the Nyquist frequency, $\omega_s/2$), the spectrum of $x_s(t)$ is identical to that of $x(t)$ but repeated with a period of ω_s. Consequently, no information is lost (Shannon's theorem). Otherwise, the spectra overlap and information is lost (Figure 8.21).

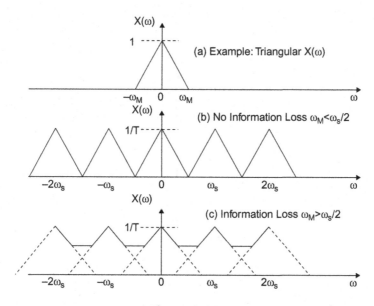

Figure 8.21:
The effect of sampling frequency on data loss

One artifact that can occur in both the real world and model cases is "aliasing." If there is a frequency component in the input signal above the Nyquist frequency at ω_M, it will fold back into the baseband at $\omega_s - \omega_M$. This is called aliasing. To prevent aliasing an ADC is usually preceded by an anti-aliasing filter, which is a low-pass filter with a cut-off frequency of $\omega_s/2$.

8.1.5.3 Digitization

In a Z-domain model, it is often the case that we define a specific number of bits, particularly if the model is to be used to specify a DSP or software code that has a limit of a number of bits. This can lead to errors and noise due to the inherent digitization of the data. Care must be taken when a model is written that this is also not inherent digitization in the simulator itself, or even in the data stored, as this may cause an incorrect assumption about noise levels.

For example, the sampling may not result in any information loss (in an ideal world), but the digitizing will as only a limited number of bits are used to represent the analog amplitude signal. This error manifests itself as noise and can

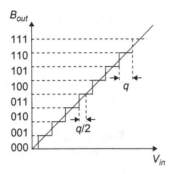

Figure 8.22:
The effect of quantization on the sampled data signal accuracy

be treated as white noise in many cases. The maximum quantization error is $\pm\, q/2$ (Figure 8.22).

The quantization noise can be calculated if we assume that the quantized noise is distributed uniformly across this range (q) (Figure 8.23):

$$e^2_{qMs} = \frac{1}{q} \int\limits_{-q/2}^{q/2} e^2 de = \frac{q^2}{12} \tag{8.4}$$

And the resulting RMS noise can be calculated as:

$$e_{qRMS} = \frac{q}{\sqrt{12}} \tag{8.5}$$

For a high number of bits, the error is uncorrelated to the input signal ($N > 5$). In the frequency domain, the error appears as white over the Nyquist range. This noise limits the signal-to-noise (S/N) ratio of the digital system analogous to thermal noise in an analog system. The importance of this noise is seen clearly when we look at the analog signal being sampled. If the peak value of a full scale sine wave (that is one whose peak-to-peak amplitude spans the whole range of the ADC) is given by $2^N q/2$, the RMS value for the sine wave is $V_{RMS} = 2^N q/2\sqrt{2}$. The signal-to-quantization noise ratio (SQNR) is given by:

$$SQNR = \frac{\left(2^N q/2\sqrt{2}\right)^2}{q^2/12} = \frac{3}{2} \times 2^{2N} \tag{8.6}$$

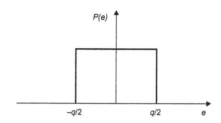

Figure 8.23:
Quantization noise distribution

Equation (8.6) can also be expressed conveniently in dB as SQNR = $(6.02N + 1.76)$ dB. Thus, each bit increases the SQNR by approximately 6 dB (e.g., for a 16-bit system SQNR = 98 dB). This is useful to see, as often a model is defined to target a 16-bit DSP, and the model may predict a noise floor of 120 dB, but, as can been seen in this case, if the number of bits is limited to 16, then the maximum noise floor is 98 dB at best. This can also be manifest in a digital simulator where the maximum resolution of the data stored is 16 bits, and again the noise floor will be too high (i.e., 98 dB maximum).

8.1.5.4 Case Study: Sigma Delta Modulator

8.1.5.4.1 Introduction

Sigma delta modulators are also called, collectively, "oversampled" converters. They turn an analog signal into a bit stream corresponding to the input with the oversampling used to partly shift the noise from the baseband into a higher end of the spectrum (more details given later in this example). Sigma delta modulators consist of, at the very minimum, a sampler, quantizer, filter function, and feedback. They can be of any order from first upward; however, it is useful to begin with simple first-order designs and then extend those into a more complex order.

8.1.5.4.2 Delta Modulation

The simplest form of this type of converter is called delta modulation. The basic concept is that the output "tracks" the input signal by either adding $+q$ or $-q$ to the signal based on the comparison (Figure 8.24).

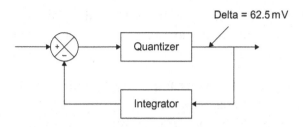

Figure 8.24:
Delta modulator block diagram

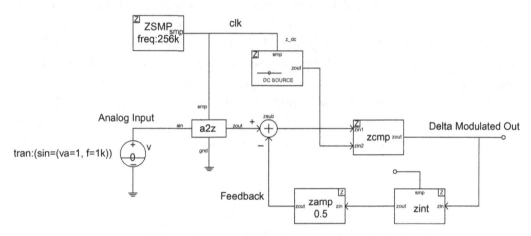

Figure 8.25:
Delta−Sigma model

Essentially, the quantization is arbitrary but defines the ability of the value fed back from the integrator to compensate for changes in the input value and differences between the input and output. In the limit, this would indicate the maximum error for a DC input with a ripple on the output.

This is a classic case of a mixed-signal model, where the analog input is sampled, and from then on the model is event-based, in this case Z-domain rather than digital logic (Figure 8.25).

The model first uses a sample function to create an event whenever the "sample" clock event occurs, and the quantizer compares the output of the difference function (which changes when there is an event change on either the sampled input or feedback value) with a static value. The integrator is a digital integrator that is clocked from the same sample clock source as the analog sampling function.

We can test the model with an example input with a sampling frequency of $F_s = 4$ kHz and the input analog waveform being a 1 kHz sinusoid. The quantizer uses a value of 62.5 mV. The resulting waveforms can be seen in Figure 8.26.

Clearly, it can be seen that the delta modulator struggles to keep up with the analog input, as would be expected from a system where the oversampling ratio (OSR) is only four. The zout signal is the actual output of the delta modulator, and, as can be seen, it is the simplest possible representation of a sinusoid – a pulse! This is also clearly not good enough for accurate data conversion so, in practice, the OSR would be much higher, and, in fact, if we set this to 256, so the sampling frequency is now 256 kHz, then the behavior of the circuit is much more convincing. The results of this simulation are shown in Figure 8.27.

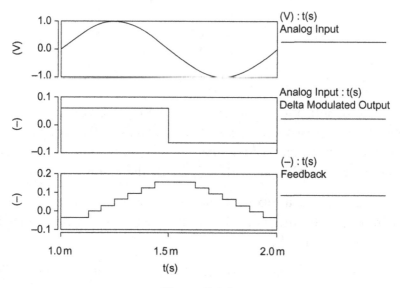

Figure 8.26:
Delta modulator with $F_s = 4$ kHz and $F_{in} = 1$ kHz, $q = 62.5$ mV

The delta modulator can now keep up with the input (delta $V = 62.5$ mV) and `zout` is a more detailed digital representation of the input waveform.

8.1.5.4.3 Sigma Delta Modulation

The delta modulator is a useful starting point in order to understand the operation of sigma delta modulators; however, if we move the integrator from the feedback path to the signal path, then we have a much more interesting functional block called a sigma delta modulator, as shown in Figure 8.28.

So, why move the integrator? The integrator is a high-pass filter for the noise and a low-pass filter for the signal, so putting it in the signal path automatically moves the noise to higher frequencies. We can demonstrate this by using a Laplace equivalent model of the sigma delta and simplify the circuit to its ideal elements, as shown in Figure 8.29, and add in the quantization noise explicitly, as shown in Figure 8.30.

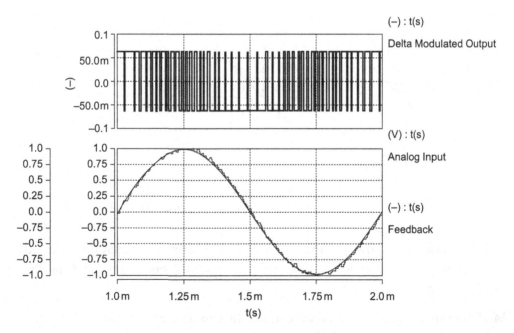

Figure 8.27:
Delta modulator with $F_s = 256$ kHz and $F_{in} = 1$ kHz, $q = 62.5$ mV

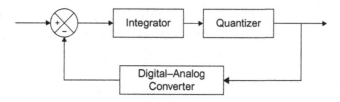

Figure 8.28:
Sigma delta modulator block diagram

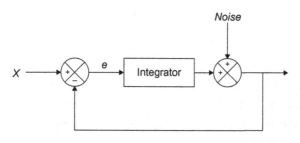

Figure 8.29:
Laplace model of sigma delta modulator

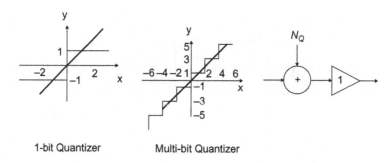

Figure 8.30:
Laplace equivalent of a single bit quantizer

A 1-bit quantizer is always linear as the gain is arbitrary; therefore, the quantizer can be modeled by a (quantization) noise source and a gain of 1, as shown in Figure 8.30.

Modeling this mathematically we can say that in discrete terms:

$$y[n] = x[n] + e[n-1] \tag{8.7}$$

where $e[n-1]$ is the quantization noise from the previous clock, and we can control its frequency response using a gain term in the feedback loop:

$$y[n] = x[n] + A_1 \times e[n-1] \tag{8.8}$$

Using the linear equivalent, we can also write Eq. (8.7) in a similar way:

$$e = x - y \tag{8.9}$$

$$y = \text{noise} + \frac{e}{s} = \text{noise} + \frac{x-y}{s} \tag{8.10}$$

$$y \times s = \text{noise} \times s + (x-y) \tag{8.11}$$

$$y \times (s+1) = \text{noise} \times s + x \tag{8.12}$$

$$y = \text{noise} \times \frac{s}{s+1} + x \times \frac{1}{s+1} \tag{8.13}$$

Now, from Eq. (8.13), we can see that y is now a function of the noise $(s/s+1)$, which is a high-pass filter, and the signal x is filtered using a low-pass filter. We can evaluate this using a linear control systems model of the type we investigated earlier in this book (Figure 8.31).

Using this model, we can measure the noise to output characteristic, and this is given in Figure 8.32.

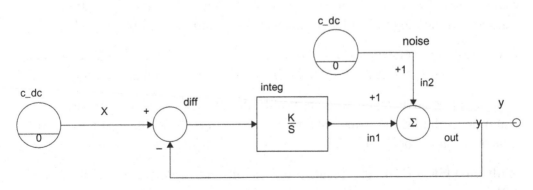

Figure 8.31:
Laplace equivalent model of sigma delta modulator for noise analysis

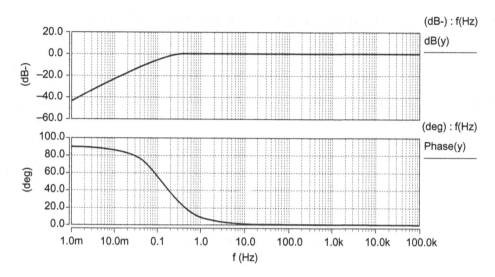

Figure 8.32:
Laplace model of sigma delta modulator — high-pass noise-to-output function

And, using the same model, the signal-to-output function is shown in Figure 8.33.

This is a graphical illustration of the rationale for using a sigma delta modulator rather than a "conventional" ADC, such as successive approximation, for example. We can also see this if we look at the noise spectrum of a typical ADC; we can see that the noise "floor" is essentially white noise and is relatively constant across the range from DC to the sampling frequency (Figure 8.34).

In contrast, an oversampled converter, the type of which the sigma delta modulators belongs, exhibits a completely different noise performance (Figure 8.35).

With all these advantages, we can modify the previous model of the delta modulator by bringing the same integrator model into the signal path to create a first-order sigma delta modulator, as shown in Figure 8.36.

With an OSR of 64, the results look a bit like a pulse width modulation (PWM) modulator where the pulse width is proportional to the signal value (Figure 8.37).

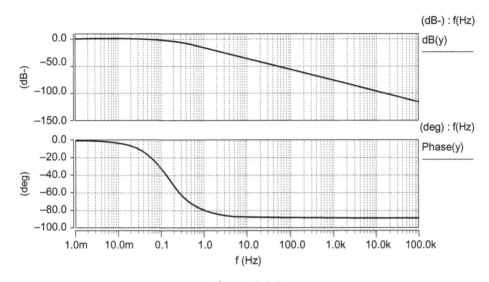

Figure 8.33:
Laplace model of sigma delta modulator —low-pass signal-to-output function

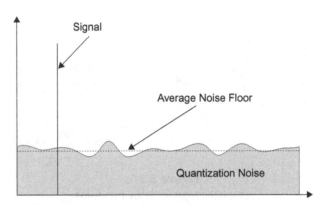

Figure 8.34:
Noise spectrum of a conventional analog-to-digital converter

Increasing the OSR to 256 (as we did with the delta modulator) increases the resolution of the modulator and gives even more accuracy than with the 64 (Figure 8.38).

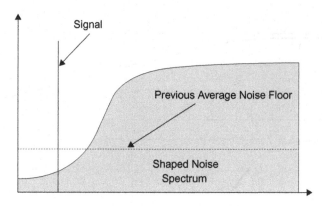

Figure 8.35:
Noise shaping of an oversampled converter

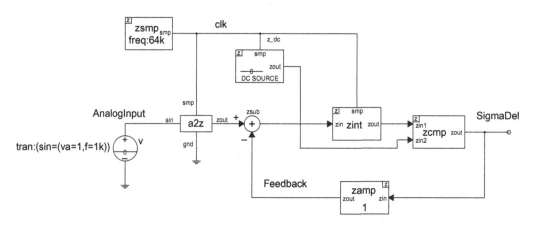

Figure 8.36:
First-order sigma delta modulator

The results of the noise performance when the model is analyzed can be seen clearly if a Fourier analysis is carried out on the output digital signal (Figure 8.39).

If better noise shaping is required, a second-order modulator can be used. A second-order modulator can be analyzed in exactly the same way as a first-order modulator (Figure 8.40).

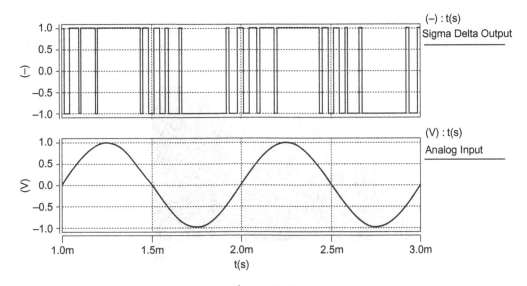

Figure 8.37:
Sigma delta modulator simulation with oversampling ratio = 64

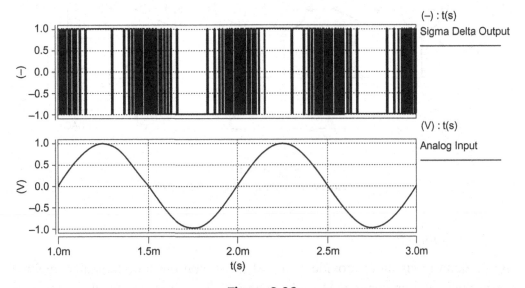

Figure 8.38:
Sigma delta modulator simulation with OSR = 256

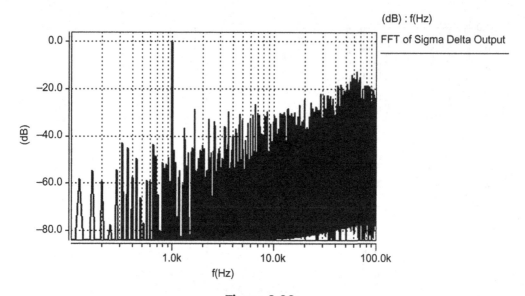

Figure 8.39:
Fourier analysis of the digital output sequence

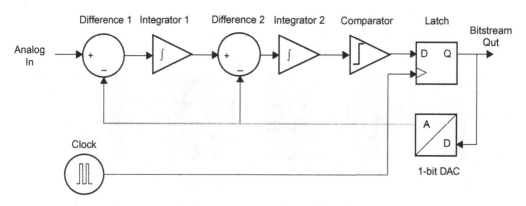

Figure 8.40:
Second-order sigma delta modulator

8.1.5.4.4 *Summary*

Sigma delta modulators provide a digital bit stream out automatically, making them excellent for conversion and streaming. They are a typical example of where event-based modeling can give a significant simulation speedup and practically useful data. They lead to a naturally simple CMOS implementation, where integrators and quantizers are easy to fabricate, and, as has been demonstrated, they are very scaleable, with the order and OSR easy to change.

Example 8.1 Logic Modeling — Combinatorial and Sequential

In this example, we will create a simple digital buffer using the event-based logic approach introduced in the first part of this chapter. If we follow the basic procedure of model creation we can define the majority of the model as we have done for other forms of model, defining the model name as buf1, an input port din, and an output port dout. For now, we will define the delay of the buffer using a single parameter td. The ports need to have a specific digital definition, and so we need to use the "signal" type and then choose an appropriate type for the logic. As we have seen already in this chapter, we could use a simple two level "bit" definition; however, the standard IEEE 1164 std_logic type is more commonly used in most simulators, so we can set this as the type. Finally, as the signals are digital, we should set the direction (IN and OUT), respectively.

The complete definition of the ports in ModLyng is as shown in Figure 8.41.

At this point we can now define the behavior of the model, and for simple combinatorial logic such as this (in fact, this is probably the simplest logic definition of all), the event assignment is the approach to use in ModLyng. What we need to do is to create a new "events" code fragment in which we can define the behavior of the buffer as follows:

```
dout <= transport din after td
```

The signal assignment (using the symbol "< = ") is equivalent to "assign the value of din to dout after a time delay of td". Using this simple assignment, the model is now complete and we can test it by creating a simple test bench where the buffer is driven

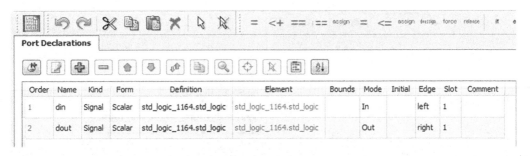

Figure 8.41:
Simple buffer port definitions

by a simple clock running at 1 MHz. The test schematic created in ModLyng is shown in Figure 8.42.

When this model was exported to VHDL-AMS in this instance and simulated, the clock and buffer signals can be seen clearly to match up in Figure 8.43, with the default delay of 0.

If we change the td parameter to 200 ns, then the result is a simple delay addition, which is basically "transport" delay as defined earlier in this chapter where the output signal always follows the input, except it is delayed by the specified amount. This is shown in Figure 8.44.

As we have seen, the default type of delay is transport, but what about inertial delay? If we want to have a buffer that ignores pulse widths of less than the specified delay, we can change the expression for the assignment as follows:

dout <= din after td

Observant readers will note that this is identical to the previous definition except the keyword "transport" has been removed. This is now an inertial delay model, and so if we increase the td parameter to 1500 ns, which is 1.5x the period of the pulse, the model should ignore the pulses on the input, and, in fact, we can see this happening with the simulation response as shown in Figure 8.45.

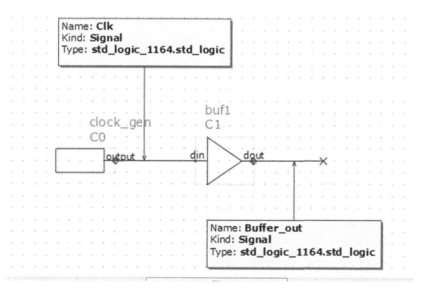

Figure 8.42:
Simple buffer test circuit

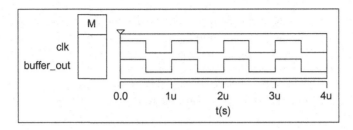

Figure 8.43:
Simple buffer simulation with td = 0 ns

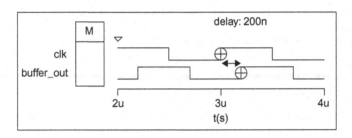

Figure 8.44:
Simple buffer with 200 ns transport delay

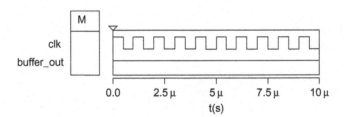

Figure 8.45:
Simple buffer with 1500 ns inertial delay

So far we have looked at simple combinatorial models, but, of course, we also need to consider synchronous or clocked models, and the simplest one of this type is a counter, so in order to demonstrate the concepts involved, a simple 4-bit counter with reset and clock inputs (rst and clk), and an integer output count (equivalent to 4 bits, with a maximum value of 15 (1111)) is shown.

As we discussed previously in this chapter, this requires a process to be defined, and the definition in ModLyng is as shown in Figure 8.46. As can be seen the sensitivity list is the two inputs `rst` and `clk`, so the reset is asynchronous as described previously in this chapter, but one minor, yet noticeable, difference is that rather than define the output variable `count` directly, an internal signal is used and then a separate signal assignment is required to set the output variable to this internal signal:

`count <= icount`

When this model is exported (in this case using VHDL) using the test circuit as shown in Figure 8.47 and simulated, the results in Figure 8.48 show the counting starting when the reset signal goes high and then reach 15 and restart at 0 once more.

This example introduces the reader to some of the important concepts in basic digital modeling, and the second example will now extend into the sampled data system or *Z*-domain realm.

Example 8.2 Sample Data Systems

As we have discussed in this chapter, the first key concept in a sampled data system otherwise known as a *Z*-domain system, is sampling. In many ways, this is the first step in translating "real-world" data or signals into a digitized form for processing, particularly DSP, where digital, rather than analog, algorithms are undertaken.

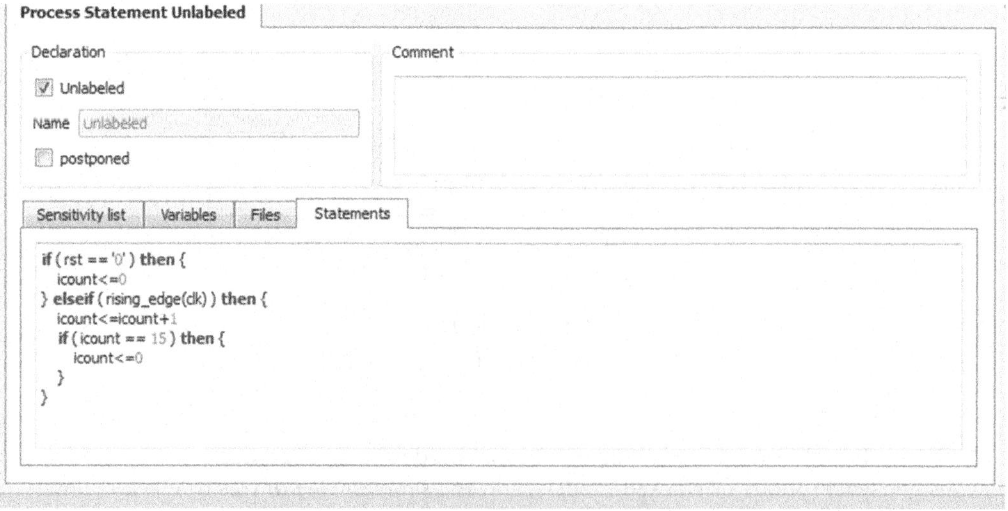

Figure 8.46:
Simple counter process definition.

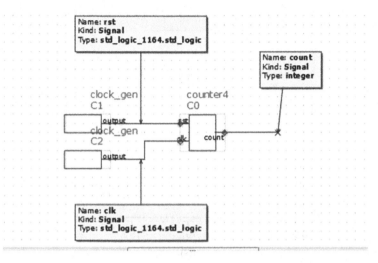

Figure 8.47:
Counter test circuit in ModLyng

With that in mind, how do we go about taking an analog signal, for example a voltage signal, and sampling it, which converts the analog signal (basically a real number) continuous in the time domain to an equivalent real number that is discrete in the time domain? As we have already seen how to obtain analog signals in the chapter on conserved systems and also seen how to implement clock signals in this chapter (see Example 8.1), the key aspect is expressing the discrete real number. In ModLyng, we can define ports as conservative or signal, and, in this case, we can define the output signal (vout) as a signal with type real and direction out.

The next issue is how to define the sampling action itself and, as we have seen already, we can simply assign the signal vout to the value of vin on an event occurring on the clk signal. In ModLyng, what we need to do is define a process, in the same manner as for the counter in Example 8.1, with clk in the sensitivity list and then, in this case, it simply has to contain the assignment:

vout <= v

where, in the model, the input branch between vin and an internal reference has the across variable defined with the name v to represent the input voltage.

Testing the model is relatively simple with the basic test circuit shown in Figure 8.49, where the input is a simple sinusoidal voltage source and the clock is the same clock source used in the digital example.

The results of sampling a 50 kHz signal using a 1 MHz clock can be seen in Figure 8.50, and if the time axis is zoomed into, as shown in Figure 8.51, and the input and output

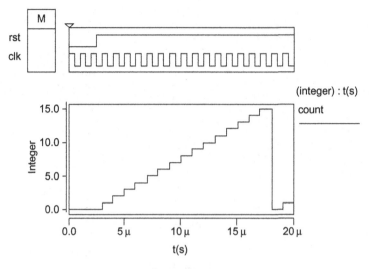

Figure 8.48:
Simple counter simulation results

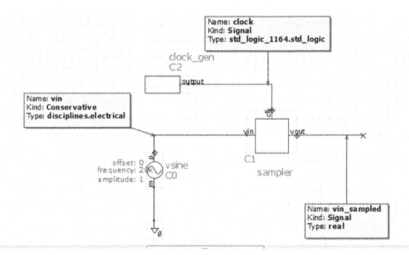

Figure 8.49:
Sampler test circuit

signals superimposed, the sampling can be seen clearly to take place, and is, in fact, similar to the operation of an ideal "sample and hold" circuit.

The advantage of this sampler model is that we now have digital equivalent (discrete) data in a form that we can carry out digital processing on. This could be in the form of

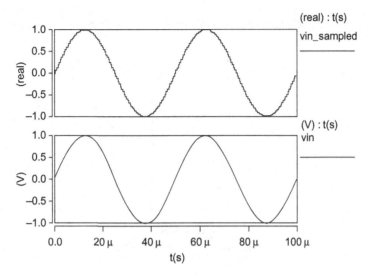

Figure 8.50:
Input and sampled data signal

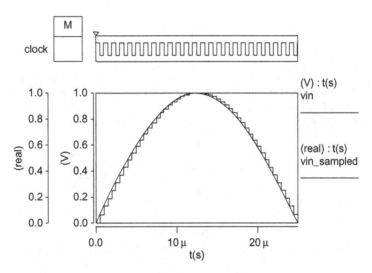

Figure 8.51:
Input and sampled data signal – detail

DSP algorithms or in digital filtering. One of the massive advantages from a designer's point of view is that this kind of model can be run in "standard" digital simulators and also can run incredibly fast as there is no nonlinear solver required when the model has been transferred into the digital domain.

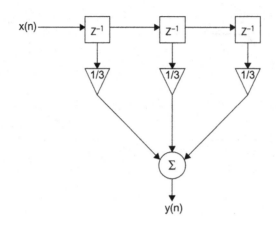

Figure 8.52:
Moving average FIR filter

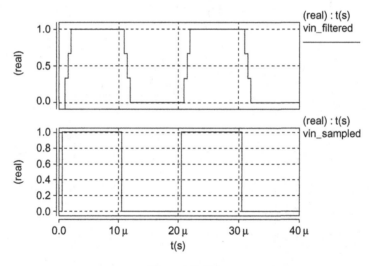

Figure 8.53:
Moving average filter simulation of a 50 kHz pulse input

For example, we could implement a simple digital filter using a series of delays of the form of a simple moving average finite impulse response (FIR) filter, as shown in Figure 8.52, and then observe the effect of this on an input signal. The FIR filter itself consists of processes to provide a delay of the input (one delay statement for each stage of the filter — in this case two delays for a third-order delay) and the summation process to calculate the filtered output.

The result of applying a 50-kHz pulse to the moving average filter is as shown in Figure 8.53. The effect of the three delay stages can be seen in the filtering of the input (nearly ideal) edges into a much slower transition time.

Conclusion

In this chapter, we have introduced the important concepts of event-based modeling. Both "standard" logic and sampled data system modeling techniques have been reviewed and demonstrated in some detail. At this point, the reader should be able to take on the task of modeling either of this type of model with confidence, and there is also the supplementary point that these modeling approaches have implications for mixed-signal modeling and also advanced techniques such as "fast analog", which we will introduce later in this book.

CHAPTER 9

Fast Analog Modeling

9.1 Introduction

As has been described previously in this book, there are two major categories of modeling for complex electronic systems: event-based or conserved equations. For the former approach, the signals are defined using states, with events triggering state changes, where time is discrete (although the values can be continuous — even floating point or real). In the latter, the models are defined using floating point, differential, algebraic equations requiring algorithms, such as Newton-Raphson or similar, to obtain a solution. In most cases the event-based simulations are significantly faster than the conserved system equivalent, as we shall see in this chapter.

In addition, there are a number of other techniques to provide faster simulations of analog systems such as state space averaging or averaged models, which can be used to model switching systems linearly, and thereby gain a massive simulation time improvement, without significant loss of accuracy. In fact, by using a linearized model, this can sometimes provide greater insight into a system's behavior than the switched version by virtue of the fact that a linear system can be analyzed using Bode plots or similar system-level analysis techniques, not available to switched systems.

To skip to the punch line of this chapter (and actually one of the main points of this book) differing representations of the design are needed to enable the acquisition of these data, depending on the information needed by the designer. These fast analog techniques offer time-proven methods of gaining insight into design performance that would be unwieldy or impossible if the designer was forced to use basic low-level transistor simulations. This chapter serves to reinforce one of the fundamental principles of model-based design and engineering: use the right tool for the job. Different methods or techniques of modeling as have been described in previous chapters, as well as this one, are simply different tools to be applied at the right time in the design process to gain the needed insight into ensuring the design performs as required to meet the specifications.

The first section of this chapter describes the highly useful averaged modeling technique for switching systems. The second section focuses on finite-difference approximations to differential equation-based models typically used for continuous-time simulation of analog circuits. This allows the analog components being modeled to be represented in a form suitable for digital simulators, thus opening up a mixed analog–digital verification approach executed exclusively in the digital simulator. This improves throughput to a point that full-chip verification of mixed-signal circuitry is realizable.

9.2 Averaged Modeling

9.2.1 Introduction

Averaged modeling is a technique often applied to switching systems, the most common being that of switched mode power supplies (SMPS). An SMPS converter uses a high-speed switched transistor(s) to switch the input power through to the output. The basic topologies of SMPS are Buck (step down), Boost (step up), and Buck-Boost (combination of the two). There are some other advanced and varied topologies, but, in many respects, they are just derivatives of these fundamental types. In order to illustrate the thought process involved in modeling such switched systems using different approaches, leading towards an averaged model, we will take the example of a simple "Buck" power supply and analyze different modeling approaches.

9.2.2 An Example Switching Power Supply: The Buck Converter

The basic topology of a Buck converter is shown in Figure 9.1. The idea behind its operation is to switch the input across the load, and when it disconnects, the current will be held by the inductor.

The Buck converter is a "step down" converter and so the output voltage will always be less than or equal to the input voltage. The duty cycle of the switch

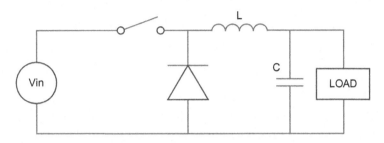

Figure 9.1:
Basic schematic of a Buck power converter

control signal controls this ratio directly, as shown in Eq. (9.1). The switching frequency will usually be several orders of magnitude higher than the change in the system voltages:

$$V_{out} = Duty \times V_{in} \qquad (9.1)$$

For example, in a typical power converter that converted a rectified 24 V dc signal obtained from the mains AC supply down to 18 V for use, in, say, a laptop power supply, the switch would perhaps be running at a frequency of 25 kHz or more to achieve a smooth output voltage with minimal ripple (and the same for the current). The converter would then have two states, one when the switch is closed, and the second when the switch is open. In the closed state, as shown in Figure 9.2, the diode is effectively not relevant in the circuit, as all the current flows through the switch and inductor. The voltage across the inductor is equal to the difference between the input voltage and load voltage, which is always positive, so the inductor current will increase. In addition, the maximum voltage possible is the input voltage.

When the switch is opened, the circuit now takes the form shown in Figure 9.3, where the voltage across the inductor is now equal to the inverse of the load voltage ($-V_L$) minus the voltage drop across the now forward biased diode, and so the inductor current will decrease.

The general concept with any switch mode power supply is that the switch (or switches) is alternated rapidly between the open and closed states to provide an "average" output voltage and current. For example, if the Buck converter is alternated between the switch "open" and "closed" states, then the basic

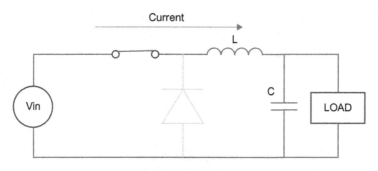

Figure 9.2:
Buck converter, switch closed

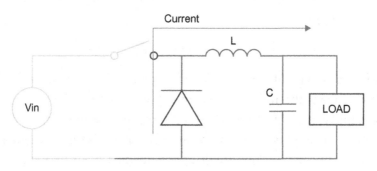

Figure 9.3:
Buck converter, switch open

electrical behavior of the inductor current and output voltage will be as shown in Figure 9.4.

The key issue when modeling systems that incorporate a switch, such as a Buck converter, is that the switching frequency tends to be several orders of magnitude higher than the overall response time of the circuit itself. The response time of the filter may be of the order of 100 Hz, but the switching frequency is often 20 kHz or even higher! As a result, a simulation time is usually required on the order of hundreds of milliseconds; however, a single switching cycle is of the order of microseconds. The effect on simulation time can be dramatic, with hours of computer simulation required to obtain a single result of a full startup.

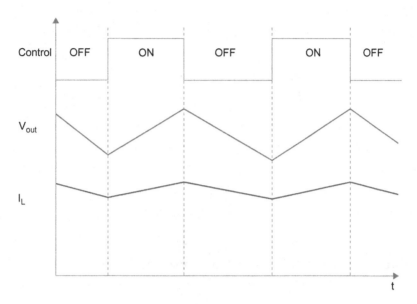

Figure 9.4:
Buck converter switching waveforms

Another problem with modeling a circuit that includes a switching element in this way is that the circuit is either in one state or another. This means that it is impossible to carry out a linear analysis to measure the stability directly. This is a particular problem for power supplies, where there is a closed loop control system to ensure stability and yet a fast response. We can illustrate these issues by first modeling the converter using a MOSFET switch model.

9.2.3 Modeling a Buck Converter Using a "real" MOSFET Model

In the case of a switching power supply, the most common switching element is, in fact, a MOSFET device, and if we were to define an example circuit, this could be simply modeled using the circuit components used in practice in a direct implementation of the power supply schematic. If we create a simple circuit using some example components, based around a practical characterized MOSFET model, as shown in Figure 9.5, then we can assess the effect on performance and accuracy as a result.

This model will simulate using a standard Newton-Raphson solver, and using a typical circuit simulator will require, in this case, 25 equation variables

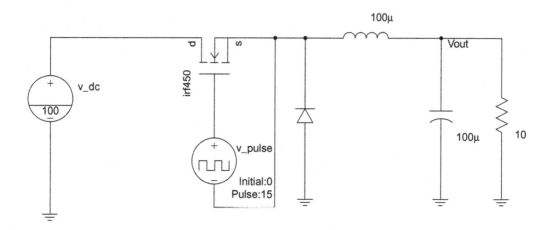

Figure 9.5:
Buck converter with a MOSFET switch and open loop control (duty cycle = 50%)

(as discussed in Chapter 3, these are simultaneous equation variables to be solved in the matrix), and, as a result, the simulation of a 10-ms analysis requires 15,446 time steps to complete. The simulation will only take a few seconds for such a simple circuit, but for more complex circuits will become time consuming very quickly.

The results of the simulation are shown in Figure 9.6. In this simple open loop model, the output voltage overshoots, with a high inrush current initially and then settles down to approximately 50 V (roughly 50% of the input, with the duty cycle set to 50%).

This is useful in understanding the cycle-by-cycle behavior on a detailed level, for example if we need to analyze the detailed turn-on or turn-off behavior of the MOSFET and evaluate its effect on the circuit performance, we are able to do that with this model. To observe the turn-on behavior of the MOSFET in this configuration, perhaps with a view to designing a suitable gate drive circuit, it is possible to do that with this model, as shown in Figure 9.7.

In many cases, it is not necessary to have this level of detail for a switching circuit in an overall system simulation, so in order to understand the overall system performance, we can replace the detailed MOSFET model with an

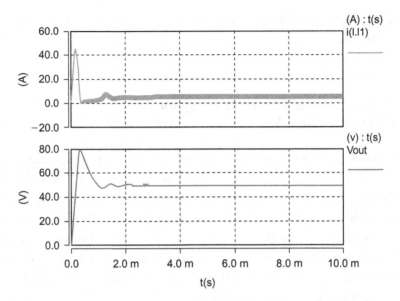

Figure 9.6:
Open loop buck converter simulation — MOSFET switch model

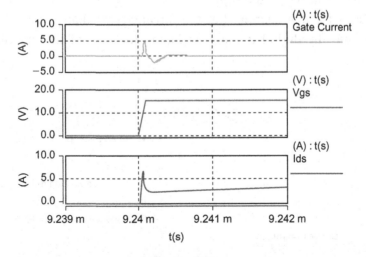

Figure 9.7:
Turn-on behavior of MOSFET based buck converter model

equivalent "switch" model. This could be an "analog"-only model (voltage-controlled impedance); however, an even more effective approach is to use a digitally controlled switch.

9.2.4 Modeling a Buck Converter Using an "ideal" MOSFET Model — Switch

The idea behind the switch model shown in Figure 9.8 is to have an impedance that has either a low on resistance (Ron) or a relatively high off resistance (Roff), and the value of the switch model impedance will be either Ron or Roff depending on the state of the digital switch input (C). Clearly, as we have discussed in Chapter 3, it is important to ensure that discontinuities are avoided in the model, so when the digital event occurs on the control input, the resistance of the switch must transition smoothly from one value to the other in a finite time to ensure a stable and robust model.

The same circuit model can be implemented, but this time replacing the detailed MOSFET model with an equivalent switch model — as shown in Figure 9.9.

This model will still simulate using a standard Newton-Raphson solver; however, with the addition of the digitally controlled switch, will also require an event simulator for that section, but the advantage is that part of the circuit is now moved into the digital domain, reducing the size of the analog matrix to

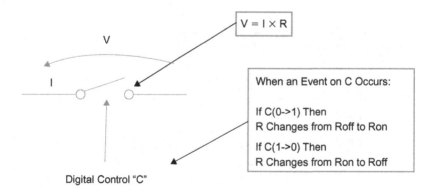

Figure 9.8:
Digitally controlled switch model

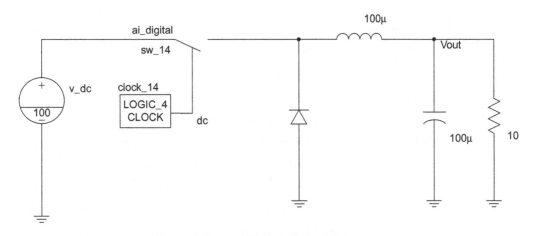

Figure 9.9:
Buck converter with an ideal switch and open loop control (duty cycle = 50%)

be solved. Using a typical circuit simulator this will require, in this case, ten equation variables (as discussed in Chapter 3, these are simultaneous equation variables to be solved in the matrix), and, as a result, the simulation of a 10-ms analysis requires 10,569 time steps to complete – which is a reduction of approximately 30% over the detailed circuit model. The reader may legitimately ask why the decrease is not even better than this. A significant portion of the circuit is still analog, including the device model of a power diode. The effect on simulation performance is manifest not so much in time steps, but in the individual Newton-Raphson iterations required to achieve a solution and, in fact, these reduce from more than 37,000 to just over 12,000, which is a reduction of nearly 70% over the device-level model. The simulation will still only take a few seconds for such a simple circuit, but clearly for more complex circuits the advantage can become significant very quickly. In addition, it is the experience of the authors that where the circuit contains multiple devices, the addition of ideal switches can have an even greater effect on the speed of the simulation.

But what about the simulation results? Are they accurate enough? Of course, we do not now have the internal detailed behavior of the device to analyze, but the overall system behavior is still available. If we plot the same variables (output voltage and inductor current) as for the device-level simulation, we can compare the two (Figure 9.10).

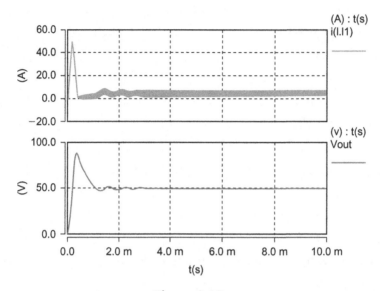

Figure 9.10:
Open loop buck converter simulation — ideal switch model

The results of the ideal switch Buck converter are remarkably similar to the device-level model, with an identical initial inrush and resulting overshoot, and then settling down to a steady state output voltage and inductor current as the device-level model. This is an interesting effect considering we have 15 fewer equation variables than before, and highlights an important design concept for advanced models, which is that the model should be only as complex as it needs to be for the task being undertaken. In this case, if the overall system behavior is required, then it certainly appears that a detailed device model is not necessary to achieve that. While these results are encouraging and also a practical step that we can take to model the system more efficiently, the simulation times for large switching power designs can still be extremely long. Another problem is the inability to analyze switching systems in the frequency domain. To achieve this we need to take a different approach, which is called "averaged" modeling.

9.2.5 Modeling a Buck Converter Using State Space Modeling Techniques

If we take a closer look at the basic BUCK converter topology in Figure 9.11, we can see that there are two important switching elements, the main switch (usually a MOSFET) and the diode. If we consider them to be effectively two

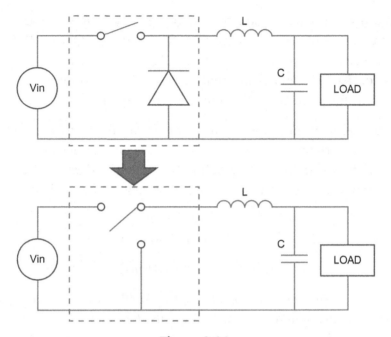

Figure 9.11:
Buck topology with switch elements identified

switches that are controlled in anti-phase (i.e. when one is ON the other is OFF), then we can represent the switching part of the circuit using the equivalent circuit also shown in Figure 9.11, where the switch and diode have been replaced with an idealized combined switch model.

This idea of there being one switch model that can be represented in two different phases is crucial to our implementation of an averaged model. One of the common approaches to modeling power supplies was the well-known approach called state space modeling [1] where the idea is to create two sets of linear differential equations and then create a combined "averaged" set of equations that enables a combined analysis. In the case of the Buck converter there are two states to consider: ON and OFF. In fact, this is a simplification. (It is beyond the scope of this book to discuss this in detail; however, power supplies, such as the simple buck supply may operate in continuous conduction mode — where the inductor current in steady state operation does not fall to zero — or discontinuous conduction mode — where the current *does* fall to zero. In the discontinuous conduction mode, the analysis becomes more complex as there are, in fact, three states, ON, OFF, and ZERO_CURRENT to consider.) As we have described earlier in this chapter, in each of these two states, the circuit consists of slightly different topologies, and therefore we can derive the expressions for these states from the circuit diagram.

The general approach for modeling a system using the state space approach is to have a set of variables (called the state variables), that when combined with the input variables can completely define the output variables. The mathematical representation of the state space approach can be defined using Eqs (9.2) and (9.3):

$$\dot{x}(t) = A(t)x(t) + B(t)u(t) \tag{9.2}$$

$$y(t) = C(t)x(t) + D(t)u(t) \tag{9.3}$$

where $u(t)$ is the input vector, $x(t)$ are the state variables, and $y(t)$ is the output vector. For example, the system may have a inputs, b outputs, and c state variables. This form is termed a continuous time-variant model as the parameters of the system defined in matrices A, B, C, and D may be dependent on time. If that is not the case, then they simply reduce to constants (or matrices of constants).

The principle of the state space *averaging* technique is to take the two different states (in this case ON and OFF) and first describe the two state space systems, and then combine them in some average manner to establish an average output response. If we do this for the Buck converter we can establish the state space equations for the overall system, where the model will therefore depend on the control signal, i.e., the duty cycle to determine the overall response.

If we consider the case in Figure 9.2, where the Buck converter switch is closed and the load is a simple resistor (R), we can use Kirchhoff's laws ($\Sigma V = 0$ and $\Sigma I = 0$) to create the system equations. This is what we have defined as the ON state.

We can say that using Kirchhoff's voltage law the network will give the following equation:

$$L\frac{di_L}{dt} + v_C = V_{in} \tag{9.4}$$

And that Kirchhoff's current law will also give an equation for the sum of the currents as follows:

$$i_L = i_R + i_C = \frac{v_C}{R} + C\frac{dV_C}{dt} \tag{9.5}$$

We can also define the output voltage and current as follows:

$$V_{out} = v_C \tag{9.6}$$

$$I_{out} = \frac{v_C}{R} \tag{9.7}$$

At this point we can begin to construct our state space description by noting that we have two outputs v_{out} and i_{out}, two state variables i_L and v_c, and, finally, one input, V_{in}. Remembering the basic state space format of Eqs (9.2) and (9.3), we can therefore arrange these expressions into the matrix form of the state space definition:

$$\begin{bmatrix} \dot{i_L} \\ \dot{v_C} \end{bmatrix} = \begin{bmatrix} 0 & -\frac{1}{L} \\ \frac{1}{C} & -\frac{1}{RC} \end{bmatrix} \begin{bmatrix} i_L \\ v_C \end{bmatrix} + \begin{bmatrix} 1 \\ 0 \end{bmatrix} V_{in} \tag{9.8}$$

$$\begin{bmatrix} V_{out} \\ i_{out} \end{bmatrix} = \begin{bmatrix} 0 & 1 \\ 0 & \frac{1}{R} \end{bmatrix} \begin{bmatrix} i_L \\ v_C \end{bmatrix} + \begin{bmatrix} 0 \\ 0 \end{bmatrix} V_{in} \tag{9.9}$$

We can repeat this analysis for the alternative OFF state given in Figure 9.3, which has the simplified equations (ignoring diode voltage drop) as follows:

$$L\frac{di_L}{dt} + v_C = 0 \tag{9.10}$$

$$i_L = i_R + i_C = \frac{v_C}{R} + C\frac{dV_C}{dt} \tag{9.11}$$

$$V_{out} = v_C \tag{9.12}$$

$$I_{out} = \frac{v_C}{R} \tag{9.13}$$

We can then do exactly the same state space matrix construction as for the ON state, which gives the state space formulation of the OFF state, as shown in Eqs (9.14) and (9.15)

$$\begin{bmatrix} \dot{i_L} \\ \dot{v_C} \end{bmatrix} = \begin{bmatrix} 0 & -\dfrac{1}{L} \\ \dfrac{1}{C} & -\dfrac{1}{RC} \end{bmatrix} \begin{bmatrix} i_L \\ v_C \end{bmatrix} + \begin{bmatrix} 0 \\ 0 \end{bmatrix} V_{in} \qquad (9.14)$$

$$\begin{bmatrix} V_{out} \\ i_{out} \end{bmatrix} = \begin{bmatrix} 0 & 1 \\ 0 & \dfrac{1}{R} \end{bmatrix} \begin{bmatrix} i_L \\ v_C \end{bmatrix} + \begin{bmatrix} 0 \\ 0 \end{bmatrix} V_{in} \qquad (9.15)$$

The averaging process takes place over one cycle, with the assumption that the switching frequency is much higher than the system response, where, as we have shown in Figure 9.4, the relative time of the ON and OFF states is controlled by the duty cycle. We can also say that if the switching period is T_s, then the duty cycle is in the range 0 to T_s, and the time of the ON and OFF state combined must be no greater than T_s.

$$T_s = \frac{1}{F_s} \qquad (9.16)$$

$$T_s = T_{ON} + T_{OFF} \qquad (9.17)$$

If we take the time per period as defined by Eq. (9.17), then to translate this into duty ratio we need to divide by the period to normalize the expression:

$$\frac{T_s}{T_s} = \frac{T_{ON}}{T_s} + \frac{T_{OFF}}{T_s} \qquad (9.18)$$

$$1 = D_{ON} + D_{OFF} \qquad (9.19)$$

We can therefore define the OFF state duty cycle in terms of the ON state duty cycle as follows:

$$D_{OFF} = 1 - D_{ON} \qquad (9.20)$$

Using the definition of duty cycle for the ON and OFF phases as being D and $1-D$, respectively (where D is the duty ratio, equivalent to D_{ON}), we can create an averaged model of each of the matrices A, B, C, and D, by simply taking an average of each, multiplied by D and $1-D$ respectively:

$$
\begin{aligned}
A_{ave} &= A_{ON}D + A_{OFF}(1-D) \\
B_{ave} &= B_{ON}D + B_{OFF}(1-D) \\
C_{ave} &= C_{ON}D + C_{OFF}(1-D) \\
D_{ave} &= D_{ON}D + D_{OFF}(1-D)
\end{aligned}
\tag{9.21}
$$

Interestingly, when we look at the matrices A, B, C, and D, we can see that, in fact, the calculation is much simpler as $A_{ON} = A_{OFF}$, $C_{ON} = C_{OFF}$, $D_{ON} = D_{OFF}$, and $B_{OFF} = 0$, so the only matrix that needs to be scaled is B, where:

$$
B_{ave} = DB_{ON}
\tag{9.22}
$$

The complete definition of the state space matrices for this system are therefore:

$$
\begin{aligned}
A_{ave} &= A_{ON} \\
B_{ave} &= B_{ON}D \\
C_{ave} &= C_{ON} \\
D_{ave} &= D_{ON}
\end{aligned}
\tag{9.23}
$$

Which is a system as follows:

$$
\begin{bmatrix} \ddot{i}_L \\ \dot{v}_C \end{bmatrix} = \begin{bmatrix} 0 & -\dfrac{1}{L} \\ \dfrac{1}{C} & -\dfrac{1}{RC} \end{bmatrix} \begin{bmatrix} i_L \\ v_C \end{bmatrix} + \begin{bmatrix} D \\ 0 \end{bmatrix} V_{in}
\tag{9.24}
$$

$$
\begin{bmatrix} V_{out} \\ i_{out} \end{bmatrix} = \begin{bmatrix} 0 & 1 \\ 0 & \dfrac{1}{R} \end{bmatrix} \begin{bmatrix} i_L \\ v_C \end{bmatrix} + \begin{bmatrix} 0 \\ 0 \end{bmatrix} V_{in}
\tag{9.25}
$$

And this system can now be used to evaluate the small-signal linear response to changes in the input or to perturbations in the duty cycle to establish the loop response. If we simulate this model in the time domain we can still obtain the step response; however; first it is necessary to translate this into a Laplace equivalent, which is given in Eq. (9.26).

$$H(s) = \left(\frac{1}{LC}\right)\left(\frac{RCs + 1}{s^2 + \frac{1}{RC}s + \frac{1}{LC}}\right)\left(\frac{V_{in}}{LC}\right) \tag{9.26}$$

With the resulting response as shown in Figure 9.12, it is interesting to note that even though most of the components are ideal in the circuit model, the resistances of the switch and diode are not modeled in this ideal version, and so the overshoot is much higher as a result. In order to more accurately reflect the real model it is important to add in the parasitic resistances to the ideal model. In spite of this difference, it is possible to see that the model also correctly reflects the effect of the duty cycle in the output voltage settling at 50 V as before.

What about the simulation efficiency? Recall that from the circuit models we were looking at around 12,000 non-linear iterations and several thousand time steps; however; in this model we have only 6 analog variables, with a linear system, so only 200 time steps were required with a simulation time of a fraction of a second. In addition to this vastly improved simulation speed, we also have the ability, with a linear model, to carry out loop response analysis to ensure stability of the power supply. If we analyze the frequency response of

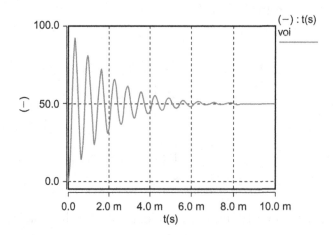

Figure 9.12:
Open loop buck converter simulation — ideal state space-derived model

the state space model we can see the gain peak, the very narrow phase margin, and also how the model loop could be closed to implement duty cycle control (Figure 9.13).

If we close the loop of the duty control by implementing a simple voltage control feedback, we can see the effect that this has on the loop response. We can implement a simple compensation scheme to increase the phase margin and provide a more realizable practical implementation, as shown in Figure 9.14.

If the loop is closed with no filter, then the response is much faster, as can be seen in Figure 9.15, but it still exhibits significant oscillation on the input and the frequency response shows a very poor phase margin.

In practice, there is a need to add a loop filter; however, this will add another pole into the system, which will further degrade the loop response, to the point that the power supply could become unstable. It is therefore necessary to ensure that a zero is added to the system to compensate for this and improve the phase margin. By adding a lead-lag filter, the necessary zero can add phase

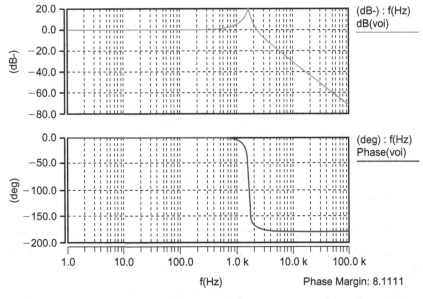

Figure 9.13:
Open loop buck converter simulation — ideal state space Bode plot

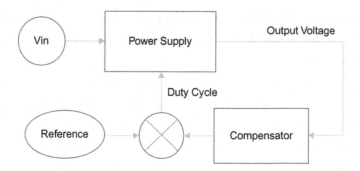

Figure 9.14:
Closed loop buck converter system model

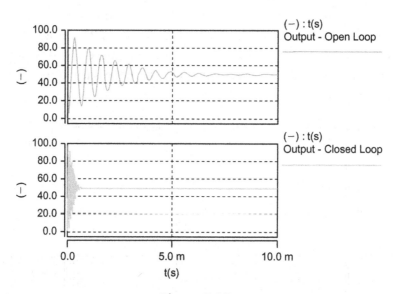

Figure 9.15:
Buck converter open and closed loop time domain response

margin, while still providing the necessary filtering; the results are shown in Figures 9.16, 9.17.

While these results are very good, there is one serious drawback with the state space approach, that is whenever a change is made to the topology of the

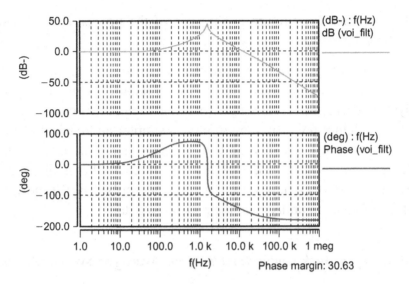

Figure 9.16:
Buck converter compensated Bode plot

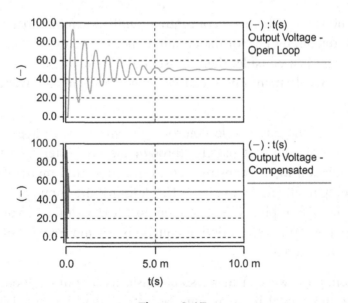

Figure 9.17:
Buck converter open and compensated loop time domain response

circuit, the designer must return to the start of the process and regenerate all the matrices and equations from scratch.

In this case, for example, we have neglected the series resistance in the inductor and capacitor, and therefore to add this in would require a complete cycle of circuit analysis to derive the correct equations, and then model creation. It is desirable from a modeling point of view to have a simpler approach, and we can do this by analyzing just the switch part of the converter. This is a specific technique called *average switch modeling* and we will cover that in the next section. There are automated techniques, such as those described in [3]; however, it is very useful to be able to understand the process in detail.

9.2.6 Modeling a Buck Converter Using an Averaged Switch Model

As we have seen, the state space technique is useful; however, it does have some drawbacks. The biggest drawback is that it is dependent on the topology remaining fixed. If the design is changed, or the topology changes then the whole process must be repeated to create the state space matrices and Laplace equations for analysis. We can often use a different approach to create an averaged model, however, where we concentrate on the switching part of the circuit and leave the remaining part of the circuit alone. If we revisit the Buck switching elements, and just consider them in isolation, then we can see that there are two switches, which themselves can be considered in an averaged manner (Figure 9.18).

As we have seen, the duty cycle controls not only the average voltage value, but also the average current passing through the converter, and therefore we can use this relationship to construct an averaged switch model. If we redraw the switch components as in Figure 9.19, defining the input pin as the active port (a), and the output pin as the common port (c), with the ground connection as the passive port (p), we can define equations for the transfer of voltage and current across the model.

Using this definition we can make some basic assumptions about the voltage and current in this model if we assume, as before, that the model is in continuous conduction mode in the steady state (i.e., that the inductor current does not

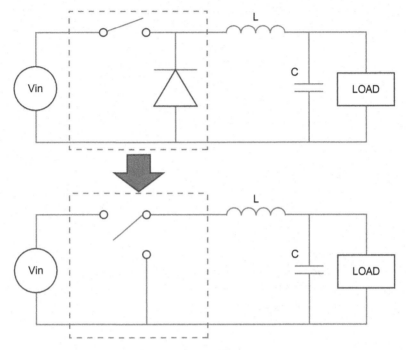

Figure 9.18:
Buck converter with average switch model

reach zero). How can we check this assumption? Well, we can calculate the average current and then, using the switching frequency and inductance, calculate the current ripple, and therefore check whether the current will reach zero or not. In this case we have already done a circuit-level simulation and clearly the inductor current is above zero after a short startup transient, so we can make that assumption safely. The first step in the definition of the averaged pulse width modulation (PWM) switch model is to define the input and output voltages. These can be defined as follows:

$$Input\ \ Voltage = v(a) - v(p) = v_{ap} \tag{9.27}$$

$$Output\ \ Voltage = v(c) - v(p) = v_{cp} \tag{9.28}$$

We can also define the input and output currents in the same manner:

$$Input\ \ Current = i(a) \rightarrow i(p) = i_{ap} \tag{9.29}$$

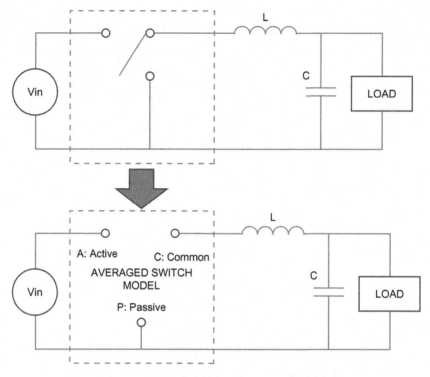

Figure 9.19:
Averaged PWM switch model

$$Output\ Current = i(c) \rightarrow i(p) = i_{cp} \tag{9.30}$$

All we need to do now is define a relationship between the inputs and outputs in terms of the duty cycle. We could define the duty cycle as a model parameter for fixed duty cycles, but obviously in most cases we would like this to be a control node in the model, so there will be an extra port on the model, d, to apply the duty cycle control. In the model itself we also need to take care that the value of duty cycle applied to any equations is strictly limited to the range $0 \rightarrow 1$; otherwise, non-physical and spurious results will arise.

If we look at the model, the switch behaves a bit like a duty controlled transformer, where the output voltage of the switch will, on average, be the input

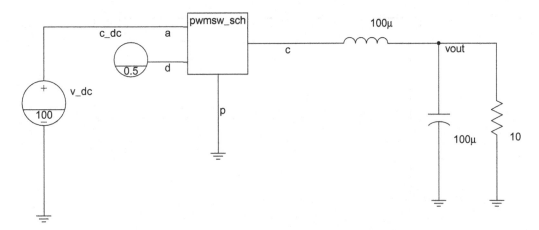

Figure 9.20:
Buck converter with averaged switch model

voltage multiplied by the duty ratio. We can therefore define the basic voltage transfer function of the averaged switch model as follows:

$$v_{cp} = D \times v_{ap} \tag{9.31}$$

Conversely, the current apparent at the input will be the output current multiplied by the duty ratio as only when the main switch is on will the current flow through the switch — the remainder of the time the current flows through the diode.

$$i_{ap} = D \times i_{cp} \tag{9.32}$$

These two equations are all that is required to model the averaged switch behavior in its most basic form. We can implement this simple model using Eqs (9.31) and (9.32), and the resulting circuit will be as shown in Figure 9.20.

If we then simulate the model with the time domain transient analysis, using the same startup time as the previous models, we can see that the voltage and current behavior closely matches the switched model equivalent (Figure 9.21).

This becomes particularly interesting when we directly compare a switched model, i.e., a model of the circuit with a switching MOSFET that has cycle-by-cycle switching *without averaging*, with the averaged model. It becomes clear

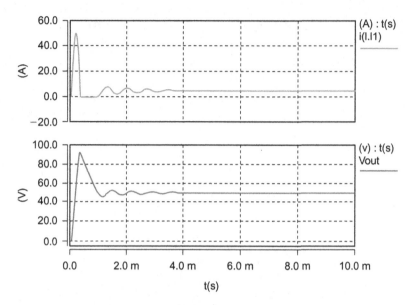

Figure 9.21:
Open loop buck converter simulation — averaged switch model

that because we have averaged the switch only, the remainder of the circuit can be simulated as it was in the original circuit, and so if we make changes to the model, then this will be reflected in the overall system response. Furthermore, even though the model does not have as few analog variables (11) as the state space model (6), it takes even fewer analog time steps (94) to complete the same simulation, so is, in fact, even more efficient. It takes around 1% of the simulation time of the switched MOSFET equivalent to complete the same time domain analysis (Figure 9.22).

If we look at the waveforms the averaged model tends to overestimate the inrush current using this simple voltage transfer model, and, in practice, we would extend the model to include some of the important switch parasitic values, such as voltage drop across the diode and switch ON resistance. However, it is interesting to note the close correlation of the waveforms and their transient behavior. If we zoom into the voltage and current waveforms more closely we can see that the averaged model bisects the switch level model (in this case we have added the on resistance of the switch to the averaged model (about 0.1 Ω) to ensure the correct output voltage) (Figure 9.23).

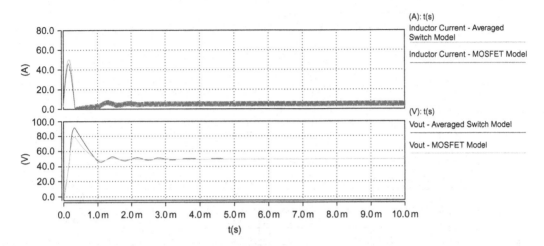

Figure 9.22:
Buck converter — comparison of MOSFET and averaged switch models

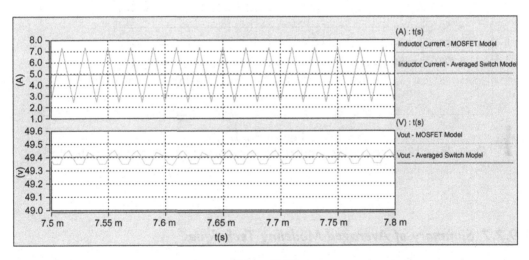

Figure 9.23:
Buck converter — detailed comparison of MOSFET and averaged switch models

This averaged model can also be used for frequency analysis, as in the case of the state space model, and we can see that if we show a Bode plot of the analysis of the Buck converter with averaged switch model. These results are almost identical to the state space approach, and the same closed loop analysis could be carried out using this model to ensure stability and performance (Figure 9.24).

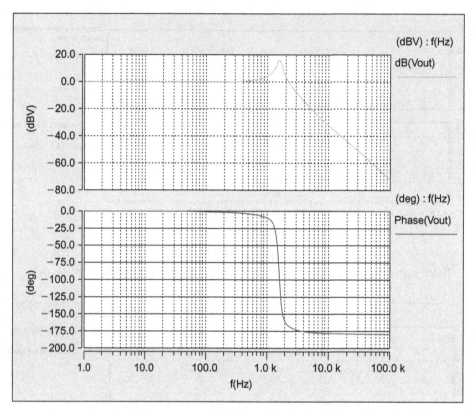

Figure 9.24:
Open loop buck converter simulation — averaged switch model Bode plot

9.2.7 Summary of Averaged Modeling Techniques

In this section we have shown how to derive averaged models for switching circuits using state space and averaged modeling techniques. The results indicate both methods have strengths and weaknesses; however, the basic principle applies in both cases that an averaged model will result that can predict the transient and frequency domain behavior accurately and faster than a detailed switch-level model. We have also demonstrated that it is useful to reduce the complexity of the switch model, which will also result in substantial efficiency improvements in time domain simulation.

9.3 Fast Analog Modeling

9.3.1 Introduction

This chapter has highlighted two distinct approaches to improving the simulation efficiency for large analog systems, one being to use averaged techniques for switching systems and the other is to use an event-based analog approach. Both methods sacrifice instantaneous accuracy for overall system understanding, with the benefit of vastly reduced simulation times. As ever with any modeling technique, the "trick" is to employ these approaches to solve the problem in hand, thus, for example, if the problem is at the system level, then is it really necessary to model *every* transistor to the nth degree? The answer in many cases is probably not.

A viable alternative to modeling analog variables using a conventional analog solver (such as Newton-Raphson) is to implement an event-based approach instead. If we were to compare a simple amplifier, as shown in Figure 9.25, where there is a corresponding equation that defines its behavior (also shown in Figure 9.25) we have two alternative routes to modeling this item.

The first approach is as we have discussed previously in this book, where we can define a conserved analog system, with an input branch and an output branch to create analog variables, or perhaps a simpler route is to still use analog variables, but to define them in a "control system" form, with no conservation of energy. In this simple linear model, the resulting simulation would be very fast; however, what if we were to implement a non-linear function, such as the output saturating to a power supply voltage? In that instance, the non-linearity would provoke the use of a non-linear solver, such as Newton-Raphson, and the resulting effect, as we have seen, will be to reduce the simulation performance as the model is solved using concurrent matrices.

Figure 9.25:
Simple amplifier

We could take an alternative approach, which, instead of analog variables, would be to implement an event-based model that would change output state only once, on the change in an input state value. This implies that the input and output variables would be defined as real in the amplitude axis, but discretely in the time axis. We have already encountered this type of modeling in Chapter 8, event-based modeling, for that is precisely what this is. Clearly, in such a simple system, the benefits will be marginal; however, if the element is in any way non-linear, then the benefits will become magnified quickly. In addition, this also enables such systems to be executed in event-based simulators which are pervasive, rather than the very few high-performance analog simulators available.

In this section of the book on fast analog modeling, we will therefore concentrate on the possibility of using event-based approaches for *analog* systems, how they work, where they are appropriate and provide examples of practical situations where they can be applied.

9.3.2 Rationale — Why Would We Do This?

In a simple electronic design, most of the circuits can be modeled and simulated using either an analog simulator (such as Spice or Saber) and the digital part in a VHDL or Verilog simulator, and even mixed-signal designs can be simulated using mixed-signal simulators based on VHDL-AMS or Verilog-AMS. However, in vastly complex systems, such as in a high-density mixed-signal System on Chip (SoC), the problem can become so massive that it is impossible to model and simulate all the different aspects of the design in any single approach using analog or mixed-signal models.

Another key issue is that of chip- or system-level verification. If you have a mixture of analog and digital devices in the design, how is it possible to adequately test that the design will work using a single unified approach? One answer to this question is the use of event-based models, where the same graphical description of a design can be targeted to a digital modeling language (such as VHDL or Verilog) and simulated in the same simulator. The verification can be handled using "assertions", which, although digitally modeled, can also handle "analog" behavior, such as thresholds or measurements.

The remainder of this chapter will recap how we can model analog blocks using event-based approaches, how a single model description can be used to generate multiple models, assertions (including analog measurement type assertions), and system validation with examples.

9.3.3 Event-based Analog Modeling

As the same suggests, the idea behind event-based analog modeling is to implement analog behavior using a fundamentally discrete implementation. For example, in the previous example of the amplifier, the governing equation was simply defined using Eq. (9.33)

$$Output = A \times Input \tag{9.33}$$

The equivalent event-based model first must establish that an event has occurred on the input and then if that is indeed the case, then assign the output variable the correct value. The resulting model behavior is therefore of the following form:

$$\begin{aligned} event(Input) &= TRUE: \\ Output &\Leftarrow Input \times A \end{aligned} \tag{9.34}$$

This is a simple and yet powerful modeling approach and can be correctly simulated with a digital simulator; however, the same issues apply with this type of modeling as we have already discussed with digital modeling more generally, i.e., what about zero delay feedback? If we have a system which is of the form shown in Figure 9.26 then the simulator will oscillate as the output will

Figure 9.26:
Amplifier with zero delay feedback

try to set the input to the inverse value in zero time. Most simulators will have a maximum oscillation threshold and then flag an error in such situations.

Clearly, the issue of feedback becomes a serious item to be resolved. One obvious question is "how does an analog solver handle this case?", and in this isolated example, the answer would be to find a value which satisfies the equation, which would be Input = 0 = Output. In the digital case, however, there is no solution mechanism to handle this case, unless we artificially set the Output initial condition to "0".

This does highlight an issue with the use of this type of approach in typical analog systems, where if we intend to utilize analog event-based techniques, there is often much more initialization required at the *model* level, rather than in the system (as we would probably normally do in a typical analog system).

The second issue with this type of modeling is how often do we generate events for accuracy? This is similar to the fixed time step analog simulation approach, except we are now dealing with events rather than time steps, and, again, this is very much defined at the model level. In contrast to a variable time step simulator that would normally vary the time step to ensure a large number of time steps on rapidly changing edges, an event-based simulator does not have this facility. It is therefore contingent on the modeler to implement sources or models that include an appropriate number of event changes to achieve an accurate solution.

9.3.4 Non-Linear Modeling

As we have seen in the pure analog domain, modeling non-linear elements is fraught with danger. We need to be careful to ensure that not only are signals continuous, but that the first derivatives are also continuous to avoid problems for the analog solver. If we use a digital, event-based approach, then this problem does not arise. Take the example of a saturation block (this would often be required if in a system we wish to ensure that an amplifier, similar to our simple example in this chapter, does not exceed the supply voltage range) we can implement the model graphically as follows (Figure 9.27).

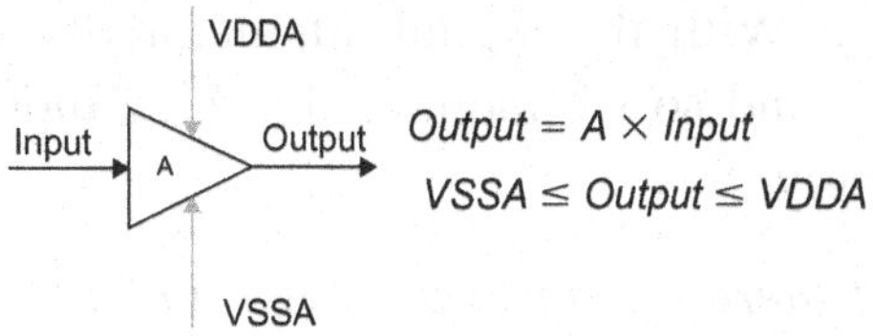

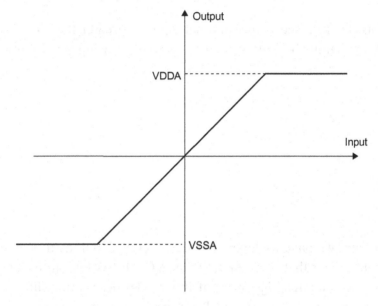

Figure 9.27:
Graphical model of a saturating amplifier

Figure 9.28:
Saturating amplifier transfer function

And the resulting transfer function will be as shown in Figure 9.28.

If we were to model this using a conserved analog model, then clearly we would require several equations and, in particular, we would need to ensure that the transitions when the output voltage reached the value of positive analog supply rail (VDDA) were handled to avoid convergence problems with the analog solver.

In the digital event-based model, we do not need to do this. We have three inputs to consider: VDDA, negative analog supply rail (VSSA), and the input (we can call this VIN). With the digital simulator, the model is *sensitive* to changes or events only, and so we monitor the three inputs and then apply the two equations accordingly.

First and *if and only if there has been a change in VIN, VDDA, and VSSA* we calculate the output (we can call this VOUT) applying the gain (A) of the model to the input.

$$VOUT \text{ (provisional)} = VIN^*A \qquad (9.35)$$

Then we can check to see if this internal value is greater than VDDA or lower than VSSA and limit the output value accordingly using the following model *pseudo-code*.

```
IF (VOUT (provisional) > VDDA) THEN
    VOUT = VDDA
ELSE IF (VOUT (provisional) < VSSA) THEN
    VOUT = VSSA
ELSE
    VOUT = VOUT (provisional)
END IF
```

In this approach, we have assigned a value to VOUT, but, as it is an event, it is more accurate to say that the output signal VOUT has had an event scheduled to take place. The interesting aspect of this model description, however, which is a massive benefit of the graphical approach, is that the intrinsic equations could be modeled either in an analog form or in a digital form, *with no changes*. This means that although we may need to add extra code for the analog model, the intrinsic behavior is consistent between the two approaches; meaning only one graphical model needs to be created (Figure 9.29).

If we create a simple test circuit to test this model, we will first define a test that uses an analog voltage input, analog supplies and simulates 10 cycles of a 1-kHz sinusoid with an amplitude of 10 V (peak), where VDDA = 3.3 V and VSSA = −3.3 V. The results are shown in Figure 9.30.

If the model is modified to be exported from the graphical modeling tool as an event-based model (which could be implemented in VHDL or Verilog in practice), and simulated, the results can be seen in Figure 9.31.

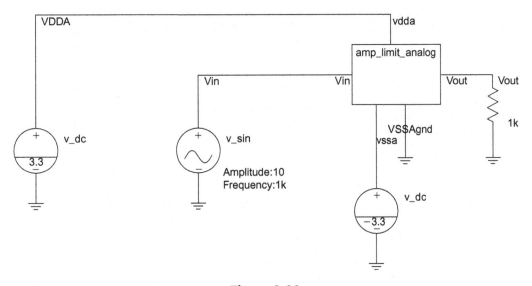

Figure 9.29:
Saturating amplifier graphical model of test circuit

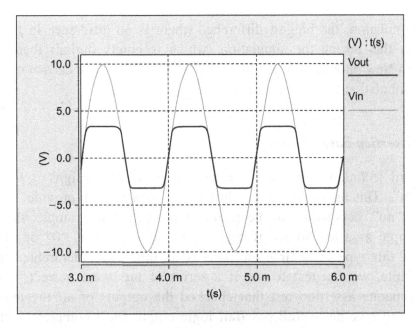

Figure 9.30:
Saturating amplifier modeled using analog equations

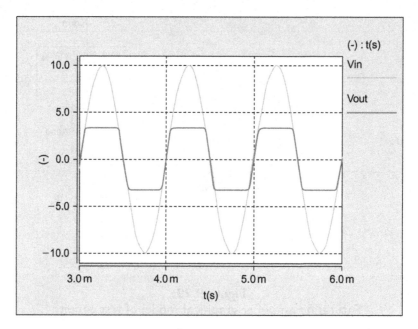

Figure 9.31:
Saturating amplifier modeled using digital equations

In this simulation, the biggest difference (there is no difference in the signal values) is that during the simulation (which is purely digital) there are, of course, no Newton-Raphson non-linear solver iterations to be carried out, therefore the simulation time is very fast.

9.3.5 Assertion-based Testing

A standard technique for functional testing in digital designs is to use an "assertion". This is simply a condition test, which will provide a logical "yes" or "no", depending on the result of the test. For example, this could be as simple as a test to see if a particular logic bit was "0" or "1". The beauty of this type of testing approach is its inherently hierarchical nature. For example, we can test to see if a series of bits were correct, and then have a separate assertion test that checked the outputs of all the individual assertions to give an overall pass/fail logic output for a complete section of the design.

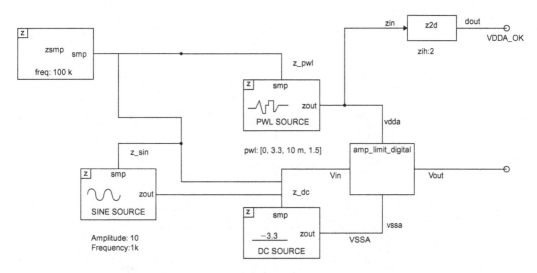

Figure 9.32:
Assertion test on VDDA (OK if VDDA > 2.0 V)

We can extend this for our complete system design when we use these fast "digital" implementations of analog models in exactly the same way. For example, we could have an assertion that gave an output if the supply voltage dropped below a minimum level, such as 2 V for the example we have used in the limiting amplifier, where the nominal VDDA is 3.3 V.

We can illustrate this by taking the digital equivalent model of the test circuit for our amplifier and adding an assertion on the VDDA supply to check that it is always above 2 V. In this case, we will change the VDDA supply from a steady state 3.3 V DC to a ramp function that starts at 3.3 V; however, it decreases to less than the 2 V minimum within a few milliseconds (Figure 9.32).

When we simulate this model, the effect on simulation time is negligible as there is only going to be one event as the VDDA crosses the threshold. The resulting behavior can be seen below in Figure 9.33. As the VDDA drops below 2 V, the assertion logic signal (VDDA_OK) changes from 1 to 0 accordingly.

One of the helpful aspects of the model is that this type of assertion-based testing can be implemented in either an analog or digital implementation of the basic model; however, the performance benefit is manifest the most in a digital model (Figure 9.33).

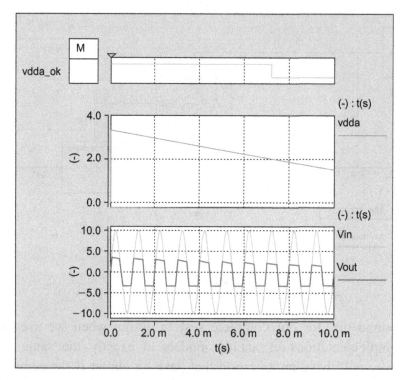

Figure 9.33:
The effect of a decreasing VDDA on an assertion test

One obvious implication of this type of modeling is that not only can these assertions be used for a testing function, but they can also lead to advanced implementation features in a digital model. For example, in a typical digital model there is not a check to see if the supplies are connected to model, but, as we have seen in this simple example, we can check that this connectivity is in place and verify that both the VDD and VDDA (digital and analog supplies) are correctly implemented.

9.4 Finite-Difference Modeling

9.4.1 Introduction

A second and extremely valuable method of fast analog modeling is referred to as finite-difference (FD) modeling. In this approach the differential equations that are normally derived for the governing equations of an analog model are approximated with finite difference equations, thus allowing the model to be implemented using event-driven techniques and made compatible for digital simulators. The advantage here is that the new model, an approximation of the continuous-time version, can typically be simulated three orders of magnitude faster by virtue of the fact that it will execute in the digital simulator. The upshot is that now models in the analog domain can be mapped over to a compatible format where an entire mixed-signal design can be simulated in a digital simulator. While this will not work for studying detailed performances of analog circuits, it will indeed prove to be one of the most valuable chip-level approaches for verifying chip-level and multi-chip functionality. This approach has actually been demonstrated by a multi-organizational team of collaborators designing a two-chip solution for sensor interface applications for extreme environment applications [2, 4, 5].

9.4.2 Description of Approach

The finite-difference approach is rooted in sampled systems theory. While a complete treatment is beyond the scope of this book, it is instructive to step back and refresh our memories on the fundamentals and then to address the applicability of this approach (i.e., when is it the right choice). The basic approach is that given a system that can be represented with differential, algebraic equations that govern its behavior, these equations are approximated through a finite-difference method. This implies a sampling rate in time to better improve the approximation. It also implies that there may be situations where the approximation fails to yield suitable accuracy, which we will address later. Lastly, implementation of FD models will: (a) require the internal definition of a sampling rate or delta time, and (b) that these models can be executed in a digital simulator for improved speed.

The FD model is achieved by replacing each differential operator in a differential equation with a discrete differential. Take, for example, a differential equation that has dV_1/dt as one of its terms. This would be replaced by

$$\frac{dV_1}{dt} \approx \frac{\Delta V_1}{\Delta t} = \frac{V_1(n) - V_1(n-1)}{\Delta t} \tag{9.36}$$

where $V_1(n)$ is the present value of the voltage V_1 and $V_1(n-1)$ is the previous value of V_1. The derivative is approximated by the simple finite difference between the most recent past and present values. As Δt approaches zero the approximation gets better. From this approximation and others like it, the governing equations are transformed into an FD set.

From an implementation perspective, the FD model will be event-driven and have an internal timer based on the chosen value of Δt, which can vary from model to model as desired. This makes the implementation compatible with digital solvers. At this point we will illustrate the approach with a simple example.

9.4.3 Example 9.1

An RC low-pass filter (Figure 9.34) is a straightforward model to understand. The behavior of this simple circuit as a low-pass filter is explained by the changing impedance of the capacitor with frequency. As the frequency increases, then impedance decreases. If the frequency increases enough, then the output is shunted to the common node and nothing passes. Therefore, the high-frequency components are filtered.

The differential equation that governs this filter characteristic can be derived via KCL at the output node in the Laplace domain and is given as

$$sCV_{out} + \frac{V_{out} - V_{in}}{R} = 0 \tag{9.37}$$

or in differential equation form

$$RC\frac{dV_{out}}{dt} + V_{out} = V_{in} \tag{9.38}$$

Applying the FD approximation to the differential term in Eq. (9.38) and rearranging yields

$$(\tau + \Delta t) \cdot V_{out}(n) = \tau \cdot V_{out}(n-1) + \Delta t \cdot V_{in}(n) \tag{9.39}$$

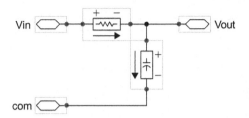

Figure 9.34:
Simple low-pass RC filter for investigating the merits of finite-difference modeling

```
1 use ieee.ELECTRICAL_SYSTEMS.all;
2 entity RC is
3 generic (
4 constant R : in real := 1.0e3;
5 constant C : in real := 1.0e-12
6 );
7 port (
8 terminal Vin : ELECTRICAL;
9 terminal Vout : ELECTRICAL;
10 terminal com : ELECTRICAL
11 );
12 end entity RC;
13 architecture consv_elec of RC is
14 quantity V_cap across I_cap through Vout to com;
15 quantity V_res across I_res through Vin to Vout;
16 begin
17 I_cap == C * V_cap'DOT;
18 V_res == I_res * R;
19 end architecture consv_elec;
```

```
1 Entity ed_RC is
2 generic (
3 constant R : in real := 1.0e3;
4 constant C : in real := 1.0e-6;
5 --Time step used to calculate RC
6 --delay
7 constant DeltaTime : in real :=
8 1.0e-6
9 );
10 port (
11 signal Vin : in real := 0.0;
12 signal Vout : out real := 0.0
13 );
14 end entity edms_RC;
15 architecture edms of edms_RC is
16 constant DeltaTimeT : Time := real2time(DeltaTime);
17 begin
18 RCDelay: process
19 constant Tau : real := R * C;
20 variable PrevVout : real := 0.0;
21 variable VoutInt : real := 0.0;
22 begin
23 VoutInt := (PrevVout*Tau+Vin*DeltaTime)/(Tau+DeltaTime);
24 Vout <= VoutInt;
25 wait for DeltaTimeT;
26 PrevVout := VoutInt;
27 end process RCDelay;
28 end architecture edms;
```

Figure 9.35:
(a) VHDL code generated from ModLyng for the continuous-time RC filter
(macromodel), (b) VHDL code generated from ModLyng for the finite-difference
version of the RC filter model

where $\tau = RC$. The two implementations are given in Figures 9.35(a) and (b), respectively. The waveforms in Figure 9.36 show the responses of each of these models.

The creation of the FD model is as easy as the macro model given that the graphical modeling tool has the built-in primitives to assemble the model. This assembly is summarized in Figure 9.37.

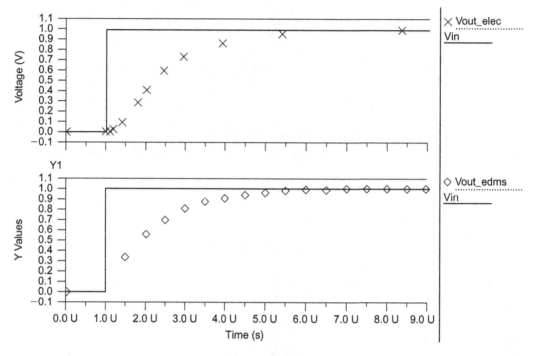

Figure 9.36:
Top curve is the continuous-time filter response while the bottom curve is the FD
model's response

The limits to which the FD modeling technique can be applied depend largely on the mapping of simultaneous $I-V$ relationships to a difference equation. Avoiding the use of simultaneous equations involves choosing the right level of abstraction for model implementation. Modeling blocks that will be operated in tight feedback configurations are perhaps not the best situation for FD modeling techniques, where race conditions could emerge. However, often times ascending the design hierarchy enables one to model that entire feedback circuit block as a block itself and this can provide a lot of advantages at the chip-level.

Conclusion

In this chapter we have introduced two important techniques for rapid and accurate simulation of complex systems. Averaged modeling is particularly

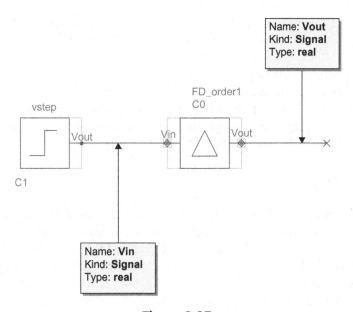

Figure 9.37:
Block diagram indicating the assembly of the RC finite-difference model using the
difference equation effect

well suited to switching systems, and has therefore been effectively used in practical systems for many years, especially in areas such as switching power supplies and also switched capacitor circuits. The use of a digital equivalent model in place of a detailed analog model is a relatively new innovation; however, the potential benefits in simulation time and integration with a rigorous verification scheme as used routinely in the digital world offers the possibility of a complete system level verification approach for complex electronic systems.

References

[1] R. W. Berger, R. Garbos, J. D. Cressler, M. M. Mojarradi, L. Peltz, B. Blalock, et al., A miniaturized data acquisition system for extreme temperature environments in space, Proc. IEEE Aerospace Conference 2008, Big Sky, MT (2008) 1–12.

[2] J. D. Cressler, M. Mojarradi, B. Blalock, W. Johnson, G. Niu, F. Dai, et al. Silicon-germanium integrated electronics for extreme environments, Proc. Government Microcircuit Applications and Critical Technology Conference (GOMAC) 2007, Orlando, FL (2007) 4.

[3] A. Merdassi, L. Gerbaud, S. Bacha, Automatic generation of average models for power electronics systems, J. Model. Simulat. Syst. 1(3) (2010) 176–186.

[4] C. Webber, J. Holmes, M. Francis, R. Berger, A. Mantooth, A. Arthurs, et al., Event driven mixed-signal modeling techniques for system-in-package functional verification, IEEE Aerospace Conference, Big Sky, MT (2010) 1–16.

[5] R. M. Diestelhorst, S. Finn, L. Najafizadeh, D. Ma, P. Xi, C. Ulaganathan, et al., A monolithic, wide-temperature, charge amplification channel for extreme environments, IEEE Aerospace Conference, Big Sky, MT (2010) 1–10.

Further Reading

[1] C. Ulaganathan, N. Nambiar, B. Prothro, R. Greenwell, S. Chen, B. J. Blalock, et al., A SiGe BiCMOS instrumentation channel for extreme environment applications, Proceedings of 51st Midwest Symposium on Circuit and Systems (2008) 217–220.

[2] J. J. DiStefano, A. R. Stubberud, I. J. Williams, Feedback and Control Systems, McGraw Hill, 1990.

[3] R. E. Harr, A. G. Stanculescu., Applications of VHDL to Circuit Design, Kluwer Academic Publishers, 1991.

CHAPTER 10

Model-Based Optimization Techniques

10.1 Introduction

Optimization methods applied to fit simulation models to measured data have been investigated by researchers such as Schmidt and Güldner [1] and Lederer et al. [2] using the well-known simulated annealing approach. Genetic algorithms provide an alternative approach to optimization that may have some advantages, especially when considering the more complex problem of fitting several loops simultaneously. The use of genetic, or evolutionary, algorithms to solve difficult engineering problems is a relatively recent innovation. Holland [3] and Goldberg [4] are two of the pioneers of this technique, and the last 10 years have seen a plethora of applications for genetic algorithms from systems design to topology analysis [5]. The fundamental difference between genetic algorithms and conventional optimization techniques, such as simulated annealing [6], is that in certain problems the computational effort involved in a standard exhaustive search method would be prohibitive. The random nature of genetic algorithms may not always find the absolute optimum solution, but this type of stochastic approach often has a greater chance of finding a relatively good solution, quickly, for difficult problems. This type of approach also works well for problems with chaotic or ill-defined behavior that is sometimes difficult to classify and also those problems with local maxima or minima that would, perhaps, trap a conventional search algorithm.

10.2 Overview of Optimization Methods

Optimization techniques have been applied to the solution of engineering and mathematical problems for many years, and a plethora of methods exist, with varying degrees of usefulness. Pierre [7] gives a good grounding in the broad spectrum of approaches, including the *univariate search method*, while

Laarhoven and Aarts [6] provide details for the *simulated annealing* approach and some practical examples. One exciting new area of progress is the development of *evolutionary* or *genetic* algorithms, and a key text in this area is Goldberg [4]. Conventionally, the simulated annealing approach has been applied to the optimization of model parameters [1, 2], but comparisons in this book have been made between all the methods. The genetic algorithm has been applied successfully to topological optimization of transformers [8], but no published literature exists for the application of the method to material model parameters. This approach is described in this chapter.

10.2.1 Univariate Search Methods

Univariate search methods [7] are probably the simplest form of optimization algorithm to implement, and require that a set of parameters are varied, in turn, until the solution cannot be improved. Once all the parameters have been varied, the whole process is repeated until the solution goal has been achieved or a maximum number of iterations has been reached. The number of times a parameter is varied before moving on to the next one indicates the *depth* of the search, while the overall number of loops through the set of parameters is the *breadth* of the search. A flowchart of the univariate search method is given in Figure 10.1.

10.2.2 Simulated Annealing

The simulated annealing method [6] uses a mathematical analogy of the cooling of liquids into solid form to provide an optimal solution for complex problems. The method operates on the principle that annealed solids will find the lowest energy point at thermal equilibrium, and this is analogous to the optimal solution in a mathematical problem. The equation for the energy probability used is defined by the Boltzmann distribution given by Eq. (10.1).

$$P(E) = \frac{1}{Z(T)} \cdot \exp\left(-\frac{E}{k_B T}\right) \tag{10.1}$$

where $Z(T)$ is the partition function, which is a normalization factor dependent on the temperature T, k_B is the Boltzmann constant, and E is the energy.

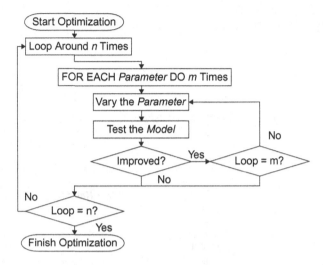

Figure 10.1:
Flowchart of the univariate search method

This equation is modified into a more general form, as given by Eq. (10.2), for use in the simulated annealing algorithm.

$$P(E) = \frac{1}{Q(c)} \cdot \exp\left(-\frac{C(i)}{c}\right) \tag{10.2}$$

where $Q(c)$ is a general normalization constant, with a control parameter c, which is analogous to temperature in Eq. (10.1). $C(i)$ is the cost function used, which is analogous to the energy in Eq. (10.1). The parameters to be optimized are perturbed randomly, within a distribution, and the model tested for improvement. This is repeated with the control parameter decreased to provide a more stable solution. Once the solution approaches equilibrium, then the algorithm can cease. A flowchart of the full algorithm applied is given in Figure 10.2.

10.2.3 Genetic Algorithms

The use of genetic, or evolutionary, algorithms to solve difficult engineering problems is a relatively recent innovation. Holland [3] and Goldberg [4] are two of the pioneers of this technique and the last 10 years have seen an abundance of applications for genetic algorithms from systems design to topology analysis [5].

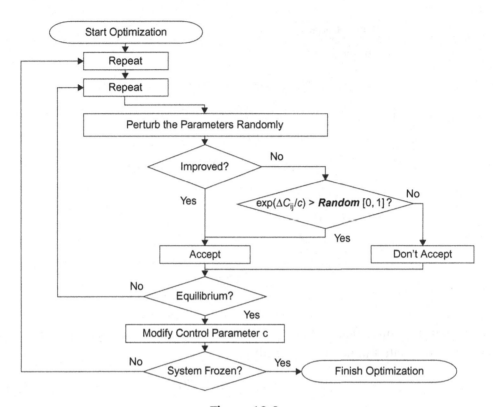

Figure 10.2:
Flowchart of the simulated annealing method

The fundamental difference between genetic algorithms and conventional optimization techniques, such as simulated annealing, is that in certain problems the computational effort involved in a standard exhaustive search method would be prohibitive. The random nature of genetic algorithms may not find the absolute best solution, but they have a greater chance of finding a good solution, quickly, for difficult problems. This randomness also works well for problems with chaotic or ill-defined behavior difficult to classify, and those problems with local maxima or minima that would, perhaps, trap a conventional search algorithm.

The first step in the genetic algorithm process (Figure 10.3) is to define an initial population of individuals, which becomes the first generation for the algorithm. For each generation there will be a number of children created by combining the characteristics of two parent individuals, determined by random

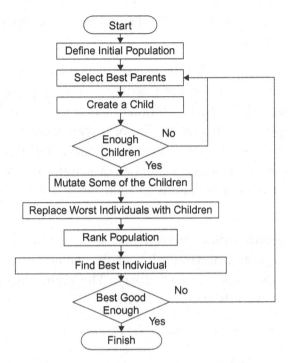

Figure 10.3:
Flowchart of the genetic algorithm

and selective methods. Once the two parents have been selected, then their characteristics can be combined to create a new individual (child). The selection of characteristics from the two parents is random, and there is a risk of small mutations. Once the required number of children have been created the population as a whole is adjusted by replacing the worst individuals in the population by the newly created children.

10.2.4 Multi-Objective Optimization

A circuit's performance is obviously a function of its designable parameters. The design goal is always to find a parameter set that meets all the performance functions and any imposed constraints. Optimization methods are important in finding the optimal solutions for a design problem. When an optimization problem involves only one objective function, this task is referred to as single-objective optimization. This is the kind of optimization, where there often exists more than one objective that needs to be optimized, known as multi-objective optimization.

In a multi-objective scenario, different solutions may produce trade-offs among competing objectives. It is often not possible to find a single solution without compromising other objectives. In a problem where such a trade-off exists, usually none of the optimal solutions are the best with respect to all of the objectives. Thus, in problems with more than one conflicting objective, there is no single optimum solution. Instead, a number of solutions may exist that are all optimal. This set of optimal solutions in multi-objective optimization is called the Pareto front [2].

A multi-objective optimization problem has a number of objective functions that are to be minimized or maximized. Usually, the optimization has a number of constraints that the solution must satisfy. The multi-objective optimization formulation can be generally stated as follows:

$$Minimise/Maximise \ f_m(x), \ m = 1, 2 \ldots M$$

$$Subject \ to \ g_j(x) \geq 0, \ j = 1, 2 \ldots J; \tag{10.3}$$

$$h_k(x) = 0, k = 1, 2, \ldots K;$$

$$x_i^{(L)} \leq x_i \leq x_i^{(U)}, i = 1, 2, \ldots n$$

where $f_m(x)$ is the set of M performance or objective functions that constitute a multi-dimensional space called the *objective space*. In Eq. (10.3), there are several constraints associated with the problem formulation. J and K are inequality and equality constraints that are defined by the $g_j(x)$ and $h_k(x)$ functions, whereas x_i is a set of boundary constraints that restrict each decision variable to take a value between a lower and upper bound. These boundary constraints represent the decision space of the design problem. A solution that satisfies none of the constraints is called an infeasible solution as opposed to a feasible solution that satisfies all constraints. For each point in the decision space, there exists a point in the objective space. Figure 10.4 shows these two spaces and a mapping between them.

In the objective space, the entire feasible region can be divided into two sets of solutions: non-dominated solutions and dominated solutions. None of the

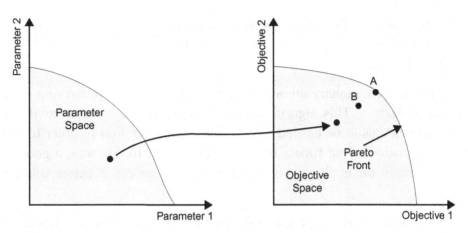

Figure 10.4:
Design space and objective space

non-dominated solutions can be said to be better than the other with respect to the objective functions, and these are referred to as the Pareto-optimal solutions. If any solution in the objective space is dominated by at least one solution from the Pareto-optimal set, then it is referred to as a non-Pareto-optimal solution. In Figure 10.4, solution B is dominated by solution A. Solution A is said to be non-dominated (Pareto-optimal solution). The thick black line in this figure that represents all the Pareto-optimal solutions is called the Pareto-front. There are several evolutionary-based algorithms that have been developed for multi-objective optimization in order to generate the optimal Pareto-front. One of the algorithms often used for this purpose is the non-dominated sorting genetic algorithm−II (NSGA-II) [2], and is explained in the next section of this chapter.

10.2.5 NSGA-II

The theory of evolutionary algorithms has been well established and widely used for solving optimization problems. Evolutionary algorithms are search and optimization procedures that are motivated by the principles of natural genetics and selection. In general, evolutionary algorithms start with a random number of individuals (design parameters) that form an initial generation. Each individual in a generation is assessed based on the specified objective and a score (fitness) value is determined. Based on the fitness score, the individuals will go through a process of selection, recombination, and mutation to generate a new set of individuals for the next generation. This process is continued iteratively until the optimization target is met and the best number of individuals with the best fitness is obtained.

NSGA-II is an evolutionary algorithm that employs an elite-preserving strategy for the optimization. This algorithm is categorized as elitist-based as it allows the elite individuals to be carried over to the next generation in order to ensure that the population's best fitness does not deteriorate. In this way, a good solution found early on in the run will never be lost unless a better solution is discovered.

The algorithm starts by creating an offspring population Q_t from a parent population, P_t. These two populations are combined together to form R_t of size $2N$ (N is the size for each population). Then, a non-dominated sorting is used to classify the entire population R_t. This step checks for non-dominated points among the individuals and sorts accordingly. The next step is to generate a new population with size N and fill this population with solutions of different non-dominated fronts from the previous sorting. The filling starts with the best non-dominated front, followed by second best and so on. As the population size is N, which is smaller than the size of R_t, which is $2N$, not all fronts can be accommodated in the new population. All fronts that cannot be accommodated in the new population are discarded. Sometimes there exists a condition where the last front has more solutions (individuals) than the available space in the population. In this case, a crowding distance metric is used to choose which members of the last front are placed in the new population. Figure 10.5 illustrates the strategy employed by NSGA-II. Once the new population is filled with all fronts, the selection, crossover, and mutation operators will be applied to this population to create new offspring, and the whole process is repeated again until the final number of generations has been reached. The step-by-step algorithm flow in NSGA-II is outlined in Figure 10.6.

10.2.6 Pareto-Based Optimization

One approach to utilizing these techniques to optimize designs under a variety of conditions is called Pareto optimization [9]. The Pareto-front resulting from multi-objective optimization is used for the performance and variation model

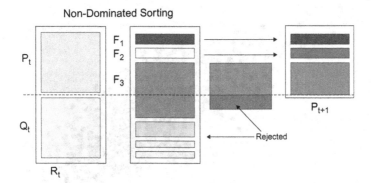

Figure 10.5:
Non-dominated sorting genetic algorithm-II procedure

NSGA Algorithm

- Generate initial random population, size N
- Create offspring population
- Combine parent and offspring population to form R_t. ($R_t = P_t \cup Q_t$)
- Perform non-dominated sorting and identify fronts, F_i (i = 1, 2...etc.)
- Set new population, P_{t+1} = 0, and fill P_{t+1} with F_i, ($P_{t+1} \cup F_i$) as long as $|P_{t+1}| + |F_i| < N$
- Perform crowding sort and place most widely spread solution in P_{t+1}
- Create offspring population Q_{t+1} from P_{t+1} and repeat until last number of generation

---

Figure 10.6:
Non-dominated sorting genetic algorithm-II (NSGA-II) algorithm

that is developed. In the Pareto-based yield optimization method, Monte Carlo simulation is used to estimate the yield of the design. This is obviously very time consuming, so in order to reduce the number of Monte Carlo runs, the simulation is only applied to a small feasible region defined by the performance specification boundaries. Because there are only a small number of solutions in the feasible region, far fewer Monte Carlo analyses are required, mitigating the computational overhead. Figure 10.7 illustrates the Pareto-based yield optimization methodology.

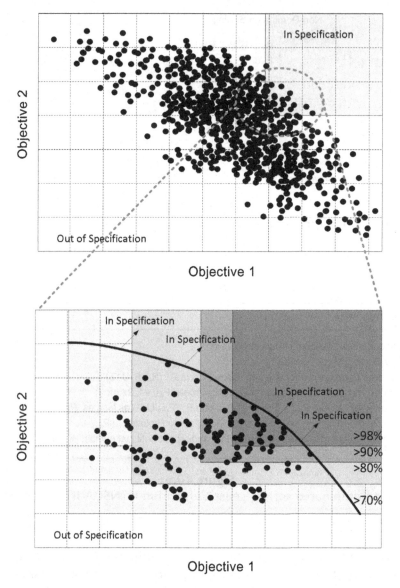

Figure 10.7:
Pareto-based yield optimization

10.2.7 Particle Swarm Optimization

Particle swarm optimization [10, 11] is similar to the genetic algorithm technique for optimization in that rather than concentrating on a single individual implementation, a population of individuals (a "swarm") is considered instead. The algorithm then, rather than moving a single individual around, will move the population around looking for a potential solution. This is an example of a heuristic approach, where there is no guarantee of an optimal solution.

Each individual in the swarm has a position and velocity defined, the algorithm looks at each case to establish the best outcome using the current swarm, and then the whole swarm moves to the new relative location.

10.2.8 Levenberg–Marquardt Algorithm

The Levenberg–Marquardt algorithm (LMA) [12, 13] is a technique that has been used for parameter extraction of semiconductor devices, and is a hybrid technique that uses both Gauss–Newton and steepest descent approaches to converge to an optimal solution. The hybrid approach is often used to trade off the best characteristics of different algorithms to solve a wider range of problems. For example, the Gauss-Newton approach is faster than LMA if the initial guess is relatively close to the optimal solution. In the event that this is not the case, the hybrid solution (LMA) uses aspects of the steepest descent approach to traverse the design space and find a potential solution area, and then find the optimum. This technique is particularly effective in solving systems of non-linear equations, and this makes it useful for semiconductor devices that have both large sets of non-linear equations and large parameter sets.

10.2.9 Summary of Optimization Techniques

So far we have described a variety of techniques for the optimization of design parameters, from the simple univariate approaches to the more sophisticated multiple objective and genetic algorithm-based methods. In order to illustrate how these can be applied in practice a case study is useful in assessing both the effectiveness of, and also how easy (or otherwise) it is to implement, these techniques. The remainder of this chapter therefore works through the optimization of a "difficult" model, a non-linear magnetic material model. This involves some complex model equations, inter-related parameters, and a difficult optimization to solve.

10.3 Case Study: Optimizing Magnetic Material Model Parameters

10.3.1 Introduction

The optimization of magnetic materials is a problem that has aspects suited to the application of genetic algorithms. Even though the Jiles-Atherton model of hysteresis [14, 15] is well understood mathematically, the parameters are inter-linked in such a way that the set of possible combinations of parameters may be large. Significant changes in the shape of the hysteresis loop may result from small parameter variations. These two aspects give a relatively high risk of local maxima or minima being found, or instabilities in a conventional algorithm. Other approaches for modeling magnetic materials have no direct link between behavior and material physical properties so are natural potential targets for the genetic algorithm approach.

10.3.2 Magnetic Material Model Optimization Procedure

The overall structure of the analysis process developed is shown in Figure 10.8. The process derives the measured field strength (H) and flux density (B) from the oscilloscope data, with the test configuration able to be customized. The Jiles–Atherton model (which is characterized by five main parameters, a,c,k,ms, and *alpha*) can be calculated using the derived field strength from the measured data or an ideal internally generated test waveform (sinusoidal or triangle) and then compared with the original test results. Metrics can be applied to both the measured and simulated results to assess the relative performance of the model. An optimization loop allows the model parameters to be modified using iteration to achieve optimum model

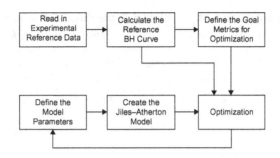

Figure 10.8:
Jiles—Atherton magnetics model optimization procedure

performance. The optimization method is a steepest descent type of approach, simulated annealing or genetic algorithm.

10.3.3 Comparison of Optimization Methods

A torroid made of the Siemens N30 material was tested to obtain some reference data and the resulting B—H loop used for optimization. The Jiles—Atherton model was optimized using the model described in [14] and [15]. The optimization was carried out using the well-understood simulated annealing method as a control and also with a genetic algorithm approach. The simulated annealing approach was carried out with a variation of 10%, a control factor of 0.001, and 2500 iterations. The genetic algorithm used 50 generations of a population including 50 individuals (50 × 50 giving a rough equivalent of 2500 iterations). Each generation produced 40 children, of whom 20 were mutated. A variation of 10% was introduced in the mutation process. In each case the fitness function used the least squares error approach.

The resulting mean errors between the simulated and measured results are summarized in Table 10.1 and Figure 10.9. Table 10.1 shows that the error is significantly reduced for the genetic algorithm. Figure 10.9 shows the error versus the number of iterations and, again, clearly the genetic algorithm is the most accurate.

Interestingly, when a combination of simulated annealing and the genetic algorithm is applied, the error was reduced further. This can be explained by the

Table 10.1: Comparison of Simulated Annealing and Genetic Algorithm
Optimization With and Without Gaussian *k*

Parameter	Genetic Algorithm	Genetic Algorithm Gaussian k	Simulated Annealing	Simulated Annealing Gaussian k
a	13.8	9.29	7.26	6.96
c	0.799	0.497	0.150	0.308
k	11.01	12.67	7.19	11.176
α	6.78e−6	29.1e−6	4.29e−6	4.54e−6
Ms	293318	268562	264500	257479
σ	−	16.55	−	15.54
Error	0.1416	0.019	0.188	0.035

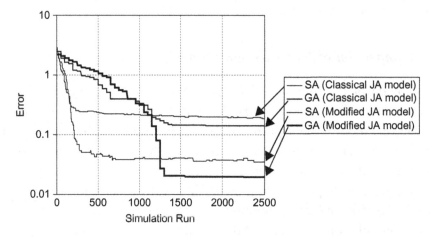

Figure 10.9:
Comparison of simulated annealing and genetic algorithm error functions

fact that the two methods have different strengths. The genetic algorithm is very good at finding the correct area of the solution, tolerant of local maxima and minima, and the simulated annealing method is excellent at refining a solution systematically to the nearest maximum or minimum. It is also worth noting the wide differences between some of the parameters obtained using different optimization approaches. The parameters in the Jiles–Atherton model are highly interdependent, and, as such, may produce similar B–H curves for widely different parameter sets.

The results of the optimization are, perhaps, best visualized by observing the resulting B–H loops. Figure 10.10 shows the optimized B–H loop using the original Jiles–Atherton model with no modification, and also the optimized curve using a combination of genetic algorithm and simulated annealing optimizations.

10.3.4 Statistical Analysis of Optimization Methods

Although the individual optimization results shown previously are encouraging, owing to the random nature of the optimization process in both simulated annealing and genetic algorithm approaches, it is appropriate to investigate the performance of the respective methods statistically. The optimizations were therefore repeated over a number of runs (20) and the resulting errors compared. The simulated annealing approach was carried out with a variation of 10%, a control factor of 0.001 and 2000 iterations used. The genetic algorithm used 40 generations of a population including 50 individuals (40×50 giving a rough equivalent of 2000 iterations). Each generation produced 40 children, of whom 20 were mutated. A variation of 20% was introduced in the mutation process. The simulated annealing approach gave a mean value for the error of 0.0219, with a standard deviation of 0.003, while the genetic algorithm gave a mean value for the error of

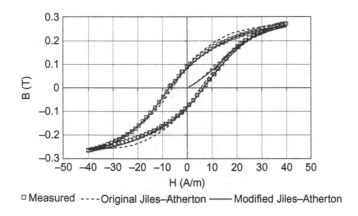

Figure 10.10:
Comparison of measured, optimized original Jiles–Atherton model and optimized modified Jiles–Atherton model B–H curves

0.0148, with a standard deviation of 0.009. Figure 10.11 shows the histogram of the respective errors for the two methods.

It is interesting to note that although the genetic algorithm has a greater standard deviation (wider spread), the average error is less than that for the simulated annealing approach. This is partly expected, as the genetic approach will search on a wider array of possible solutions, some better and some worse. The strength of the method is that, in some cases, the error will be spectacularly better than the simulated annealing approach, as shown in Figure 10.11.

10.3.5 Multiple Loop Optimization

In practice, for circuit simulation, the resulting optimized model for a magnetic material must be accurate over a wide variety of operating conditions. To ensure this is the case, the optimization goal function was extended to allow the optimization of a set of B−H loops rather than a single major loop. Each loop in the set has its own weighting, so if it is essential that the minor loop has a high level of accuracy, but the major loop is not significant, then the weighting can be increased for the minor loop accordingly. An example of this

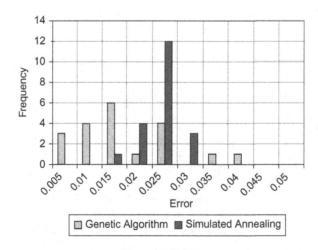

Figure 10.11:
Statistical comparison of the performance of the simulated annealing and genetic algorithm approaches

is shown in Figures 10.12–10.14. The minor loop weighting was set to 5 to improve the relative optimization for the smaller loops. The resulting family of curves shows a good match for the minor loop, a reasonable match for the major loop, but a poor match for the medium-sized loops.

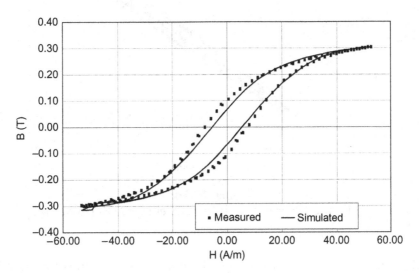

Figure 10.12:
Multiple loop optimization results for Siemens N30 (major loop)

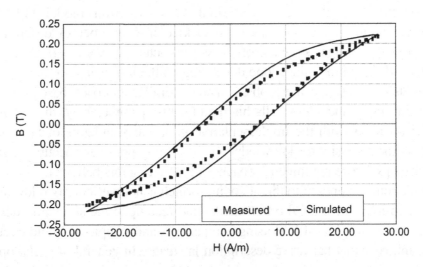

Figure 10.13:
Multiple loop optimization results for Siemens N30 (medium loop)

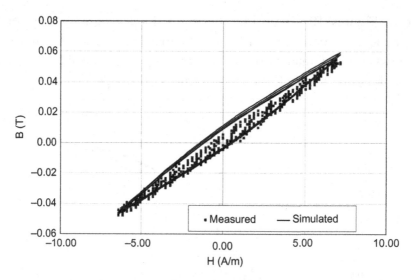

Figure 10.14:
Multiple loop optimization results for Siemens N30 (minor loop)

While the multiple loop optimization techniques described thus far are effective in the characterization of different fixed applied magnetic field strengths, a different approach is required for more general modeling of arbitrary loop shapes. The problem of modifying the original Jiles–Atherton model [14, 15] to address the minor loop issue has been tackled in a number of different ways. Carpenter [16] investigates the use of a new parameter (Λ), which effectively scales the slope and size of the minor loop depending on the last change of direction. Jiles [17] has also investigated the problem using a similar approach based on the two previous turning points (Jiles calls his modification parameter, *vf*, the *volume factor*). The difficulty with the implementation in circuit simulators of both of these methods is the necessity for knowledge of the turning points for accurate modeling of minor loops. There are three approaches to modeling this behavior for use in circuit simulation (as opposed to direct coding, which is not particularly relevant here). These are to implement a C/Fortran model and recompile the simulator, develop a complex macromodel in SPICE using comparators and sample-and-hold circuits, or to use a mixed-signal hardware description language. In general, the first option is not practical for anyone other than a seasoned simulator writer and is therefore not suitable for general purpose application.

10.3.6 Outline of Minor Loop Modeling using Turning Points

The key to both of the methods described by Carpenter [16] and Jiles [17] is to monitor the changes in direction of applied magnetic field strength (*H*). Carpenter's approach is to scale the irreversible magnetization using the last turning point of *H*. The strength of this approach is that only the last turning point is required to model the behavior. Jiles uses a slightly different method, which modifies the irreversible part of the magnetization using the last two turning points. Jiles refers to the factor that varies the irreversible magnetization as the *volume factor* and Carpenter refers to the factor as Λ (in the remainder of this section we will use this notation for this modification). In both the Carpenter and Jiles' methods, the models do not require any extra parameters and are self-consistent.

10.3.7 Testing the Modified Jiles–Atherton Model Behavior

In order to test the basic behavior of the volume factor modification for the Jiles–Atherton model, the original model was modified to include the volume factor as a parameter to the model. When a fixed amplitude *H* was applied, then the effect of the parameter could be ascertained on the minor loop modeling performance. Firstly, a Jiles–Atherton model was created based on the major loop characterization results, and then a magnetic field strength of ± 30 A/m applied (about 20% of the major loop values) to create a minor loop. The value of Λ was varied from 0.2 through to 1.0 (1.0 is the same as the original unmodified model). Figure 10.15 shows the results of the measured B–H loop, the unmodified Jiles–Atherton model, and the modified model with $\Lambda = 0.4$.

It is clear from Figure 10.15 that making this modification can significantly improve the performance of the model for minor loops, without the need for different sets of parameters. The model is now extended to include the dynamic behavior of modifying Λ according to the recent *H* turning points.

Conclusion

In this chapter we have introduced several key optimization methods and some recent approaches to improve secondary parameters, such as yield, in addition

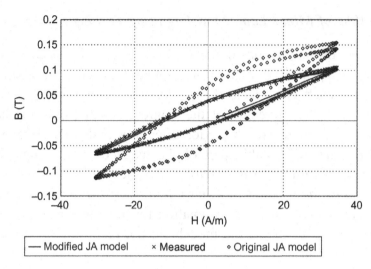

Figure 10.15:
Measured, original and Λ modified Jiles-Atherton (JA) B−H loops

to the optimization of specific design parameters. We have described how we can apply these methods using an example, and hopefully this gives an awareness of how the use of optimization can be a powerful tool in design, once the correct detail of models are in place. In the context of model-based engineering, optimization methods are relevant in that most often the systems we are modeling have to be fit to existing data to improve the fidelity of the simulations. Excellent frameworks for this fitting activity exist. Optimization methods also play a key role in design centering and tolerancing during the latter stages of the design process. This chapter has shown how the designer can take advantage of these techniques to improve the accuracy of the model parameters, and achieve a better quality design.

References

[1] N. Schmidt, H. Güldner, Simple method to determine dynamic hysteresis loops of soft magnetic materials, IEEE Trans. Magn. 32(2) (1996) 489−496.

[2] D. Lederer, H. Igarashi, A. Kost, T. Honma, On the parameter identification and application of the jiles-atherton hysteresis model for numerical modeling of measured characteristics, IEEE Trans. Magn. 35(3) (1999) 1211−1214.

[3] J. H. Holland, Adaption in Natural and Artificial Systems, University of Michigan Press, 1975.

[4] D. E. Goldberg, Genetic Algorithms in Search, Optimization and Machine Learning, Addison-Wesley, 1989.

[5] Genetic Algorithms in Engineering Systems: Innovations and Applications, IEE Conference Publications, 1997.

[6] P. J. M. Laarhoven, E. H. L Aarts, Simulated Annealing: Theory and Applications, Kluwer Academic Publishers, 1989.

[7] D. A. Pierre, Optimization Theory with Applications, second ed., Dover, 1986.

[8] J. W. Nims, R. E. Smith, A. A. El-Keib, Application of a genetic algorithm to power transformer design, Elect. Mach. Power Syst. 24 (1995) 669−680.

[9] S. Ali, R. Wilcock, P. Wilson, A. Brown, Yield model characterization for analog integrated circuit using a Pareto-optimal surface, IEEE International Conference on Electronics, Circuits, and Systems, 2008.

[10] J. Kennedy, R. Eberhart, Particle Swarm Optimization, Proceedings of the IEEE International Conference on Neural Networks, 1995, pp. 1942−1948.

[11] Y. Shi, R. Eberhart, A modified particle swarm optimizer, Proceedings of IEEE International Conference on Evolutionary Computation, 1998, pp. 69−73.

[12] K. Levenberg, A method for the solution of certain non-linear problems in least squares, Q Appl. Math. 2 (1944) 164−168.

[13] D. Marquardt, An algorithm for least-squares estimation of nonlinear parameters, SIAM J. Appl. Math. 11(2) (1963) 431−441.

[14] D. C. Jiles, D. L. Atherton, Theory of ferromagnetic hysteresis (invited), J. Appl. Phys. 55(6) (1984) 2115−2120.

[15] D. C. Jiles, D. L. Atherton, Theory of ferromagnetic hysteresis, J. Magn. Magn. Mater. 61 (1986) 48−60.

[16] K. H. Carpenter, A differential equation approach to minor loops in the Jiles-Atherton hysteresis model, IEEE Trans. Magn. 27(6) (1991) 4404−4406.

[17] D. C. Jiles, A self consistent generalized model for the calculation of minor loop excursions in the theory of hysteresis, IEEE Trans. Magn. 28(5) (1992) 2602−2604.

CHAPTER 11

Statistical and Stochastic Modeling

11.1 Introduction

If we consider a typical design process, as we have described so far in this book, there is a development of the design from an initial specification, through conceptual design, to detailed design and finally optimization. This is worthy and necessary; however, in the "real world" we do not usually have quite such a deterministic situation. In practice, there is an inherent variability in both the parameters and intrinsic behavior of components that must be considered. For example, we have already used the simple example of a resistor that obeys Ohm's law ($V = I \cdot R$) that in practice may not be an entirely realistic model. We have already discussed in Chapter 7 how the thermal interaction with the environment can modify model behavior. In addition to this behavioral change there is also a need for stochastic behavior to be implemented. This takes two forms — noise and basic statistical variation. If we consider resistor noise there will be stochastic noise as a result of Brownian motion, which is termed thermal noise, that follows the characteristic defined in Eq. (11.1).

$$S = 4kRT \qquad (11.1)$$

where S is the noise power, k is Boltzmann's constant (1.38×10^{-23}), and T is the temperature in degrees Kelvin.

Therefore, for a specific value of resistance, at a specific temperature, there will be random noise behavior that will have a noise power of S, and a root mean square (RMS) noise power of N. This is a specific type of stochastic behavior that we will develop further in this chapter.

The second kind of stochastic behavior is the statistical variation from one resistance value to another. If we specify a 1 kΩ resistor, then the actual value of the resistance will deviate from this nominal value depending on the tolerance of the component. For example, if the tolerance is specified as $\pm 10\%$, then the value of any particular resistance will be within the nominal value $\pm 10\%$, which, in this case, would be from $900\,\Omega$ to $1100\,\Omega$. This can have a profound and sometimes catastrophic effect on a design's performance, and so this concept will also be described in detail in this chapter.

11.2 Fundamentals of Noise

While a complete discussion of all of the sources of noise and the effects on design are beyond the scope of this book, the interested reader can see a useful discussion about noise in Horowitz and Hill [1] and also the Circuit Designer's Companion [2], which are both written from a practical designer's perspective. Before embarking on a detailed discussion of modelling noise it is important that some fundamental concepts are defined, and these are now introduced.

11.2.1 Definitions

Two important definitions for noise calculations are the mean value and the RMS value. For any noise signal based on random values we can calculate the mean value using the integration over a time period as shown in Eq. (11.2), where, over time, it will eventually average out to be zero.

$$\overline{v_n} = \frac{1}{T}\int_0^T v_n(t)\ dt = 0 \tag{11.2}$$

The RMS value is the square root of the integration of the square of the instantaneous values as given in Eq. (11.3), where, unlike the average calculation, it will result in a non-zero value.

$$v_{n(RMS)} = \sqrt{\frac{1}{T}\int_0^T v_n(t)^2\ dt} \neq 0 \tag{11.3}$$

Using these definitions we can make some important statements about noise signals as follows.

- The instantaneous value of a noise signal is undetermined.
- To characterize a noise signal, the *mean* value, *mean square,* and *root mean square value* are used.

- The standard deviation is equal to the RMS; the variance is equal to the mean square value (literally the RMS^2).
- The mean square value is a measure of the normalized noise power of the signal.

If we consider the RMS noise, then this can be considered as a voltage or current noise. The power dissipation by a 1 Ω resistor with a DC voltage of x V applied across it is equivalent to a noise source with a RMS voltage of x V$_{RMS}$. In a similar manner we can say that the power dissipation by a 1 Ω resistor with a DC current of A applied across it is equivalent to a noise source with a RMS current of y V$_{RMS}$.

11.2.2 Calculating the Effect of Noise in a Circuit

Typically, a circuit contains many noise generators. For circuit analysis we usually sum all noise sources into a single noise source (either at the output or the input of the circuit) and treat the functional part of the circuit as noiseless.

Summing j noise sources (where j is the number of noise contributors in the circuit) is done by adding their mean square values as shown in Eq. (11.4)

$$V_{n(RMS)}^2 = V_{n1(RMS)}^2 + V_{n2(RMS)}^2 \cdots + V_{nj(RMS)}^2 \qquad (11.4)$$

This is only valid provided that the noise sources are independent of each other (uncorrelated). For circuit analysis this is usually the case.

Using this approach we can therefore calculate the effective total noise power in a circuit, by summing the individual noise powers. For example, if we have two noise sources in a circuit with values of RMS noise voltage of 10 μV and 20 μV, respectively, the total noise power can be calculated using the approach given in Eq. (11.4) and the specific result in this case is given in Eq. (11.5).

$$V_{total(RMS)}^2 = (10\mu V)^2 + (20\mu V)^2 = 500(\mu V)^2 \qquad (11.5)$$

This results in an RMS noise voltage of the square root of 500 $(\mu V)^2$, which is 22.36 μV. At this point it is worth considering precisely what we have achieved in this analysis. We have calculated that for two components with uncorrelated noise of 20 μV and 10 μV, the combined overall noise voltage is 22.36 μV, which is the RMS of the noise signal.

11.2.3 Power Spectral Density of Noise

We have considered the mathematical nature of noise signals using RMS calculations; however, it is important to note that the noise is always spread out over the frequency spectrum. We must therefore consider not only the RMS noise, but also the way that the noise is distributed over the spectrum and how we can possibly mitigate the effects of noise on our design. The way that we think about noise is that the overall spectrum is divided into 1 Hz bandwidth "slices" so the noise power *density* is the mean square noise within that 1 Hz bandwidth.

For example, if we consider a total noise power in terms of noise voltage squared, then the power spectral density in our 1 Hz bandwidth is therefore defined in terms of Volts Squared per Hertz (V^2/Hz). In a similar manner, if we have a noise *voltage*, then this is in terms of the square root of the power spectral density, and therefore we can define the equivalent voltage noise spectral density in terms of $V/\sqrt{Hz}$.

If we calculate the autocorrelation function of the time domain noise function, we can express the noise spectral density in the frequency domain using an integration of the individual spectral harmonics, as defined by Eq. (11.6)

$$\overline{v_{n(RMS)}}^2 = \int_0^\infty S(f) \ df \qquad (11.6)$$

where $S(f)$ is the autocorrelation function. It is beyond the scope of this book to go much further into the description of how this is derived; however, it can be calculated using the Wiener-Khinchin theorem, which states that the spectral density is the Fourier transform of the autocorrelation function of a random signal (further details can be found in many signal processing texts such as [3]).

11.2.4 Types of Noise

We have two main types of noise to consider in practice for real components — thermal noise and flicker noise. "White" noise has a flat (e.g., constant) spectral density, i.e. $S(f)$ is a constant and is produced by thermal noise generators (e.g. Johnson, Boltzmann). Examples of sources of such noise are resistors, bipolar transistors, diodes, and MOSFET transistors. While, of course, the noise may not be exactly white up to infinity, we can assume the noise is effectively white into the THz range. Flicker noise has a different characteristic where the noise spectral density is proportional to $1/f$ (where f is the frequency). Flicker noise is particularly important in MOS transistors, especially at low frequencies, as we shall see. It is important to remember that MOS transistors have both flicker noise and white noise.

If we take an example of a MOS transistor and plot its noise spectral density as in Figure 11.1, then we can see that the flicker noise will dominate the behavior at low frequencies, but that the white noise continues to very high frequencies.

If we consider a white noise source, in contrast, and we define a white noise source with a power spectral density (PSD) of $10 \, \mu V^2/Hz$, this means that the level of noise power will be $10 \, \mu V^2$ *across the entire spectrum*. We can see this graphically if we plot the PSD for a single white noise source as shown in Figure 11.2.

In the next sections of this chapter we will discuss these types of noise in more detail.

11.2.5 Thermal Noise

As we have seen, thermal noise is a form of white noise – uniform across the entire frequency spectrum. It is present in resistors and, as we have seen, has the following basic equation that describes the noise power in V^2/Hz:

$$S = 4kRT \qquad (11.7)$$

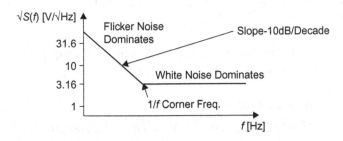

Figure 11.1:
Typical noise spectral density of a MOS transistor

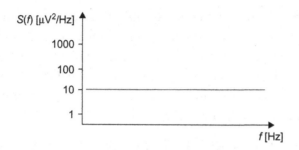

Figure 11.2:
White noise source example

where S is the noise power, k is Boltzmann's constant (1.38×10^{-23}), and T is the temperature in degrees Kelvin.

We can also take the square root of Eq. (11.7) and calculate the thermal noise voltage, which is shown in Eq. (11.8).

$$N = \sqrt{4kRT} \tag{11.8}$$

If we consider Eq. (11.8) in more detail, it is clear that it is *not* dependent on frequency, but *is* dependent on the resistance and temperature, so this explains why the characteristic of the noise is "white", i.e., uniform across the frequency range. As a designer, it is important to understand the effect that this will have a on a circuit's performance and we can illustrate that with an example. Consider a 100 kΩ resistor, where the temperature is 27°C, then we can calculate the RMS noise power (S) using Eq. (11.7):

$$\begin{aligned} S &= 4kTR \quad V^2/Hz \\ S &= 4 \times 1.38 \times 10^{-23} \times (273 + 27) \times 100 \times 10^3 \quad V^2/Hz \\ S &= 1.66 \times 10^{-15} \quad V^2/Hz \end{aligned} \tag{11.9}$$

and from this we can calculate the noise voltage by simply taking the square root of the noise power as given in Eq. (11.9), shown in Eq. (11.10)

$$\begin{aligned} N &= \sqrt{4kTR} \quad V/\sqrt{Hz} \\ N &= \sqrt{1.66 \times 10^{-15}} \quad V/\sqrt{Hz} = 4.07 \times 10^{-8} \end{aligned} \tag{11.10}$$

If we look at Eq. (11.10) in more detail we can see that the noise voltage RMS value is of the order of 40 nV/$\sqrt{\text{Hz}}$, and so as we have seen in Figure 11.2, this means that across the entire spectrum there will be 40 nV (RMS) of noise voltage.

So, what are the implications of this for our circuit designer? Consider the situation where we have a circuit that is designed to provide amplification for a signal that will eventually go into an analog-to-digital converter (ADC). If we consider the case where the converter is 16 bits, then we can state that the number of quantization levels for a N bit converter is 2^N and the resolution is given by $V_{FS}/(2^N-1)$; (where V_{FS} is the full scale voltage). This is equivalent to the smallest increment level (or step size) q of that converter.

If we have a converter of 16 bits, and a voltage supply of 3.3 V, giving a $V_{FS} = 3.3$ V, then we can estimate the resolution as being of the order of 50 μV, and so the RMS noise from this single resistor will be of the order of 0.16% of the resolution of this circuit.

We can see how this can quickly increase if we start to build circuits using multiple components, and if we take a simple example of a voltage divider, ideal amplifier, and an RC low pass filter, we can investigate the noise behavior of the complete circuit. Now, as we have already calculated the noise for the 100 kΩ resistor we can use that in our calculations. As the amplifier is ideal, we do not need to consider its noise performance (although, in practice, of course, that would also need to be included in the calculation). Finally, we can assume that ideal capacitors have negligible thermal noise, although, again, in practice, there will probably be a figure due to parasitic elements that needs to be included, although this will be very small (Figure 11.3).

As we have already calculated, the thermal noise power in each resistor of the potential divider where $R_1 = R_2 = 100$ kΩ, and this was calculated to be 1.66×10^{-15}V^2/Hz. As $R_1 = R_2$ we know that the voltage gain (A) of the voltage divider is 0.5, and therefore the combined noise power contribution will be multiplied by the square of the voltage gain to give the effective contribution due to the voltage divider resistors.

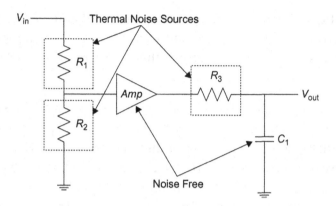

Figure 11.3:
Simple circuit for the illustration of thermal noise calculation

So, given that the noise in R_1 is 1.66×10^{-15} V²/Hz, the noise after the voltage divider *contributed from R_1* will be

$$S_o = (0.5)^2 \times 1.66 \times 10^{-15} = 4.14 \times 10^{-16} \ V^2/Hz \qquad (11.11)$$

and there will be the same contribution from R_2. The filter resistor is only 100 Ω; however, it will also make a contribution to the overall noise in the system.

$$\begin{aligned} S &= 4 \times 1.38 \times 10^{-23} \times (273 + 27) \times 100 \ V^2/Hz \\ S &= 1.66 \times 10^{-18} \ V^2/Hz \end{aligned} \qquad (11.12)$$

As the amplifier has been considered to be ideal, and the capacitor is also ideal (with no noise contribution), we can therefore calculate the overall noise of the system as follows:

$$\begin{aligned} S_{total} &= S_{R1} + S_{R2} + S_{LP} \\ S &= 4.14 \times 10^{-16} + 4.14 \times 10^{-16} + 1.66 \times 10^{-18} \ V^2/Hz \\ S &= 8.30 \times 10^{-16} \ V^2/Hz \end{aligned} \qquad (11.13)$$

We can therefore calculate the noise voltage:

$$\begin{aligned} N &= \sqrt{S} \quad V/\sqrt{Hz} \\ N &= 2.88 \times 10^{-8} \ V/\sqrt{Hz} \\ N &= 28.8 \ nV/\sqrt{Hz} \end{aligned} \qquad (11.14)$$

This is an interesting result as it shows that even though the overall noise contribution of an individual resistor is 40 nV/$\sqrt{\text{Hz}}$, due to the gain of the circuit, the overall noise on the output will be less than a single resistor, even though we have three resistors in the circuit. It is also an illustration that we need to be careful in assumptions relating to noise calculations, and that it will quickly become difficult to correctly predict the noise contribution for anything other than the simplest circuits.

11.2.6 Modeling and Simulation of Noise

Luckily, we have the option to carry out noise analysis using simulations, and we have two choices to consider when we do that. The first option is to add random noise sources that have the correct RMS value of noise and complete time domain simulations, and while this is very accurate, it is also extremely time consuming to accomplish. The second option is to work in the frequency domain, and in this way the noise analysis in a simulator works like a small-signal frequency analysis (AC) by sweeping the frequency and plotting the output. In the noise analysis, however, unlike the conventional frequency analysis, the sum of the individual noise contributions is calculated over the frequency range specified by the designer.

This is particularly useful in situations such as the simple amplifier example we have just considered, where there is a frequency dependence in the characteristic, and as such it would be a laborious task to calculate the overall frequency behavior, whereas in a simulator a single analysis will give that response.

If we implement the same amplifier example circuit in a circuit simulator (in this case Saber) as shown in Figure 11.3, then the noise voltage spectral response can be simulated and the results are shown in Figure 11.4.

As we know from the circuit diagram, there is a low pass filter on the output, and the noise is subject to the same filtering as the signal, so it is a useful check to see that the noise response exhibits the same low pass response. If we compare the overall noise at a low frequency we can see that the simulator predicts a value for the noise voltage RMS of 28.818 nV/$\sqrt{\text{Hz}}$, which is consistent with our previously calculated value.

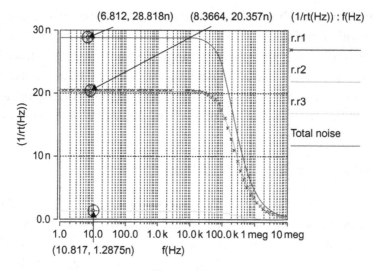

Figure 11.4:
Noise voltage spectral response using a simulated noise analysis

So, how is this noise behavior implemented in a model? If we consider one of the components, such as a resistor, then the noise voltage must be added to the model. Of course, this is in addition to the functional governing characteristics, and is only valid during a noise analysis (if we use a small-signal noise figure). We can add the noise source as a Thevenin voltage source, or as a Norton current source as shown in Figure 11.5.

11.2.7 Summary of Noise Modeling

As we have seen, the implementation of stochastic noise is relatively straight-forward with the addition of noise sources added to the functional models. The noise can be added in the AC domain for noise analysis or as a randomly vary-ing signal in the time domain. While this (particularly in the time domain) is stochastic behavior, the nominal component values are still used in the simula-tion, and so the remainder of this section will describe the case where there is a statistical variation in the model parameters to consider.

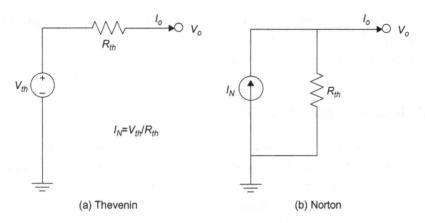

Figure 11.5:
Thevenin and Norton noise sources

11.3 Statistical Modeling

11.3.1 Introduction

Yield and reliability have been identified as one of the greatest present and future challenges associated with nanometer process technologies. This was illustrated graphically by a team of researchers at IBM investigating the effect of variability on high performance integrated circuits [4]. Aggressive scaling of process nodes has led to a significant reduction in CMOS parameter precision as a result. The outcome of this is that the problem of device variability becomes a first order limitation for integrated circuit designers rather than a minor secondary design issue [5].

The causes of variability can be seen broadly in terms of spatial or temporal effects. Spatial effects include die-to-die parameter mean shifts, on-chip layout induced variations and device-to-device mismatch caused by atomistic dopant variations, line edge roughness, and parameter standard deviation; all leading to variability in circuit behavior. Temporal effects refer to time dependent changes in performance and reliability, such as dielectric breakdown (DB), hot carrier injection (HCI) and negative bias temperature instability (NBTI). The difference with temporal effects is that they may cause significant changes in a circuit's performance over its lifetime [6]. In the case of analog circuits the effect of variability can be complex owing to a large number of performance specifications; therefore, it is important to use a model-based approach to predict the effect of variability on circuit performance as a result.

11.3.2 Basic Statistical Behavior

If we consider any parameter of a device or circuit as being defined by a single number, then this is defined as the nominal value. It is easy, particularly when an optimization has been used, to consider this an exact and precise definition of the model behavior. This is, however, an error in many cases, as, in fact, the nominal value is often simply the *mean* value obtained from multiple tests with a complete batch of components to establish the values.

For example, consider a batch of 1000 resistors that have been manufactured to have a designed value of 100 Ω. Each device is measured on a resistance bridge to establish its actual value, and if we plot the results we would in fact see a "spread" of results due to the intrinsic variability of the manufacturing process (Figure 11.6).

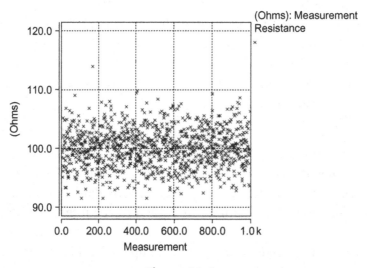

Figure 11.6:
Random spread of resistance measurements

Clearly, we can see that that the results of the measurement show a spread around the desired value of 100 Ω, but that, in fact, most of the results are not exactly the correct value and some results are quite some distance from the nominal value. If we take the same results and calculate a histogram of the values, then we can measure the mean value and also the standard deviation.

As we have seen in the discussion of noise earlier in this chapter, the calculation of the mean value for a continuous signal is the integration over a time period; however, in this case, we are looking at a number of discrete values, so the integration becomes a summation as defined in Eq. (11.15).

$$\mu = \frac{1}{n}\sum_{i=1}^{n} x[i] \tag{11.15}$$

where n is the number of samples, $x[i]$ is the individual sample and μ is the mean value.

We also need to calculate the variability of the measurement, i.e., how much do the measurements deviate from the nominal value? To do this we need to calculate the variance from the nominal value using the expression defined in Eq. (11.16).

$$V = \frac{1}{n}\sum_{i=1}^{n} (x[i] - \mu)^2 \tag{11.16}$$

The variance is not a particularly useful number as it is scaled (squared, in fact) from the original units, and so in order to get a measure of the variability we take the square root of the variance to obtain a measure of the typical deviation of a sample from the mean value. This is called the standard deviation (σ), as defined in Eq. (11.17).

$$\sigma = \sqrt{V} = \sqrt{\frac{1}{n}\sum_{i=1}^{n} (x[i] - \mu)^2} \tag{11.17}$$

If we take the results shown in Figure 11.6 and calculate the mean and standard deviation, we can superimpose them on the statistical "count" plot in

Figure 11.7. We can see that the mean value is measured as 99.949 Ω and the standard deviation is measured as 3.3854 Ω.

A valid question that the circuit designer could ask at this point is "what do these numbers actually represent?". Clearly, the mean value is useful as it demonstrates that, on average, the device is very close to our specified value of 100 Ω; however, in many cases the results deviate from this. The standard deviation is a useful measure in that it can be used to estimate the proportion of devices within a certain deviation from the nominal value. We can only make this assumption, however, if we have a well-defined statistical variation behavior. In practice we can say that a truly random variation will follow a statistical function called a "normal" distribution. This has, if there are enough samples, a bell curve shape with the probability density function (PDF) as defined by Eq. (11.18).

$$PDF = \frac{1}{\sigma\sqrt{2\pi}}\exp\left(-\frac{[x-\mu]^2}{2\sigma^2}\right)$$ (11.18)

If we plot this graphically, we can see the symmetrical and smooth nature of the PDF of the normal distribution.

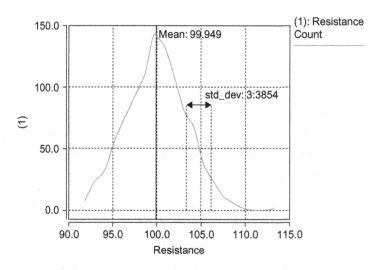

Figure 11.7:
Mean value and standard deviation of the resistance

A key significant aspect of the PDF shown in Figure 11.8 is that we can quantify using this function proportionally by how much of the values will be within one standard deviation of the mean, two standard deviations and so on. As we can see from Figure 11.9, 68.2% of the samples are within one standard deviation of the mean value, 95.45% are within two standard deviations and 99.73% are within three standard deviations. In fact, most engineering designers will work to a tolerance of three standard deviations, which leads to the term "six-sigma design", which is to plus and minus three standard deviations (σ-sigma).

If we return briefly to our previous example of the resistor, where we have obtained a mean value of 99.949 Ω, and a standard deviation of 3.3854 Ω, we can therefore say that if we assume that the variation follows a normal distribution that more than 99% of the values will fall in the range $\mu \pm 3\sigma$, which is 99.949 $\Omega \pm 10.1562$ Ω. This indicates that we will see a spread of values mostly in the range 90 Ω to 110 Ω, and if we return to Figure 11.6, we can see that this is indeed the case. In our 1000 samples, we can see that there is only one measurement that falls outside this range.

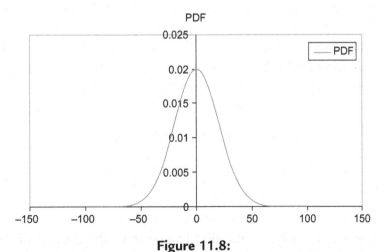

Figure 11.8:
Probability density function (PDF) of ideal normal distribution with mean value of 0 and standard deviation of 20

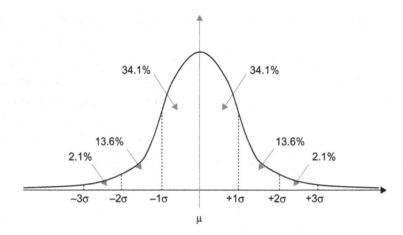

Figure 11.9:
Proportions of samples with a normal probability density function relative to the
standard deviation

11.3.3 Modeling Distributions

We can implement a probability distribution function on a parameter by applying a "normal" distribution directly to the parameter and passing this to the
model. For example, if the resistance parameter is normally passed to the
model using the mean value (in our example $r = 100\ \Omega$), we can, instead, pass
the output from a normal distribution function (with the same mean value, but
also with a "tolerance" value, which is the same as the 3σ value we calculated
in the previous section). In this example we could therefore round the measured
values to a mean value of $100\ \Omega$ and a tolerance (3σ) value of $10\ \Omega$. This is
implemented in the model as a normal function, as shown in Eq. (11.19).

$$r = normal(100,\ 90,\ 110) \tag{11.19}$$

where in this case the *normal* function parameters are the mean value, mean
-3σ, and mean $+3\sigma$.

We do have another type of distribution that we can consider, which is called a
"uniform" distribution. One of the disadvantages of the normal distribution is
that clearly the results are skewed such that most of the samples are within the
$\pm\sigma$ range, and so outliers (samples far from the mean) will occur more rarely.

If we have a case where it is important to concentrate more on the periphery results and have more samples further from the mean value, then we could use a uniform distribution, where the values are distributed evenly across the range we define. For example, taking the case of the resistor, we could define a uniform distribution and implement the function using the same approach as the normal function and implement the same tolerance as in Eq. (11.19):

$$r = uniform(100, 90, 110) \tag{11.20}$$

The probability density function of the uniform distribution is different from the normal distribution function, and this is illustrated in Figure 11.10.

If we implement this function in a model, using the expression for resistance defined by Eq. (11.20), the resulting samples in 1000 different simulations can be plotted and these can be seen in Figure 11.11. The contrast with the normal distribution is apparent immediately, with the samples evenly spread over the range of values and, interestingly, no outliers.

If we look at the distribution of values across the range of values in Figure 11.12, then we can see that there is no evidence of a normal variation as we saw for the measured results, and the values are reasonably evenly spread across the range of possible values from 90 Ω to 110 Ω.

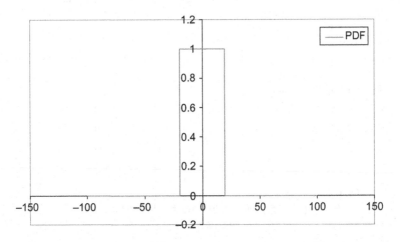

Figure 11.10:
Probability density function (PDF) of a uniform distribution

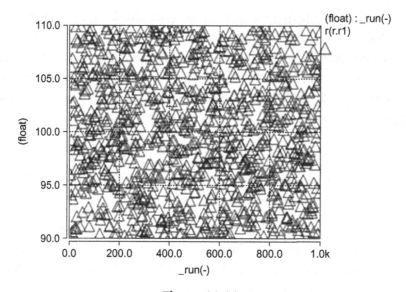

Figure 11.11:
Uniform spread of resistance simulated values

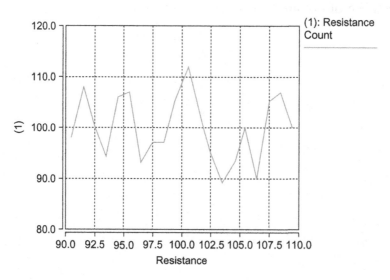

Figure 11.12:
Distribution of resistance values using a uniform distribution

11.3.4 How to Interpret Variation in Models

So far we have looked at the specific case of applying a variation to a specific design parameter, but we can also briefly consider the case where a model parameter is dependent on the intrinsic variability of the real devices to produce the model parameter in the first place. For example, consider the input stage of a typical operational amplifier, as shown in Figure 11.13.

If we model and simulate this circuit using transistors, a good question to ask is "how can we establish the input offset voltage of this circuit?" As circuit designers, we will either use a data sheet value or make an estimate based on experience, but how can we establish this from a simulation? In fact, we can only do this if we have statistical information for the two transistors Q1 and Q2 such that we can run multiple simulations to calculate the *statistical* voltage offset value based on a number of simulations. Once we have the value and its standard deviation, then, as we have seen, we can implement the model, but not before.

11.3.5 Statistical Simulation Methods – Monte Carlo

We have talked quite a bit in this chapter about statistical modeling, but how is this implemented in a simulator? In circuit simulators, there is usually a

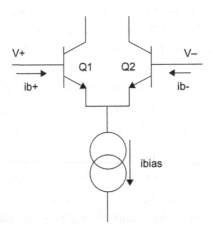

Figure 11.13:
Typical operational amplifier differential input stage

built-in method of analysis called a "Monte Carlo" simulation (sometimes abbreviated to MC analysis). The idea behind the analysis is very simple. The designer specifies a number of simulations to be carried out (say a steady-state analysis, followed by a time domain analysis), and then this same analysis is repeated for a set number of iterations, again chosen by the designer. For each iteration of the loop, any parameters that have been defined using statistical distributions will be varied and the values applied in each simulation of the loop. Once the analysis has been completed, then instead of a single time domain analysis, for example, there will be a *set* of time domain analyses, where the statistical variations have been applied in each one.

After the analyses have been completed, then all the results can be plotted and analyzed to establish the statistical behavior of the circuit overall.

11.3.6 Random Numbers and "Seed"

For any random process, at the heart of the variation is a random number generator. In fact, for a computer system there is no such thing as a truly random number and, actually, a *pseudo* random number generator is used to provide the variation in simulations. In simulators, the random functions can be defined with a known "value", which is used to start the randomization process. The semantics for this approach are described by the algorithm published by Pierre L'Ecuyer in [7]. The method is based on the combination of two multiplicative linear congruential generators for 32-bit platforms. Before the first call to the function, the seed values have to be initialized to values in the range (1, 2147483562). The seed values are modified after each call to the random function. This random number generator is portable for 32-bit computers, and it has a period of $\sim 2.30584*(10**18)$ for each set of seed values.

From a practical perspective, the seed value is important as, owing to the pseudo random nature of the random number generator, if we choose the same seed, then we can repeat the same statistical analysis. This is important, as we can recreate a simulation scenario, and all we need to store are the models and the seed to be used for the analysis and we can rebuild the complete set of results.

11.3.7 Practical Statistical Simulation

As we have seen in this chapter, the effect of statistical variation on model parameters can be implemented fairly easily; however, the real benefit for a circuit designer is in the ability to assess the effect of the model parameter variations on the circuit performance. If we take a simple example, as shown in Figure 11.14, of an operational amplifier gain stage, with a built-in low pass filter, then we effectively have two design parameters of critical concern – the gain and the filter cut-off frequency.

Taking this circuit, we can carry out a frequency response analysis to measure the response of the circuit, which will give the DC gain and the −3 dB cut-off frequency; when this is carried out, the result obtained is shown in Figure 11.15. As expected the DC gain is approximately 0 dB (with a DC gain of the amplifier designed to be 1.0) and the frequency cut-off measured at 1588 Hz, which is consistent with the expected values with the feedback capacitor of 10 nF.

When a signal of 10 kHz and 200 mV (peak-to-peak) is applied to the input, at low frequencies the transient output will also match the input; however, we would expect to see attenuation above the cut-off frequency. If we choose a frequency of 10 kHz, then the AC model predicts an attenuation of −16 dB, and when the signal is applied to the time domain model we can observe the

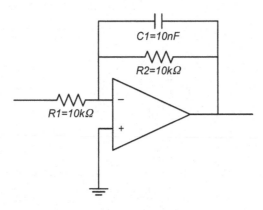

Figure 11.14:
Operational amplifier circuit – nominal case

behavior as seen in Figure 11.16, where the output voltage is measured to have a peak-to-peak voltage of 0.3093 V, which is −16.2 dB. Therefore, we can say that the time domain analysis confirms the frequency response predicted behavior.

If we look at the circuit we can add in statistical variability to the nominal parameters of each of the passive components, and also to the operational amplifier characteristics, such as open loop gain, offset voltage, input and output impedance, and pole frequency. In this case, we will add normal distributions of the form defined previously in this chapter.

The resulting AC analysis when 100 simulations are carried out can be seen in Figure 11.17. Clearly, there is a variation in both the DC gain and also the frequency cut-off; however, it is difficult to quantify from these graphs alone. The approach, therefore, is to use measurements of the key design parameters

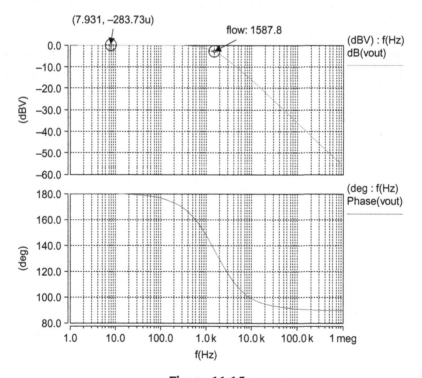

Figure 11.15:
Operational amplifier circuit nominal frequency response

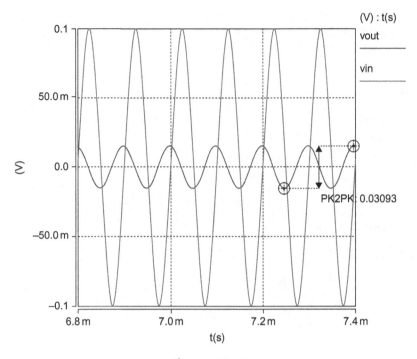

Figure 11.16:
Operational amplifier circuit nominal time domain response at f = 10 kHz

(DC gain and cut-off frequency) and extract those results from the 100 simulation runs.

One question that often arises at this point is "how many simulation runs are required?" This is a slightly holistic issue in the sense that there needs to be enough simulations to be statistically significant. There is a practical issue with simulation time, so usually a few hundred runs would be adequate, and less than a hundred would probably not be enough in most cases.

If we apply the measurement of DC gain at 1 Hz, then we can plot the 100 individual measurements, and then calculate both the mean value and also the mean value $\pm 3\sigma$ range. This can be seen in Figure 11.18, and we can observe that the mean value of the DC gain is 0 dB (within a small margin) and the $\pm 3\sigma$ range is between -1.27 dB and $+1.24$ dB. What this means in practical

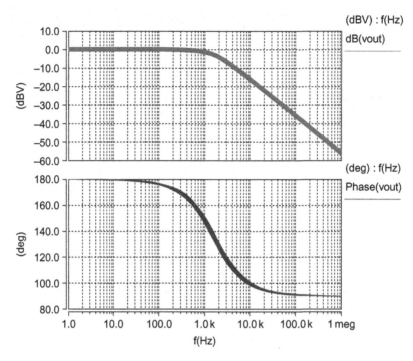

Figure 11.17:
Operational amplifier circuit statistical frequency responses: 100 runs

terms is that the amplitude of the DC gain could be up to 15% variant from the specification value of 1.0 (unity).

If we carry out the same analysis for the cut-off frequency, then we can see that the resulting variation is as shown in Figure 11.19, where the mean value is measured at 1598 Hz (which is very close to the nominal value), and the $+3\sigma$ range is 1812 Hz and the -3σ range is 1385 Hz. This is an illustration that the effect on the frequency cut-off is such that the value can range over 500 Hz.

Now that we have established the basic circuit behavior we can understand clearly how individual tolerances can affect the behavior over a realistic practical scenario, taking into account the potential component tolerances.

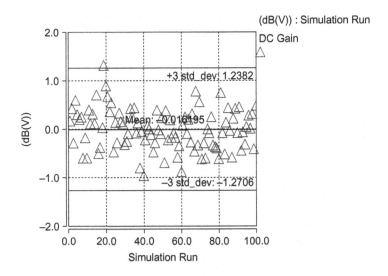

Figure 11.18:
DC gain measurements for operational amplifier circuit

11.3.8 Establishing the Relationship Between Component and Performance Variation

Now that we know that individual component tolerances have a profound effect on the overall circuit performance, we can investigate how the individual components relate directly to output performance measures. For example, we know that there are several components in our operational amplifier circuit, but how do those individual components directly influence the circuit performance?

Given that we have simulated the circuit with a statistically varying set of component parameters, we can carry out a correlation analysis and establish which parameter is the most significant in our design. For example, if we plot the relationship between the passive components and the DC gain performance, we should be able to establish which is the most important. If we plot the scatter plot of the dc gain against the values for C_1, R_1 and R_2, we can carry out a basic regression analysis to see if there is any relationship between the parameters and the design performance, and we can see the results of doing this in Figure 11.20. We can see that there is a positive relationship between R_2 and the gain, and a negative relationship of about the same between R_1 and the

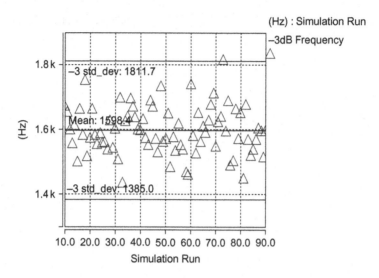

Figure 11.19:
Variation in -3 dB frequency cut-off over 100 simulation runs

gain, which is entirely what we would expect in this operational amplifier circuit. We can also see that the capacitor C_1 has almost no effect, although, as we are only approximating the DC gain by measuring at 1 Hz, there is a very small correlation.

If we consider the frequency response and carry out the same analysis, then we can observe that for this design parameter (Figure 11.21), there is a correlation between R_2, C_1 and the cut-off frequency, but that R_1 on this occasion has almost no effect.

We can use this technique to ensure that if we need to improve the statistical performance of the circuit overall, we can target the appropriate components based on the correlated results.

11.3.9 Improving the Circuit Yield Based on Simulation

So far we have implemented statistical model parameters, demonstrated how to measure the statistical performance, and understand the relationship between component parameters and overall circuit performance. From a circuit designer's perspective, of course, we need to also be able to predict how well

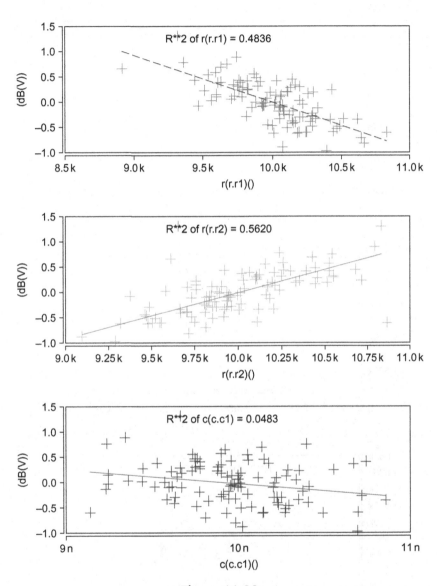

Figure 11.20:
Correlation between DC gain and R_1, R_2, and C_1

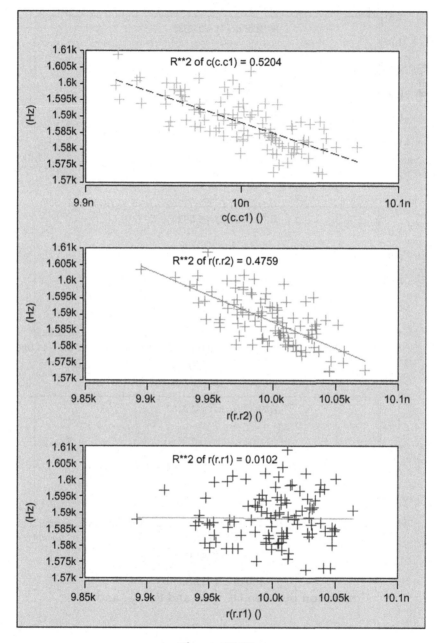

Figure 11.21:
Correlation between cut-off frequency and R_1, R_2, and C_1

our circuit meets the specification. Clearly, we can make some predictions of the range of possible circuit performance, as we have seen previously in this chapter, but, in most cases, that is not how the design process works in practice. As has been discussed in Chapter 2, the design starts from a set of requirements that lead to a detailed design specification, and this will define the acceptable performance limits of the circuit and not the other way around. For example, if our operational amplifier circuit has a specification as shown in Figure 11.22, then how do we establish if our circuit will meet this performance under all conditions?

In many respects we have already done the hard work in that we have completed the statistical analysis to establish the potential range of performance of the circuit, which this has resulted in a nominal value, with $\pm 3\sigma$ values so that we can state that the design will meet those figures to within 0.1% of the time. Another way of expressing this is to apply the specification limits to the statistical analysis and to calculate what proportion of the designs will meet the specification, which will give a figure called yield. For example, if all the designs meet the specification, then the yield will be 100%; however, if only half the designs meet the specification then the yield will be 50%.

If we apply the specification limits to the analysis, we can see that the current design will have 97% yield, as shown in Figure 11.23. If we apply the frequency cut-off specification, we can see that the current circuit will have a yield of 94% using the current component tolerances, as shown in Figure 11.24.

If we wish to improve the yield, we now have the intrinsic knowledge to make an informed choice as to which components to target to improve the results. As we have seen, the only component that directly affects the two performance

Performance parameter	Nominal value	Lower limit	Upper limit
dc Gain	1.0	0.9	1.1
Cut-off frequency	1580 Hz	1400 Hz	1700 Hz

Figure 11.22:
Specification of circuit performance

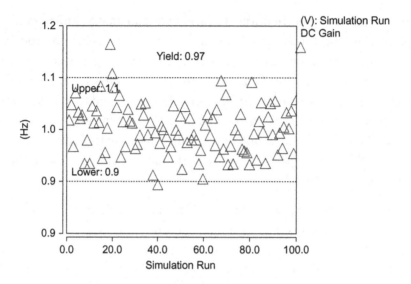

Figure 11.23:
DC gain yield calculation − tolerances = 10%

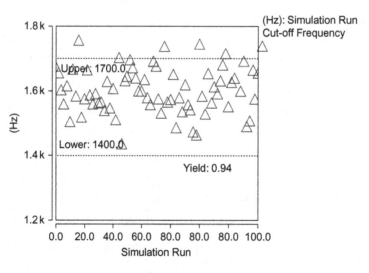

Figure 11.24:
Frequency cut-off yield calculation − tolerances = 10%

parameters is R_2; therefore, if we reduce the component tolerance from 10% to 1% we should be able to improve both performance parameters.

As we now have the circuit model in place it becomes a simple matter to change the tolerance of R_2 to 1% and re-evaluate the yield, which, for the DC gain, becomes 99%, and for the frequency cut-off becomes 96%. This is a clear improvement; however, it is still not yet at the 100% we require and so if we reduce the component tolerances to 1%, the circuit will easily meet the specification and provide us with 100% yield on both performance parameters (Figures 11.25 and 11.26).

Conclusion

In this chapter we have seen how we can predict stochastic behavior in circuits using noise analysis, and statistical modeling and simulation. This is an incredibly powerful set of techniques and bridges the gap between a nominal and somewhat idealized view of circuit design to a more realistic and reliable set of results from which we can predict how circuits will operate in practice. While our treatment has been just an overview with some simple illustrations, it is important to note that any design process should include the steps of design

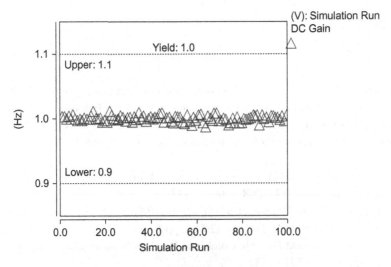

Figure 11.25:
DC gain yield calculation — tolerances = 1%

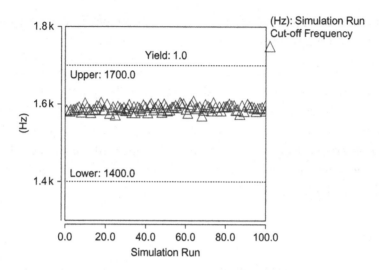

Figure 11.26:
Frequency cut-off yield calculation — tolerances = 1%

centering and tolerancing in order to achieve a design that not only meets the nominal specifications, but will hold up under environmental and other external stimuli that will cause the design parameters to shift. This level of robustness is one key to avoiding costly design recalls.

References

[1] P. Horowitz, W. Hill, The Art of Electronics, second ed., Cambridge University Press, 1989.

[2] T. Williams, P. Wilson, Circuit Designers Companion, third ed., Newnes, 2011.

[3] L.W. Couch, Digital and Analog Communications Systems, sixth ed., Prentice Hall, 2001.

[4] K. Bernstein, D.J. Frank, A.E. Gattiker, W. Haensch, B.L. Ji, S.R. Nassif, et al., High performance CMOS variability in the 65-nm regime and beyond, IBM J. Res.Dev. 50 (2006) 433–449.

[5] T. Mcconaghy, G. Gielen, Emerging yield and reliability challenges in nanometer CMOS technologies, Proc. of DATE 2008 (2008) 1322.

[6] S.V. Kumar, C.H. Kim, S. Sapatnekar, Impact of NBTI on SRAM read stability and design for reliability, Proc. of ISQED (2006), pp 6–11.

[7] P. L'Ecuyer, Efficient and Portable Combined Random Number Generators, *Commun. of the ACM* 31(6) (1988) 742–774.

Design Methods

The last section of the book does as its title implies — it brings all of the modeling techniques, representation methods, design flows, and analysis together in the form of a case study illustrating model based design and engineering. Chapter 12 describes the design flow to be followed, which indeed may vary from one company to another, but serves the point here. Chapter 13, the final chapter in this book, leads the reader through the process of designing a mixed-signal ASIC for the purpose of demonstrating how such design is done. This chip has actually been fabricated.

Design Flow

One of the engineers said, "It looks like you're just solving the problems by throwing a lot of silicon at the problem." I pointed out that the silicon is very cost-effective. It's the screw-ups that are expensive, as well as the ability to get something good enough to ship consistently. "It's bad product design that's expensive," I said.

Bob Pease, from "What's all this Bridge Amplifier Stuff, Anyhow?",
Electronic Design

12.1 Introduction

At this point in the book it is time to bring it all together and describe how all the pieces of this model-based engineering (MBE) puzzle fit. There are some things that are common to the design of systems independent of the type of system (i.e., electronic, mechanical, structural, etc.). One of these common aspects is some form of approximate or first-order design and analysis to achieve a first cut design. This design is a basic rendering of the designer's early concept of the solution to the design problem(s). Another commonality is that the designer has a set of computer-based analytical tools that he/she uses to validate the first-order design, but also subsequent refinements of the design as more detail and second-order effects are considered making hand analysis too difficult or unwieldy. Implicit in both of these steps of the design process, or flow, is the use of models to perform the design activity. The models may not be executable in the first cut, but merely equations on a piece of paper. A third commonality is verification of a design against specifications. No matter how the specifications were created, or how the design was ultimately realized, verification of the design against specification is the most important signoff step in the process. After all, if the design does not meet spec, then what's the point?

We have covered a wide array of techniques and methods of representing models so far in this book. Now we will focus our attention on electronic design and MBE of electronic and electronic-based systems. While there are many possible design flows depending upon what type of electronic system (analog, digital, RF, etc.) is being realized, we are not attempting to elaborate this entire space. We will focus on a mixed analog–digital–RF (i.e., mixed-signal) integrated circuit design flow (again, one of many possibilities) and demonstrate the MBE concepts somewhat generically in this chapter. We will add more specifics through a case study in the next chapter.

12.2 Requirements and Specifications

All design activity begins with requirements and specifications, i.e., statements of what and how the design has to perform. Strictly speaking, requirements are the top-level descriptions of what the system being designed is "required" to do. Specifications are statements that describe more detail on what the system must do in terms of performance measures. Example requirement statements are:

The solar inverter shall take a DC voltage input from the solar array and convert it to a three-phase AC voltage output.

The sensor interface chip shall be able to process analog sensor inputs and convert them to a digital representation for data collection.

Specification statements, which can also be inequalities or entries in a table, can be written as follows:

The range of allowable DC inputs on the solar inverter can be from 200 to 400 V and the output will be converted to a 480 V, three phase AC output.

The sensor interface chip must be able to process analog inputs from 5 to 50 V and convert them to 3.3 V logic with 10 bits of precision.

The synthesis of specifications from the more generic or behavioral requirements of a system is itself a process that can be achieved through MBE. This is the essence of the initial steps of system engineering – defining and refining specifications from the system requirements. This definition and refinement activity is often performed by both the systems engineer and the chip architect that will define the IC solution. The next section describes how this process leads to *executable* specifications, where the term executable refers to the fact that these specifications can be represented in a form that allows them to be analyzed and verified through computer-based analysis, such as simulation.

12.2.1 Executable Specifications

It is a common occurrence in the semiconductor industry for IC companies to employ highly skilled technical marketing engineers to work with application companies to define the requirements and specifications for the chip sets that IC companies propose to build for them. The roles that we will define for the purpose of this description are:

- *system engineer*, who works for the application company (such as a cell phone company, an airplane manufacturer, an automotive company) and who is responsible for building a working system using the chip sets that they can get from their suppliers;
- *chip architect*, who works for the IC company and who is responsible for designing and building the chip(s);
- *design engineer*, one of many that will design, analyze, and verify the various circuitry − analog, digital, RF − that comprise the chip(s);
- *software engineer*, engineer responsible for the real-time operating software executed in a digital core.

With these definitions established, we can begin to describe the design flow and activities, and who is typically taking these responsibilities.[1]

Chip Architecture

MBE is used to aid in the exploration and creation of a chip architecture to address the application company's requirements.

---

[1] We will not spend much, if any, time discussing hardware−software co-design or the software side of the design process in this book. However, it is important to point out that MBE techniques have been, and continue to be, commonly employed on the software side using the Unified Modeling Language (UML) and other methods.

The use of MBE begins with winning the IC business, or "the socket", from the application company. The chip architect and the technical marketing engineer respond to the requests and requirements put forth by the applications company by conceptualizing and modeling a potential solution to the challenges that these requirements represent. This chip-level modeling is done to: (a) convince the application company that the IC company can meet the needs and (b) convince themselves that they can build the required chips. Clearly, the chip architect and the technical marketing engineer are extrapolating from past designs, so design reuse is implicit. However, for a dramatically new product line consultation with design engineers and managers as to what can be achieved beyond current designs is an important step. This internal exchange keeps everyone calibrated as to what is possible, what the risk level in the future, and what, if anything, should be negotiated with the application company in terms of modifying requirements because of potential game-changing technology breakthroughs that the IC company may be offering. All of this is best described through system-level simulations involving a model of the chip at a very high level.

Once agreement is reached on meeting requirements, the result is the beginning of a chip architecture. The chip architect, in order to satisfy the application company requirements, conceives of an architecture that will meet or surpass the needs of the company. This architecture now needs to be taken from concept to a more firm realization. Now we are entering into a specification definition and capture phase where MBE makes its second major impact (the first being a means for winning the business!).

Specification Definition and Capture

MBE facilitates the definition and capture of chip specifications from the requirements. This leads to executable specifications, which has profound implications on verification.

For more than two decades the engineering community has desired to create tools and methods for the creation of executable specifications as defined earlier. As a result it is doubtful that the reader would question the value of executable specifications, but, nonetheless, some of the benefits of creating them are listed.

- Executable specifications create a vital link between the hardware specifications that will ultimately drive IC design activity and the system performance requirements of the application. The executable spec is a mechanism for *confirming* that requirements have been translated successfully into specifications.
- Once captured, executable specs provide a framework for hardware validation and verification. The executable specs are the pass–fail criteria against which designs are measured through the use of testbenches and assertions that are the means by which executable specs are realized. This investment makes specification tracking and signoff much more automated and straightforward.
- Executable specs also provide a means for managing risk in the design process. This is done by understanding the tolerances being achieved on specs and translating these into yield numbers for the overall chip.

The creation of VHDL was driven by the US government's desire to have a language to represent executable specifications for traceability and signoff. While hardware description languages (HDLs) are an excellent technology for modeling and executable specifications, only recently has a tool emerged that can expedite their capture in the same graphical sense as the way circuit schematics are captured for analog, RF, and some digital circuitry. The importance here is ease of use and design efficiency because design flows that employ executable specifications are not universally adopted owing to a perception that it is too inefficient or not viable.

Specifications define the performance boundaries that a circuit must achieve in order for the system to perform as desired. In order to create an executable specification for a circuit, two primary things must be captured: a test bench and the assertions that define the performance boundaries. The conditions under which the circuit must be placed (i.e., stimulus, loading, environmental) are defined by a test bench. If the circuit is placed into this context and the simulation executed, then waveforms (i.e., data) are generally produced that indicate the response of the circuit to the conditions the spec demands.

However, to ascertain whether the circuit meets the specification or not, the data must be analyzed for, or reduced to, *information*. This data reduction is accomplished by means of waveform measures coupled with pass–fail criteria to indicate automatically to the user whether the specification has been met or not. These data reduction instruments are referred to as *assertions* after the term used in VHDL. The capture of an executable specification is illustrated in Example 12.1 where both the test benches and the assertions are captured graphically in the ModLyng tool. Pay particular attention to the ease with which alternative designs can be substituted into the executable spec to determine if it meets spec. Now imagine that an entire suite of executable specifications is available, and, as a design is evolved, the entire suite is able to be executed regressively. This enables the designer to trace back to changes that might have adversely affected the design and manage the design tradeoffs more clearly and succinctly.

Example 12.1 Executable Specification Capture

In this example we will capture the executable specification for the slew rate of the operational amplifier that comprises part of our chip. The specification is that the op amp must slew at greater than or equal to ± 15 V/μs for the system to respond to large input signals fast enough to meet the overall systems specs. The first step is to define the test bench for the slew rate simulation, including the proper loading. For this particular instance of the op amp, it is not driving an external pad, but simply other internal circuitry. This loading is estimated to be 1 pF, but will be made a parameter so that it can be easily swept or modified later. Figure 12.1 shows the op amp instantiated within the executable specification, which consists of the test bench and the graphical assertion. Note: a library of test benches can be created and used in subsequent design activity therefore obviating the need to define a new test bench for the slew rate of an op amp every time one is designed. Also, the entire executable spec can be saved out into a library of op amp specifications, as well making the entire process much easier to reuse. Figure 12.2 shows the details of the test bench itself once you descend into its hierarchy. It is this schematic that you can see where the test bench was captured and then subsequently saved off into the op amp test bench library. Figure 12.3 shows the test bench instantiated along with the assertion for slew rate measurement. This represents the executable specification that, just as the test bench, can be saved off into a library of op amp specifications.

As we can see, the "model" is not just the functional behavior of the device itself, but rather, having captured the test bench, the assertions, and the

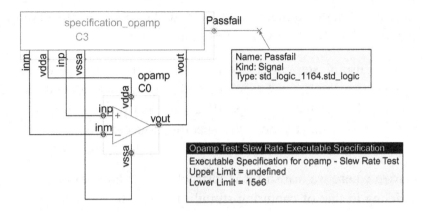

Figure 12.1:
Testbench with device under test (DUT) (op amp device instantiated)

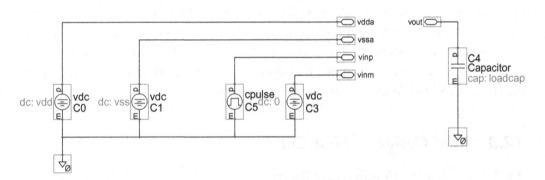

Figure 12.2:
Test bench internal components

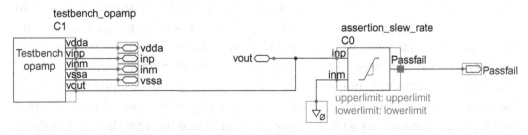

Figure 12.3:
Assertion and test bench together

combination of all these elements, there is now the scope for the definition of a complete executable specification as this combination.

This is a very useful step as it not only forces the designer to capture the design, but also to think, in detail, about the appropriate stimuli for the test bench and how that relates to the specification. That process is also forcing the designer to consider exactly how the specification can be tested, not just on paper, but in reality.

To summarize where we are so far is instructive. We have expanded the thinking of modeling to that of capturing stimuli, loading, test benches and test conditions (assertions) that go along with a design. We have also introduced putting these into libraries for reuse within design groups, across entire companies, and between companies to ensure that communication breakdowns do not occur and that silly mistakes are avoided in the design. Now that all of this information is associated with our design − our model-based design − it has evolved from concept to realization to validated, centered, and toleranced design ready for tapeout. Further, as design changes occur, this infrastructure is executable in a regressive manner to quickly determine if and where any problems arise. Wow! Why would you do it any other way?

12.3 Initial Design − First Cut

12.3.1 Design Partitioning and Reuse

As the early specifications are defined and captured as high-level test benches plus assertions, they are then instantiated with the system level model(s) of the chip to create the top-level executable specifications. The purpose of this is to now refine the chip architecture. The chip architect will begin partitioning the design into hardware and software and then into analog and/or digital and/or RF circuitry for the hardware. The executable specifications are used to evaluate alternative architectures, but at this point the architectural design activity is on the designer's shoulders. Our own experiences, along with those of many companies that we have worked with, have demonstrated that the best designs are those that benefitted from proper design partitioning in the early phases of the project.

Once the chip architecture is established and the executable specifications have been used to validate the architecture, then the hardware is further partitioned into the functional blocks of circuitry that will comprise the chip. These blocks may be reused from prior designs at the technology process node (e.g., 90 nm SOI CMOS) that the chip will be designed in. Existing blocks may have to be redesigned or altered for the chip. New blocks may have to be designed either because the IC company does not have that block or does not have that block at this process node.

Whatever the circumstance, as the MBE paradigm (which leads to executable specifications) is adopted and used the libraries of test benches, assertions, and thus executable specifications grow very rapidly, making it highly likely that new specification capture is a rare activity. The executable specifications generically exist in libraries and simply need to be given parametric information and instantiated into executable form with the design. The parametric information that the executable specification needs is the transformation from chip-level specifications to the block level specifications.

For our op amp example, this means answering the question of what *is* the specification for slew rate that the op amp must have to satisfy the higher-level performance requirements? The way to answer this is with the substitution of behavioral models into the chip architecture possessing some degree of nonideal behavior reflecting the parametric specification to be defined. For example, if a sensor interface channel that consists of a Wheatstone bridge, flying capacitor, variable gain amplifier, filter, and A/D converter is an analog channel within a chip, then a behavioral model of the op amp within the variable gain amplifier and the filter with the appropriate functionality will allow rapid analysis that will assist the designer in evaluating the minimum slew rate needed [1−4].

Design Partitioning

MBE facilitates evaluation of partitioning options the designer proposes.

This level of analysis may not be required for each and every possible parameter simply because experienced designers will be able to know from the context of a chip's functionality which specifications will be most important. However, the executable specification library makes it easy to establish a quick definition of the complete set and regressively ensure that as the design is evolved, and refined, and optimized nothing falls through the cracks. It also affords the chip architect the opportunity to play *what-if* games at a level higher than the circuits being designed to ensure that circuit blocks integrate together well.

12.4 Detailed Design

12.4.1 Second-Order Effects

As the designer completes the activity of instantiating the executable specifications around the circuit blocks that they are responsible for, the circuit design process then begins in earnest. As each design is investigated, the designer will be concerned about second order effects and nonideal behavior in analog and RF designs. Because of the large number of design procedures, we cannot generalize on how the designer overcomes these challenges. However, we can say that the same framework for executable specifications coupled with MBE will serve the purpose. For analog and RF design this typically means circuit-level simulations with semiconductor foundry-provided device models. Behavioral models may work to define specifications at the first order level, but once the detailed design activity begins it is off to the transistor level schematics to (a) verify the first-order specs are being met and (b) to tackle any second-order issues that must be dealt with.

12.4.2 Focusing on Interfaces and Design Complexity

Another very important activity that was touched on earlier is making circuit blocks work together in larger pieces of the chip and, ultimately, the overall chip. A wise man once said "problems migrate to interfaces". This is certainly true of analog–digital interfaces, digital–RF, and even circuit block to circuit block. One of the primary reasons that mixed-signal chips typically require more than one spin is owing to interfaces. The hierarchical nature of the executable specifications within MBE provides an excellent framework for managing the interfaces to ensure a working chip. Chapter 13 will provide excellent examples of analog–digital, RF–digital, and circuit–circuit interfaces and how they can be managed.

Evaluating Interfaces

MBE is extremely valuable for defining and evaluating interfaces between circuitry – where problems are most likely to occur.

Design complexity can take a variety of forms. One of these being the complexity of interfaces (i.e., word length, speed, drive capability, noise, signal level, etc.). Handling interfaces is managing design complexity and MBE, as we have described it herein, is shown to enhance the designer's ability to manage this complexity. Other forms of design complexity involve integrating more circuitry/transistors on a single chip. It has been proven that the only way to handle this situation is hierarchically. Again, this hierarchical partitioning, modeling, specification, and verification lends itself naturally to MBE techniques. Another form of design complexity is that of pushing the envelope on what a given technology process node can deliver. At a basic level there is nothing in the design of high-performance circuitry that produces any special or different requirements in design flow, so these methods apply equally as well, but do not, necessarily, offer any additional benefits. High-performing circuitry at a given technology node falls more to the designer's creativity than to the tools in use. A fourth form of design complexity involves multi-chip designs. If the millions and billions of transistor chips demanded hierarchy and greater degrees of specification management, then, certainly, multi-chip solutions do as well. Now the functions are being realized in a family of complex chips and multi-chip verification is required. The problems and their solutions (i.e., the designs) have to be partitioned and analyzed at levels the human designers and the tools can cope with.

Managing Complexity

MBE is the key to managing design complexity of many types as we span the V diagram.

12.5 Optimal Design

As we have seen previously in Chapter 10 on optimization, there are a number of standard techniques that we can employ to formally optimize a design. In the context of a complete design process, the idea of "optimal design" is more than simply identifying the best parameter set in a particular configuration. The designer has a significant role in choosing the most appropriate topology for a design, and this is integral to the ability of the design to be the most optimal for a particular specification. For example, if the design requires excellent common mode rejection, then it is unlikely that a single ended design will be particularly good.

The first stage in any exploration of possible designs is therefore to assess potential topologies and decide on which one will meet the specification or requirements the best. In many instances this can be subjective, or perhaps benefit from designer experience, and also a holistic approach can turn out to be particularly powerful. In most cases, formal methods or optimization software will be of limited value at this stage.

The second stage is to then focus on details of the topology chosen and ensure that the key specifications or requirements are clearly understood. It is often helpful in this case to rank the top requirements; even if an automated or numerical approach is to be used, at some stage a "figure of merit" will need to be defined in order to assess the effectiveness of whichever optimization approach is to be used. Another variation on this is to either rank or weight individual requirements to ensure that an intelligent assessment takes place. This is essentially developing a refined "goal function" that we can use in an optimization algorithm.

Once the topology has been defined, and the goal function specified, then we can leverage the techniques described in Chapter 10 to decide on the best design parameters to achieve the goal function. This is what is called "nominal optimization" and will ensure that the nominal design parameters are, indeed, the best fit to achieve the goal function. In practice, however, we need to ensure that not only the nominal design works, but also that every design (or at least, as many as possible) meets the specification. This approach to optimal design is also called "robust design", and, as we have seen in Chapters 10 and 11 on statistical modeling, being able to predict the variation in designs accurately is essential to understand whether all the resulting designs will meet the specification, but also in assessing the "robustness" of a design. In other words, given that the design has intrinsic variability, how tolerant is the design to those changes and how sensitive are the performance metrics of the design to the design parameters?

12.6 Chip Integration and Verification

Chapter 2 described the V diagram for design and verification. At this point in the design flow we have reached the bottom of the V and begun our ascent back up the right side. Before we reach all the way to the chip level, it is often necessary to have abstracted models of the actual design realizations that maintain significant accuracy for key circuit performances. These abstracted models can improve the verification activity of full channels of circuitry such as an analog signal processing block, a sensor interface channel before an A/D, etc. Modeling tools like those in [5–7] are handy to abstract models from the circuit descriptions of the designs. However, these are not available for all classes of circuits, so the typical approach has been to use tools and languages to create the generic behavioral models of commonly used circuits and fit their behavior to the actual circuit.

The integration of blocks together with the optimization of their performance is the re-assembly of the chip architecture and the successive verification of each of those major blocks. Ultimately, we get to the chip level and are at a point where it has become unwieldy to be performing true analog simulations for verification. The fast analog methods described in Chapter 9 must be used to represent the analog portion so that chip level throughput can be achieved. This is also true once we begin to deal with multi-chip verification. We really cannot afford to repeatedly be solving differential equations at that point of complexity.

Conclusion

This chapter has established the general MBE design flow for mixed-signal integrated circuits from chip definition and specification through verification. Key effects of the MBE approach are highlighted along with relating the techniques described in Section 2 to the flow itself. Chapter 13 will follow this flow with a specific case study of a mixed-signal chip design fully illustrating each of these concepts and effects.

References

[1] C. Ulaganathan, N. Nambiar, K. Cornett, J.A. Yager, R.L. Greenwell, B.S. Prothro, et al., A SiGe BiCMOS instrumentation channel for extreme environment applications, VLSI Design 2010 (2010) 12 Article ID 156829.

[2] R.M. Diestelhorst, S. Finn, L. Najafizadeh, D. Ma, P. Xi, C. Ulaganathan, et al., A monolithic, wide-temperature, charge amplification channel for extreme environments, IEEE Aerospace Conference, Big Sky, MT (2010) 1–10.

[3] R.W. Berger, R. Garbos, J.D. Cressler, M.M. Mojarradi, L. Peltz, B. Blalock, et al., A miniaturized data acquisition system for extreme temperature environments in space, Proc. IEEE Aerospace Conference 2008, Big Sky, MT (2008) 1–12.

[4] J.D. Cressler, M. Mojarradi, B. Blalock, W. Johnson, G. Niu, F. Dai, et al., Silicon-germanium integrated electronics for extreme environments, Proc. Government Microcircuit Applications and Critical Technology Conference (GOMAC) 2007, Orlando, FL (2007) 4.

[5] W. Zheng, Y. Feng, X. Huang, H.A. Mantooth, Ascend: automatic bottom-up behavioral modeling tool for analog circuits, IEEE Proc. Int. Symp. Circuits Syst. (2005) 5186–5189.

[6] A. Mantooth, M. Francis, V. Chaudhary, M. Vlach, Model-based design tools for extreme environments, Proc. Government Microcircuit Applications and Critical Technology Conference (GOMAC) 2005, April 2005, pp. 4.

[7] H.A. Mantooth, A.M. Francis, Y. Feng, W. Zheng, Modeling tools built upon the HDL foundation, IET Proc. of Computer and Digital Techniques 5 (1) (2007) 519–527.

CHAPTER 13

Complex Electronic System Design Example

A complex system that works is invariably found to have evolved from a simple system that worked. The inverse proposition also appears to be true: A complex system designed from scratch never works and cannot be made to work. You have to start over, beginning with a working simple system.

Gall's Law, John Gall

13.1 Introduction

Having devoted the previous chapters in this book to describing both fundamental modeling and simulation techniques and the concepts for successful design, in many respects the "proof is in the pudding", and it is now time to put these ideas into practice and develop a design. This chapter will therefore describe the design and implementation of a complete integrated circuit that has the overall function of a complete wireless sensor node integrated circuit. The idea behind the integrated circuit is to provide a complete solution for sensing multiple data channels to a high level of precision, but not necessarily at a particularly high data rate. In fact, the circuit has been designed for environmental sensing where the sensor nodes are designed to be asleep for long periods and woken up periodically to provide data to the central node.

The architecture of the system is designed to be a "star" network, as shown in Figure 13.1, where there will be a single "hub" node that talks to a number of sensor nodes (nominally 256 per hub).

Each sensor node has a number of input sensor channels that are intended to be used to measure the voltage of multiple sensors in each location. Each chip will also have a built-in temperature sensor for local node temperature measurement (on chip).

The basic operation of the device is such that once a device is activated, it will go into a "standby" mode using almost no power until its radio frequency (RF) channel detects activity and then the Receive (Rx) channel is turned on and the node will wait for a short period until it recognizes its own node ID, at which point it will download the command from the hub. This may instruct a sensor cycle, or it may require a transmission (Tx) of the stored data from a previous cycle.

The overall architecture of the chip will be as shown in Figure 13.2.

The remainder of this chapter will describe how we get from these conceptual overall ideas of what the system is intended to do and how each individual chip will function overall to specific requirements and design details that demonstrate how exactly each individual section of the design will work, and, of course, how we can use model-based engineering (MBE) to achieve a successful design.

As we have discussed in Chapter 2, the design process can be seen as following a "V" shape, and it is useful to reprise this in Figure 13.3 once again. We can see that the general philosophy is to start with high-level descriptions of the design, to ensure that the basic idea is consistent and correct, and gradually

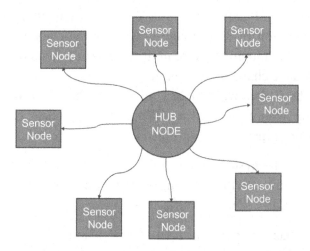

Figure 13.1:
Star network configuration of sensor network

increase the complexity of the design description (using models) through architecture and detailed design until an implementation level is reached (usually a circuit level description). Once we have reached this stage, then the second

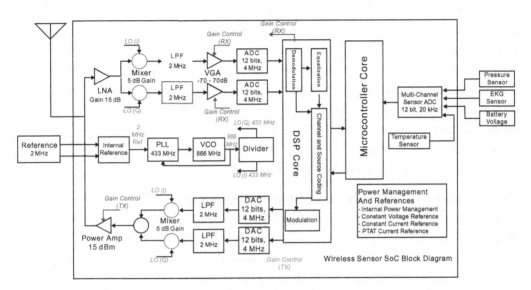

Figure 13.2:
Overall wireless sensor node integrated circuit topology

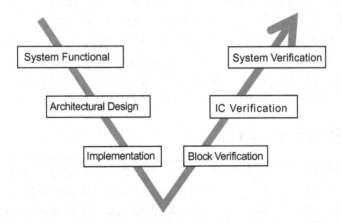

Figure 13.3:
The design process represented using a "V" diagram

phase of the design can start, which is the verification of all of the individual elements from the most detailed up to a complete system description of the design.

As we have discussed throughout this book, it is generally not tenable or useful to go straight to the implementation level and then expect to be able to validate the entire design, and also that the V diagram itself is a rather linear expression of a process that is nonlinear. One of the most effective ways to do this type of design is to use models as they arise. For example, it makes lot of sense, as we have seen in this book already, to utilize the high-level system models to ensure that the architecture is correct and the designer has not made some fundamental errors in interpreting the specification. In reality, therefore, there are multiple loops and iterations within the general framework of the V shape. Usually the more the designer does, the better the understanding of the design and specifications, and risk is reduced.

We will, therefore, use this idea of the V diagram as a framework for this final chapter and illustrate how we develop the design, manage its complexity, and solve problems along the way until we have a reasonably complete final implementation. It is, of course, impossible to show every detail of the design, reasonable space precludes this; however, we will show the key steps along the V diagram and, most importantly, demonstrate the process that takes place.

This case study was used as a class example by the University of Arkansas' Department of Electrical Engineering, and appreciation is given to Dr. Matt Francis and the Electrical Engineering students who completed the work. There was also an international aspect to the collaboration with designers from the School of Electronics and Computer Science at the University of Southampton involved in the project for a number of years, including Dr. Reuben Wilcock, Dr. Li Ke, Dr. Matthew Swabey, and Robert Rudolf.

13.2 Key Requirements

The overall requirements for the chip are that the chip should be able to monitor a number of discrete sensor interfaces, digitally process the measured responses in addition to its onboard temperature sensor, and wirelessly transmit their data to a central control node. In addition, the sensor node should be able to receive signals to configure its own behavior as a result.

The system was, therefore, designed to be a low power wireless sensor network application, with functionality over a temperature range of $-55°C$ to $125°C$ to ensure its potential application in somewhat extreme environments. While this general requirement enabled the design to be applicable for a wide range of potential applications, including biomedical, civil structure, environment, and industry applications, the initial target application field was defined to be agriculture.

Key features of the design were that the integrated circuit was to contain RF and analog blocks in addition to the digital signal processor (DSP) and digital processing. It was specified to be a fully differential signal at a 2-MHz baseband frequency. The RF specification was to use the 433 MHz RF band for wireless communication.

Multisensor analog-to-digital converters (ADCs) were to provide information on more than one property and the data were to be processed using an onboard DSP core (which was also required for the modulation and demodulation, joint source and channel coding, and Trellis-coded quantization). The microcontroller core was to be used to investigate low-power asynchronous design using the techniques of null convention logic and asynchronous delay insensitive design.

Using this overall set of requirements, the chip specification can be defined in order to provide an accurate definition for the design work to begin.

13.3 Top Level Model and Chip Architecture

13.3.1 Chip Architecture

MBE is used to aid in the exploration and creation of a chip architecture to address the application company's requirements.

The top level of the design is intended to define the broad division of the chip into the main elements of functionality, which are the RF channel (Rx and Tx), the digital core, the power management, and the baseband mixed-signal interface. The idea behind the design is to have eight input sensor channels, which are multiplexed into a set of digital registers, which are then processed for transmission to the hub system.

We can make an initial judgment as to a suitable partitioning of the design into initial blocks using this broad division of functionality and define the top level blocks for the design before we embark on any detailed modeling at this stage.

Using the overall initial chip architecture defined in Figure 13.2, a top-level symbol can also be developed which defines the overall outline of the chip. Using this architecture, the individual design tasks could be allocated and distributed amongst the design team. Table 13.1 shows the complete list of design tasks and the broad area of technology to which they belong. For example, RFC is radio frequency circuit, DIG is digital IC design, POW is power management, and finally ANA is baseband analog/mixed-signal design.

In this project, each block was designed by a separate design engineer as part of the overall project team. Even Table 13.1 shows the complexity that needs to be managed in order for the entire design to function correctly. There are four main groups of design type (RF, reference and power, digital and baseband analog) and each have their own distinct methods of design and description. This leads to a tension early in the process of how best to represent each

Table 13.1: List of Individual Design Blocks

RFC: Mixer
RFC: LPF x2
RFC: VGA
RFC: ADC
RFC: RF switch
RFC: DAC
RFC: PowerAmp
REF: PLL
REF: VCO
REF: Divider
DIG: Microcontroller
DIG: DSP modulator
DIG: DSP demodulator
DIG: DSP equalization/source coding
ANA: Sensors xN
ANA: ADC
POW: Battery management/charging

block and how to share information. In addition, there is the sheer scale of such a project — there are 18 separate designs and designers to manage in addition to channel integrators and system integrators, making this a project of more than 20 different people, all with their own deadlines to meet and challenges to overcome. The obvious place to start is with the overall system integrator, and for them it is critical to identify the key specifications and disseminate those to the individual design teams and engineers.

13.3.2 Specification Definition and Capture

MBE facilitates the definition and capture of chip specifications from the requirements. This leads to executable specifications, which has profound implications on verification and design cycle reduction.

In order to start the process of chip architecture, using the initial specification developed from the initial requirements gathering phase, the process of detailed specification and capture can begin with the development of a model of the overall chip design. This corresponds to the first stage in the V diagram, as highlighted in Figure 13.4.

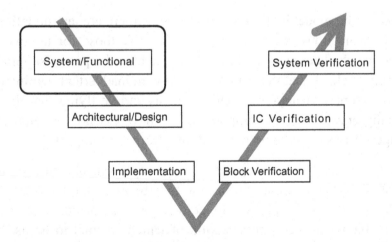

Figure 13.4:
Starting the design process at the system level

The first practical step is to define the hierarchy of the top level of the chip into the four main sections (RF, digital, power, and baseband) as described earlier. In the interest of space, we will focus on the three signal processing sections (RF, digital, and baseband) that integrate together to perform main chip functionality.

With this basic architecture we can first develop a simple test bench that is used to check that the basic connections are in place. This also helps in developing the ideas of design functionality and executable specifications. For design functionality, it is most important to define the tests required to ensure that the specification is both feasible and that the design will operate correctly in principle. For executable specifications, the appropriate tests exercise a model that is used to not only validate the design, but also to demonstrate that the specification is achieved.

So, how do we use this model to check these basic tasks? The first step is to determine if whatever model we generate will actually compile. Using the export model function in the graphical modeling software, we can generate a Verilog-A or VHDL-AMS model and check using our simulator of choice to see that the model will compile correctly, and we have connected and named each pin correctly. One advantage of using a graphical approach is that we can actually see all the connections on the schematic of the top level design, rather than as just a code-based (i.e., handwritten) model.

In addition, as the model is truly mixed-signal, if there are any undefined nodes, or missing connections, the simulator will identify those for us, even using a simple test bench. The initial test bench can include the power supplies, and so we can begin to check that there are no shorts with inadvertent naming of signals by running a simple steady-state (DC) analysis and verifying that there is zero current being drawn from the supplies (as we have essentially "empty" models at this stage, there should be no power drawn from any supply).

At this stage of the initial design, the first simple mistake in the case study design was discovered. Although the interface between the baseband and digital models was defined using a start conversion, end of conversion, and data transfer (16 bits) sequence, the selection of which channel to be used was not correctly included, and so this was added at this stage. This demonstrates the power and effectiveness of a graphical approach as the design engineer is then able to identify a potential mistake or omission much quicker than would be the case using a modeling approach where the code is handwritten.

Having investigated these first basic steps, we can now begin to think about the specifications of the design itself and look at how these could be tested. We already have some initial requirements in place that we have defined previously in this chapter and some idea of the kind of specifications we need, but now it is possible to define those in much more detail in a formal and structured manner than was possible before. It is useful before even embarking on an initial "first cut" design to review some of the major design issues involved. This is the first time we can begin to investigate using models to understand these issues in more detail, even though we may not use all of these results in the final design. In order to accomplish this, we can use our rough division into RF, baseband, and digital sections to develop these ideas and contemplate the design decisions we need to make later in the design process. Even though we have made an initial decision about the architecture of our wireless sensor chip, it is useful to consider all the implications of separate block designs early so as to make an informed choice about design details early. Therefore, as part of our preliminary systems analysis, we will use models to understand some of the fundamentals in each of the major blocks and some of the key design issues.

13.3.3 RF Section Design

13.3.3.1 Initial Requirements Definition

The first issue is what kind of RF Transceiver design to use? This is governed partly by the overall requirements for the sensor chip and also the environment in which it is to be used. A key initial decision is what kind of network is the chip to integrate with? If the network is intended to be WiFi or Bluetooth, then, obviously, the frequency will be 2.4 GHz. One of the goals of this chip is to be able to transmit over a reasonable range, and this pushes the choice of frequency down, as the lower the frequency, the better the range for a given power. If we investigate the potential options for such a network, especially if the system is to be able to operate in both the USA and Europe, then very few standard license-free bands exist for that purpose. One of the common bands is the 433 MHz range, and this was a suitable compromise between frequency and power.

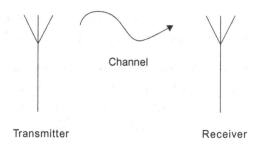

Figure 13.5:
Typical transmission channel

The second decision to be made is what kind of modulation scheme should we use. The system needs to have the transceiver designed in such a way that the chip will have both transmit and receive channels in the design and be able to communicate with a "host" or "hub" system. We have a wide variety of potential systems to choose from; however, the basic principle is that we have a transmitter, a channel, and a receiver. The key criteria are the transmit power required, the range possible, and the data rate that can be achieved. This is a classic case where we can use MBE to help assess the possible options, and understand the strengths and weaknesses of each approach.

As we can see in Figure 13.5, the transmission system can be divided into three parts: the transmitter, the channel, and the receiver. Each of the three elements can be modeled and designed to fit the requirements of our application. So, what are the RF requirements for this design? We can define these in terms of some basic criteria, such as data rate, range, and power as given in Table 13.2.

Using the data in Table 13.2, we can estimate what the effective range of this system could be. Obviously, at this stage, the range will be an estimate; however, as we refine the model this estimate can be improved. We can use the following expression to estimate the potential range of this system, which is a simplified version of the propagation of waves in free space first defined by James Clerk Maxwell [1], and while Eq. (13.1) is a simplified equation[1] that

---

[1] A more detailed link budget can be calculated that also builds in the antenna gains at either end of the link, and this can be extended to include detailed effects of amplifier gain and other factors in the system; however, this is suitable for an initial "ball park" figure for power, sensitivity, and distance as an initial starting point.

Table 13.2: Radio Frequency System Requirements

Requirement	Value
Date rate	10 kbit/s
Output power	10 dBm
Operating voltage	2.2–3.3 V

assumes unity gain for the antenna, it is a useful starting point for estimating the scale of power required and the sensitivity of the receiver.

$$d = \frac{\sqrt{30P_{\mathrm{T}}}}{E} \tag{13.1}$$

P_{T} is the transmitted power (W) and E is the sensitivity of the receiver in V/m. As we have specified a value of $P_{\mathrm{T}} = 10$ dBm, this equates to a power of 10 mW. We can then make an initial estimate of sensitivity to calculate the range (or vice versa). Using an initial estimate of 50 μV/m for E, the resulting range estimate (d) using Eq. (13.1) can be calculated as 11 km. Obviously, we can experiment with different values of range or sensitivity to obtain shorter range with higher sensitivity or longer range with lower sensitivity. As the system is intended for environmental applications, even if we only achieve a fraction of this range, then the design will be acceptable.

The next stage of the design is to consider which type of modulation scheme to use, and there are numerous choices in this area from the simplest amplitude modulation (AM) approach to the more complex forms of multiple level quadrature modulation.

13.3.3.2 Modulation Scheme Options

13.3.3.2.1 AM

The simplest approach would be to use a basic form of AM. The idea with a simple AM scheme is to multiply the date signal with the carrier (in this case at 433 MHz). The resulting signal is then transmitted over the channel, as shown in Figure 13.6. (We have shown a carrier expressed as a sine function; however, a cosine function could equally be used.)

The amplitude modulated signal is also called a "double sideband amplitude modulation" scheme (DSB-AM) as the frequency components of the modulated system are the carrier, and two sidebands containing the signal and its mirror image (as can be seen from Figure 13.6, the output modulated signal has not only the signal, but also the inverse, as the carrier is at a higher frequency than the signal itself, and, as a result, follows the inverse, as well as the main, signal), resulting in two sidebands as shown in Figure 13.7.

This approach is relatively straightforward to implement; however, it is notoriously power inefficient (about 33%, in fact), as the carrier being transmitted is essentially a waste of energy because the information is actually stored in the sidebands. A variation is an approach called on−off keying (OOK), which is essentially the idea used in Morse code transmission, where the signal is either on or off. This can be visualized if we replace the generic signal shown in Figure 13.6 with a digital signal (on's and off's) and observe the resulting waveform. As can be seen in Figure 13.8, the OOK approach is effectively the multiplication in the time domain of a series of pulses with the carrier sinusoidal waveform.

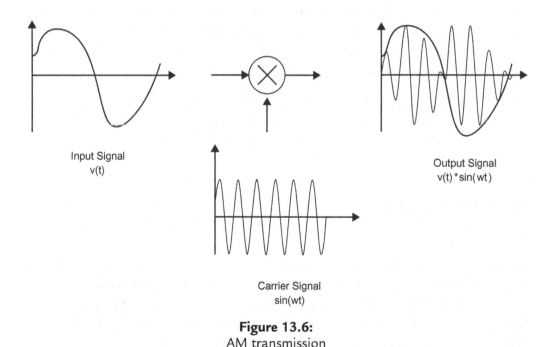

Input Signal
v(t)

Output Signal
v(t)*sin(wt)

Carrier Signal
sin(wt)

Figure 13.6:
AM transmission

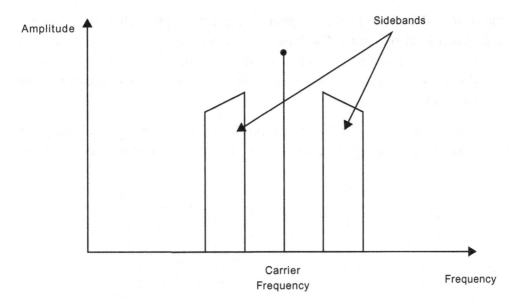

Figure 13.7:
AM spectrum

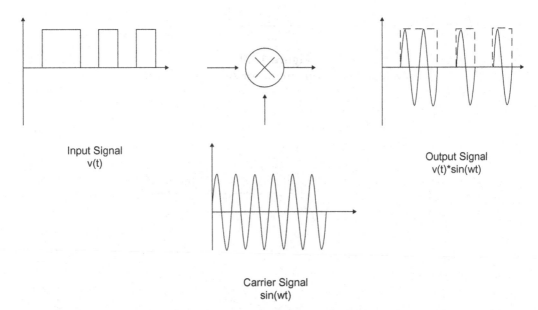

Figure 13.8:
On-off keying AM

One of the advantages of the simplicity of the AM approach is the demodulation, which is also simple. The basic idea is to rectify and filter the received signal and the resulting output will approximate the input. This type of demodulator is also called an envelope detector, with a simple circuit as shown in Figure 13.9.

The output waveforms for the envelope detector are shown in Figure 13.10, where the recovered signal is clearly a function of the filter frequency and

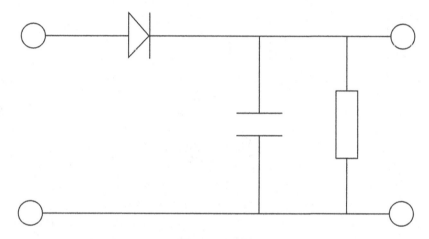

Figure 13.9:
Envelope detection circuit for AM demodulation

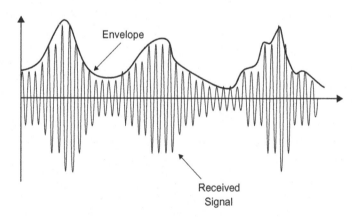

Figure 13.10:
Envelope detection waveforms for AM demodulation

received power. This simple example shows simple rectification, and full-wave rectification can also be used; however, in both cases, clearly there will be a ripple as a result of the demodulation process, which can become problematic in real systems because of the introduction of distortion and noise into the recovered signal.

An alternative approach is to multiply the received signal with a local oscillator that is the same nominal frequency as the transmitter carrier. This is called product demodulation and operates on the simple principle that if we have a signal that is a function of a pure sine or cosine signal, the multiplication of the identical signal will result in a product of double the frequency, plus a low frequency term. Using this approach, we need to use a cosine rather than sine, as we are taking advantage of the standard trigonometric identity in Eq. (13.2):

$$\cos(a)\cos(b) = \frac{1}{2}(\cos(a+b) + \cos(a-b)) \tag{13.2}$$

If $a = b$, then the resulting product is:

$$\cos(a)\cos(a) = \frac{1}{2}(\cos(2a) + \cos(0)) = \frac{1}{2}\cos(2a) + \frac{1}{2} \tag{13.3}$$

And, by filtering out the $\cos(2a)$ product term, the result is a constant multiplied by the input signal:

$$
\begin{aligned}
x(t) &= (dc + v(t))\cos(\omega t) \\
y(t) &= (dc + v(t))\cos(\omega t)\cos(\omega t) \\
y(t) &= (dc + v(t))\left(\frac{1}{2} + \frac{1}{2}\cos(2\omega t)\right)
\end{aligned}
\tag{13.4}
$$

If the product term $\cos(2\omega t)$ and dc offset are filtered out, then the result will be the original signal attenuated:

$$y(t) = \frac{1}{2}v(t) \tag{13.5}$$

At first glance this looks ideal; however, there are some serious issues with this approach. The principle relies on the local oscillator of the receiver being almost exactly the same as the transmitter; if there is any difference, the results on the recovered signal can be catastrophic. The effect on the recovered signal can be fade in/out of the amplitude, or frequency shifting. The other issue is that if there are any phase errors, then the signal will be attenuated, but the noise won't be, and so the signal-to-noise ratio will be poor, as will the resultant signal quality.

13.3.3.2.2 Frequency Modulation

Frequency modulation (FM) takes a similar approach in that a carrier signal is modulated by the input signal except, in this case, the amplitude of the modulated signal is constant, but its frequency changes. A simple example of a kind of frequency modulator could be a voltage-controlled oscillator (VCO), where the frequency of the output is controlled by the input voltage.

Mathematically, we can consider the carrier signal to be the same as for AM; however, rather than the amplitude changing, the frequency of the modulated signal changes by a factor called the frequency variation, or frequency deviation. The amount of deviation is usually specified as part of a commercial radio standard. The carrier frequency for a zero input is the nominal frequency, and the frequency deviation can be positive or negative, so the total carrier swing is twice the frequency deviation. For example, if the FM signal is assigned a 200 kHz bandwidth, this is equivalent to the carrier swing and so the frequency deviation would be 100 kHz. FM radio stations are usually assigned a frequency in the range of 88 to 108 MHz in contrast to AM radio which is in the range of 0.55 to 1.6 MHz, and this is one reason why AM radio has a longer range; however, FM radio operates better in reception areas that are closed in, such as tunnels and buildings, owing to the higher frequency and corresponding shorter bandwidth.

We can see an example of a signal that is FM modulated with a frequency of 1 kHz on a carrier of 50 kHz, with a full range sensitivity in Figure 13.11.

The massive improvement in quality inherent in FM signals over AM signals is the result of almost all of the power being contained in the modulated signal, whereas in AM, as we discussed previously, most is wasted in transmitting the carrier.

FM demodulation takes place using a "superheterodyne" demodulator as shown in Figure 13.12. This is actually quite similar to an AM demodulator often used in integrated circuits.

The major drawbacks with FM systems are the relatively wide bandwidths required per channel and, although similar, the transceiver topology complexity is greater than that of a basic AM demodulator.

13.3.3.2.3 Quadrature Amplitude Modulation

While AM and FM are both fine for the transmission of music and speech — basically narrow band audio signals — in data-logging and sensor networks, it is important to be able to send data reliably and securely. As we have seen with the AM OOK system, and the equivalent in an FM system frequency shift keying, the systems are fine for the transmission of single bit streams; however,

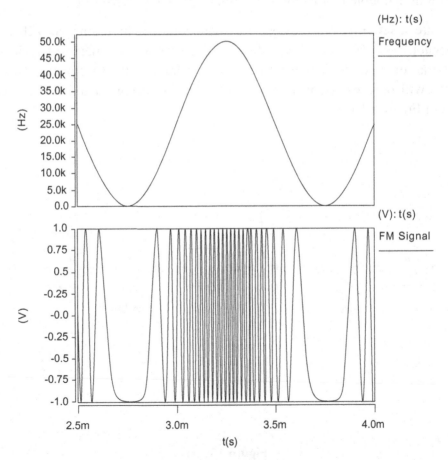

Figure 13.11:
FM modulated signals

this does lead to fairly low data rates (usually only a few kbits/s). If we need to transmit multiple bits in one step, then a different approach is required and a common technique often used in cellular communications and data communications systems is quadrature amplitude modulation (QAM).

The basic approach with QAM is to use two AM or FM channels using oscillators shifted by 90° (i.e., one $\sin(\omega t)$ and one $\cos(\omega t)$) combined to give a number of bits per transmission cycle. For example, if we say that in an AM system, where the digital inputs in each channel (incident and quadrature — otherwise known as I and Q) are implemented using OOK, then we can transmit a "symbol" that consists of two bits at a time instead of one. The effective bandwidth is double that of the single message being transmitted.

There are a few different options for demodulation, including zero-IF (direct conversion), low-IF, or high IF. Zero-IF is the simplest scheme as it does not require an extra stage of demodulation at the intermediate frequency (IF), and zero-IF will only work for an IQ system. The overview of a QAM system is given in Figure 13.13.

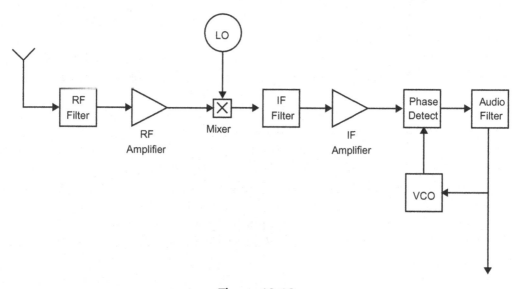

Figure 13.12:
Superheterodyne FM demodulator

Transmitting data using I and Q channels becomes a way of encapsulating magnitude and phase data if we consider the I and Q channels in the same way as complex data (real and imaginary). The data can then be described graphically using a "constellation" diagram, where the values of instantaneous data are plotted on an X−Y graph.

A simple scheme can be to use a form of OOK in each channel to provide either positive or negative I or positive or negative Q signals, as shown in Figure 13.14.

This appears simple; however, there are some issues. As can be seen from the rotation of the bits, the sequence is binary, not grey code, which makes the transitions uneven. Also, the coding depends on both bits in I and Q channels. An alternative approach is to use the technique shown in Figure 13.15. This has the grey code sequence, and the advantage of this coding scheme is that the I channel depends on one bit and the Q channel on the other.

This QAM scheme can be modified to include multiple phases or multiple amplitude levels to give more symbols per transmission phase, for example 64-QAM has 8 amplitude levels (2^3) per channel − giving a total of 6 bits transmission per symbol. The design trade-off is to trade the ability to transmit multiple bits against signal-to-noise tolerance.

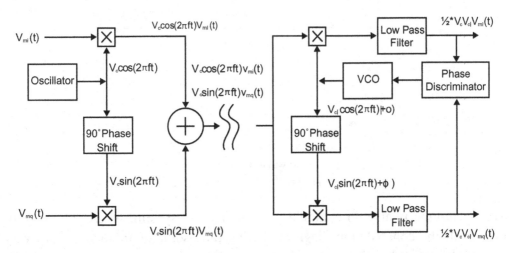

Figure 13.13:
QAM system overview

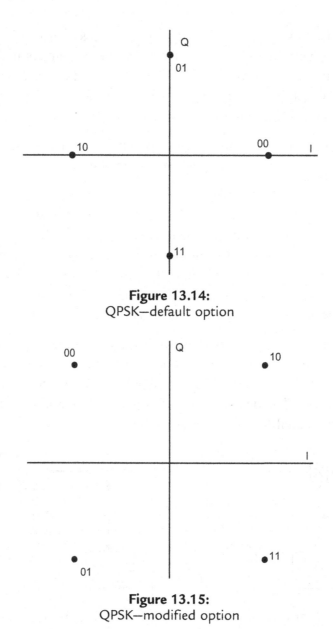

Figure 13.14:
QPSK—default option

Figure 13.15:
QPSK—modified option

This technique is also an example of a "coherent" system, where the assumption in all the constellation diagrams is that the oscillator frequencies are very accurately defined with very small errors. In practice, of course, there will be frequency and phase errors, and the result can be a rotation in the constellation diagram to reflect that error. Therefore, practical systems do not measure the absolute phase differences but rather the relative phase cycle-to-cycle.

Example 13.1 Modeling Modulators

We can use models to understand the behavior of modulation schemes at a variety of levels; however, a useful starting point is in the creation of a basic system level model of the QAM modulator described in Figure 13.13. If we start from the basis that this system is operating on continuous real numbers (a typical system-level design), we can apply simple mathematical operators, such as multiply, addition, sine, and cosine functions to create the modulated signal. As we have seen already in this book, we have options as to how we go about creating a model for system level elements such as this. If we take a general view of the modulator in terms of two inputs representing the I and Q channels and modulated output, we could use an equation to implement the modulation:

$$y(t) = i(t)^*\cos(\omega t) + q(t)^*\sin(\omega t) \tag{13.6}$$

Using this approach in graphical modeling, we simply create a top-level symbol for the "modulator" with two inputs (i and q) and a single output (y). For system-level modeling these will be defined as "quantities" and have type "real" — in other words completely generic analog signals with no units defined at all. Using ModLyng, the resulting symbol and equation can be seen in Figure 13.16.

Clearly, this is a simple approach and very efficient; however, for more complex systems, sometimes the equations can become either very complicated, or difficult to manage. In such cases, as we have seen throughout this book, it may be preferable to take a more graphical approach to the process of creating models, and use primitive elements such as multipliers, adders, and trigonometric functions, such as SINE or COSINE. Taking this approach to model the same modulator would use a schematic description and leverage the existing simple system building blocks either available within ModLyng libraries or easy to make by the user.

As shown in Figure 13.17, the modulator is made up of several simple blocks and, although in some respects looks more complicated than the equation-level model, each of the building blocks is defined once and can then be reused any number of times in different models. Also, it is very easy to then use the block diagram version of the model in reports or to explain concepts in design reviews to nonexperts, which can be very helpful. In both cases, the resulting model behaves in an identical manner.

The same approach can be taken when it comes to modeling the demodulator, and it makes a lot of sense to reuse many of the blocks used in the modulator. By adding a low pass filter (LPF) we can construct a block-level demodulator as shown in Figure 13.18. The two new blocks in the demodulator are a simple gain block ($y = x^*a$) and a LPF. The gain can be used to compensate for the inherent attenuation of 0.5 in the demodulator, as shown in Figure 13.13, by adding a gain of 2.0 in each signal path.

The LPF is a simple model of a standard Laplace first order filter with the basic equation as shown in Eq. (13.7), where ω_c is the cut-off frequency and k is the DC gain.

$$y = \frac{k}{s + \omega_c} x \qquad (13.7)$$

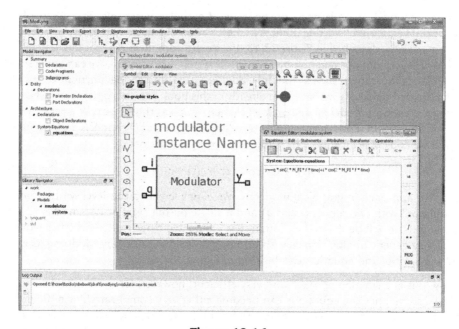

Figure 13.16:
System-level model of a modulator using equations

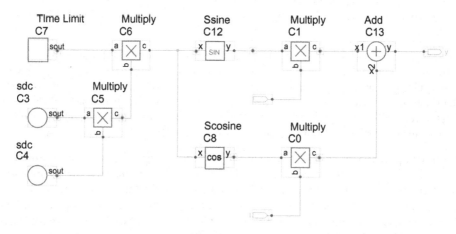

Figure 13.17:
Block-level modulator model

As we described in the chapter on system-level modeling, the way we implement a Laplace function is to consider the operator s as a "d_by_dt" function. Therefore, if we rearrange Eq. (13.7) into the form shown in Eq. (13.8) we can then apply the Laplace operator as a derivative in Eq. (13.9).

$$y(s + \omega_c) = kx$$
$$\Rightarrow ys + y\omega_c = kx$$
$$\Rightarrow y\omega_c = kx - ys \tag{13.8}$$
$$\Rightarrow y = \frac{1}{\omega_c}(kx - ys)$$

$$y = \frac{k}{\omega_c}x - \frac{1}{\omega_c}\frac{dy}{dt} \tag{13.9}$$

This can then be modeled in the graphical modeling tool to provide a LPF function. The resulting modulator and demodulator can be tested by applying two different quadrature signals (a pulse down one channel and a sinusoid down the other) and comparing the output waveforms with the input waveforms. The resulting test bench can be seen in Figure 13.19.

When the model was generated (in VHDL-AMS), the resulting simulations using an approximate signal frequency of 1 kHz and carrier frequency of 1 MHz (much lower than the design 433 MHz to reduce simulation time) were carried out. The results show that the modulation scheme generally works; however, some refinement is clearly necessary.

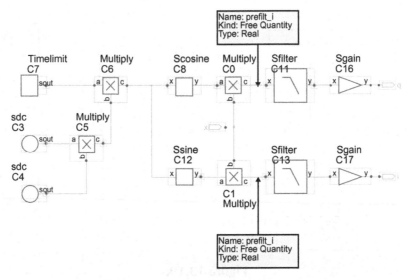

Figure 13.18:
Demodulator block diagram

One of the obvious issues is that the amplitude of the output is a little lower than it should be. So why is this happening? As we have seen, the equations do not indicate any attenuation, and the system-level model is simply an encapsulation of those equations, so we would not expect a problem with the output. The solution is a result of the use of the simulation time. As is often the case in a trigonometric or integration model, as the time (in this case) value increases, the potential for numerical noise becomes greater. This is analogous to integration where the absolute value becomes greater, and the resulting numerical noise floor also rises accordingly. A point occurs at which numerical noise begins to degrade the quality of the calculated trigonometric function (in this case a sine or cosine value) as the absolute value of time increases. In fact, in this example after about 200 cycles of the carrier fundamental, the numerical inaccuracy begins to have an effect. This can be seen clearly in Figure 13.20, where, after about 200 μs, the accuracy of the output degrades dramatically, whereas prior to this point the output tracks the input very well.

So how do we address this issue? The solution is to modify the time calculation function so that we calculate the period of the carrier and then calculate the relative time within a period, so that the maximum absolute value of the time variable is always less than or equal to the period of the carrier frequency. We can do this using a simple digital model where the resolution (number of sample per period) can be set as a parameter and the new time calculated. The resulting simulation in Figure 13.21 shows a much better correlation between the input and output, with a small filtering effect due to the LPF, apparent on the fast edges of the pulse input.

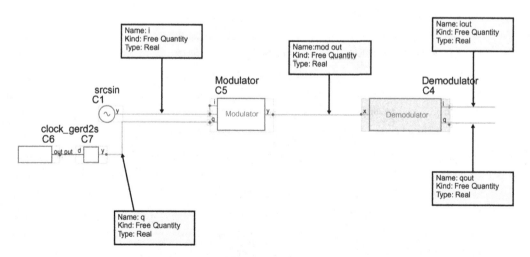

Figure 13.19:
Quadrature amplitude modulation test bench

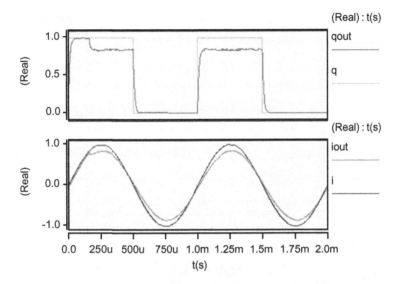

Figure 13.20:
Quadrature amplitude modulation simulation of system level design

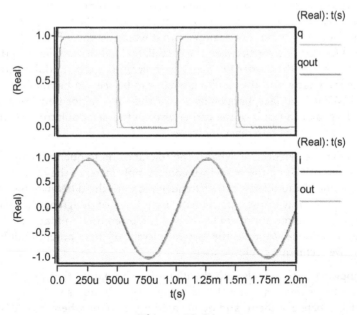

Figure 13.21:
Modified system model with relative time rather than absolute time

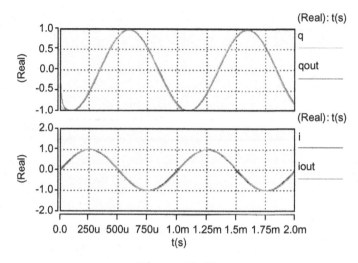

Figure 13.22:
Carrier frequency error of 1 part per million

So, where does this leave us in terms of a design? We now have a simulation model of the system that includes both the modulator and demodulator. We can change the modulation carrier frequency and can test different waveforms to analyze the behavior of the system. We can also modify key parameters, such as the LPF cut-off frequency, amplifier gain, and LPF order. For example, what would be the effect of a 1 parts per million (ppm) frequency error between the modulator and demodulator in the system? We can simply change the carrier frequency parameter in the demodulator to +1 ppm (1 Hz) of the modulator and the resulting effect can be seen to be minor in Figure 13.22. What is, in fact, happening is that there is a minor frequency error in the demodulated signal as a result of the carrier error in the demodulator, and this gets progressively worse as the difference increases.

So, from a designer's perspective, what is the tolerance on frequency before we see an effect on the output? Using the model we could simply increase the difference between the modulator and demodulator carrier frequencies until the difference became intolerable; however, this is not a particularly useful way of designing the system (although a trial and error approach like this is still often used in many examples). In practice, we would prefer to make the system tolerant of these potential differences and so an alternative technique could be used.

Instead of applying static signals with the IQ QAM modulator, if we apply a sinusoidal waveform at a lower frequency (say 1 kHz) to the I and Q inputs with a phase difference of $\pi/2$ between them, and define a coding scheme where if we wish to send a "1" then apply sine to the I input and cosine to the Q input, and vice versa for a "0", we can see how this affects the behavior of the modulator. As we saw previously,

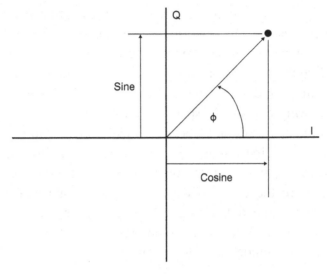

Figure 13.23:
Rotating vector of IQ signal

we could define a static value (DC) for I and Q signals and this would result in a static QAM as we saw in Figure 13.15. By applying a sine or cosine signal we produce a rotating vector as shown in Figure 13.23.

On first inspection we are no further forward than we were before; however, this has a significant advantage in that we now do not care about how fast the vector rotates, but rather which direction the rotation is taking place. This gives much more tolerance to errors in the signal and very simple signal processing can be used to estimate the rotating phase from sample to sample and therefore the value of the data being transmitted. The downside of this approach is that we trade off reliability for data rate, as this approach is by its very nature going to be relatively slow.

The other advantage of this type of approach is that the frequency error will manifest itself as a constant error in rotation (either too fast or too slow) and can therefore be filtered out using a simple moving average filter over a number of samples.

So far in this section we have explored various architectural options for the RF part of the design, which will enable a system-level exploration to take place. At this point we can investigate the individual building blocks in more detail, at least to a first cut of the design.

This process begins with capturing top-level specification evaluative criteria. For the chip-level test bench, performance measures at the output pins based on given input sequences can be designed for chip-level pass/fail criteria. Even though the underlying models are nonexistent at this point and the test would fail, spending time to define what "good" is at this point puts the focus squarely on capturing executable specifications. Then, as a subsequent step in the process. As the next section elaborates, ideal models can replace the empty models to achieve the first passing executable specs in simulation. For example, for a given analog value on one of the input pins, the performance criteria could be to compare the decoded digital equivalent and assess whether it is within the tolerance expected. Being a largely ideal model at this stage, any fundamental problems with the architecture can be identified before any significant time is spent on detailed circuit design.

At this stage of the design process, therefore, it is fundamentally important to ensure that the overall design is consistent and coherent, without necessarily spending a long time establishing that the performance is entirely achieved; rather, the goal is to minimize the complexity of both the model and the resulting design analysis. It could be said that at this point, the main aim of the project manager is to know that all the main blocks are in place, with resources to move into a more detailed design phase later.

This section of modeling has brought us to a point where we understand clearly not only the RF design options available to us, but also some potential pitfalls when we simulate our designs later. We can now move onto the baseband section, in particular the data converter, and look at some of those issues.

13.3.4 Baseband Analog Design

13.3.4.1 Interface Considerations

As we have seen thus far both in this book and in this chapter, we have decisions to make about the level of detail included in a particular model. As we saw in Figure 13.2, the overall circuit topology has a number of analog inputs (8) which are multiplexed into a single analog channel, which is then converted to a digital signal using a 16-bit ADC (Figure 13.24).

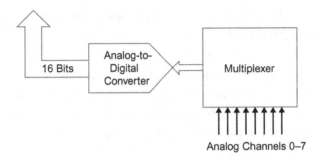

Figure 13.24:
Baseband analog section of the design

As we can see, there are two parts to the design, both mixed-signal elements, and we can take these in turn. However, this relatively conceptual system diagram does leave some important questions unanswered. For example, what types of signals are being measured? What is the dynamic range? Is it a voltage-, current-, or charge-based sensor? What is the output impedance of the sensor? We cannot simply take a random sensor and plug it into our circuit and hope that it will work. Also, we have some constraints on the signal that we can measure owing to the limits of the process technology (especially relating to supply and breakdown voltages).

13.3.4.2 Baseband Analog Design — Data Converter

With the selected analog channel in place, the signal can then be converted using an ADC. There are several choices of converter, depending on the type of signal performance and system performance we need. For a very fast conversion, a flash converter is used; however, the hardware resources required for 16 bits might be considered too much. Over-sampled converters are very useful for noise reduction, and, finally, successive approximation converters give an excellent trade-off between performance and resources required.

There are a number of key criteria that we require to understand to ensure we correctly specify the type and parameters of an ADC.

- Number of bits (typically 8–20).
- Sampling rate (typically 50 Hz to 100 MHz).

- Relative accuracy — deviation of the output from a straight line drawn through zero and full scale:
  - integral nonlinearity or linearity
  - differential linearity — measure of step size variation; ideally, each step is 1 bit, but, in practice, step sizes can vary significantly
  - usually, converters are designed so that they have a linearity better than ½ least significant bit (LSB) (if this were not the case then the LSB would be meaningless).
- Monotonicity — no missing codes (i.e., $1001 \rightarrow 1011$ is impossible).
- Signal-to-noise ratio (S/N or SNR; same as dynamic range).

Using these parameters we can make an informed decision as to the best choice of ADC and also how to specify its performance in detail. If we consider the most important stage of the conversion, the digitization phase, we can see how the number of bits available in the conversion limits the accuracy of the ADC, as shown in Figure 13.25.

The sampling did not result in any information loss (in an ideal world) in itself, but the digitizing will as only a limited number of bits is used to represent the analog amplitude signal. This error manifests itself as noise and can be treated as white noise in many cases, and the maximum quantization error is defined as $\pm q/2$.

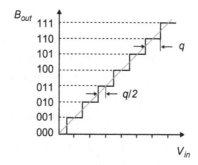

Figure 13.25:
Analog-to-digital converter digitization illustration

We can define some basic terminology as a result, most of which can be defined directly from this aspect of the ADC:

- the number of quantization levels for a N bit converter is 2^N
- the resolution is given by $V_{FS}/(2^N - 1)$ (V_{FS}: full scale voltage), which is equivalent to the smallest increment level (or step size) q
- Most significant bit (MSB) — weighting of $2^{-1}V_{FS}$
- LSB — weighting of $2^{-N}V_{FS}$
- oversampling ratio (OSR) $= f_s/2f_m$
- monotonicity — a monotonic converter is one in which the output never decreases as the input increases; for ADCs this is equivalent to saying that it does not have any *missing codes.*

If we assume that the quantization noise is uniformly distributed as shown in Figure 13.26, the mean square value of the error can be calculated as shown in the following:

$$e_{qMS}^2 = \frac{1}{q}\int_{-q/2}^{q/2} e^2 \mathrm{d}e = \frac{q^2}{12} \tag{13.10}$$

Therefore,

$$e_{qRMS} = \frac{q}{\sqrt{12}} \tag{13.11}$$

For a high number of bits, the error is uncorrelated to the input signal ($N > 5$). In the frequency domain, the error appears as white noise over the Nyquist range. This noise limits the S/N ratio of the digital system analogous to thermal noise in an analog system.

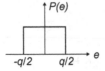

Figure 13.26:
Quantization noise

The peak value of a full-scale sine wave (i.e., one whose peak-to-peak amplitude spans the whole range of the ADC) is given by

$$2^N q/2 \tag{13.12}$$

The RMS of the sine wave is hence

$$V_{\text{RMS}} = 2^N q/2\sqrt{2} \tag{13.13}$$

The signal-to-quantization noise ratio (SQNR) is given by

$$\text{SQNR} = \frac{\left(2^N q/2\sqrt{2}\right)^2}{q^2/12} = \frac{3}{2} \times 2^{2N} \tag{13.14}$$

or, in dB,

$$\text{SQNR} = (6.02N + 1.76) \text{ dB} \tag{13.15}$$

Thus, each bit increases the SQNR by approximately 6 dB, so for an example 16-bit system the SQNR works out to be 98 dB. This equation is an extremely useful way of estimating rapidly the number of bits required to achieve a specified SQNR figure.

The previous calculation assumes that the input signal is sampled at the Nyquist rate (twice the sampling frequency), and the power spectral density of white quantization noise is given by:

$$E^2(f) = 2e_{\text{qRMS}}^2/f_s. \tag{13.16}$$

The OSR is given by

$$f_s/2f_m \tag{13.17}$$

The noise power is therefore given by

$$n_0^2 = \int_0^{f_m} E^2(f)\mathrm{d}f = e_{\text{qRMS}}^2 \left(\frac{2f_m}{f_s}\right) = \frac{e_{\text{qRMS}}^2}{\text{OSR}} \tag{13.18}$$

SQNR is then

$$SQNR = \frac{(2^N q/2\sqrt{2})^2}{q^2/12OSR} = \frac{3}{2} \times 2^{2N} \times OSR \qquad (13.19)$$

or, in dB,

$$SQNR = (6.02N + 1.76 + 10\log(OSR)) \text{ dB} \qquad (13.20)$$

Thus, doubling of the oversampling ratio increases the SQNR by approximately 3 dB, or half a bit.

A "Flash" ADC is the quickest type of converter and a brief look at its architecture shows why. Essentially, it consists of multiple individual level comparators and, when the voltage input reaches each individual level, then a subsequent bit is set by a comparator. Output logic then translates this code into a standard binary code (Figure 13.27).

The drawback with this type of approach is that the resources required to achieve a relatively low number of bits — the flash architecture requires $2^N - 1$ comparators and 2^N resistors. It is, however, the fastest converter; and a conversion can be performed in one clock cycle. The sheer number of components leads to a high

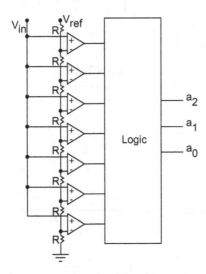

Figure 13.27:
Flash converter architecture

circuit complexity, and the converter accuracy depends on resistor matching and comparator performance (practical up to 8 bits).

A "counting" ADC is one of the simplest converters to implement and has the architecture shown in Figure 13.28. It is easy to implement; however, a major drawback is that the conversion speed depends on the difference to previous sample. It is therefore very slow for rapidly varying signals and fast for slowly varying signals. It is also referred to as a "tracking or counting A/D converter".

The successive approximation ADC is one of the most commonly used ADC types, and has an excellent compromise between efficiency and accuracy. The architecture is shown in Figure 13.29.

Unlike the "counting" ADC, it successively tries each bit in turn − so it is much faster than using a binary approach. It converts the MSB first and then

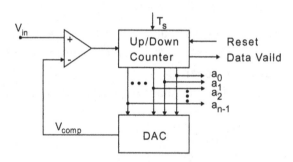

Figure 13.28:
Counting analog-to-digital converter

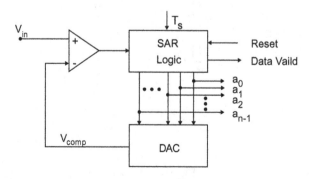

Figure 13.29:
Successive approximation analog-to-digital converter

progressively smaller bits — always taking N cycles, so has a uniform conversion time unlike the counting ADC.

An alternative approach is the "dual slope" ADC, which uses an integrator to create the slope between input values, as shown in Figure 13.30. It has relatively high resolution (up to 14 bits), is independent of exact values of R and C, and is implemented readily in CMOS; however, it is relatively slow as it depends on the time constant of the integrator.

Sigma Delta ADCs (Figure. 13.29) are in the "oversampled" converters category. They turn an analog signal into a bit stream corresponding to the input, and consist of a sampler, quantizer, filter function, and feedback functions (Figure 13.31).

If better noise shaping is required, the order can be increased from first order to any number; however, as the order increases, the potential for instability becomes much higher. If better noise shaping is required, a second order modulator can be used and the RMS noise power can be calculated using the following expression:

$$n_0 = \int_0^{f_B} \left| N_q(f) \right|^2 df = e_{\text{RMS}} \frac{\pi^2}{\sqrt{5}} \text{OSR}^{-5/2} \qquad (13.21)$$

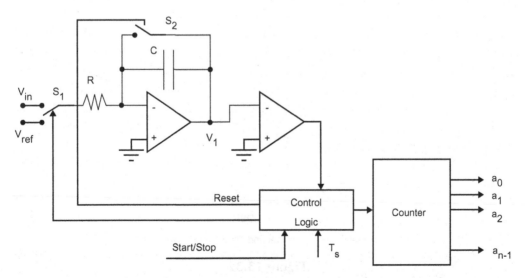

Figure 13.30:
Dual slope analog-to-digital converter

The implication is that both increasing the order and also the OSR leads to an improved SNR, and this can be seen graphically in Figure 13.32.

In summary, the advantages of a sigma delta modulator are that you get a digital bit-stream out automatically – making it an excellent choice for conversion

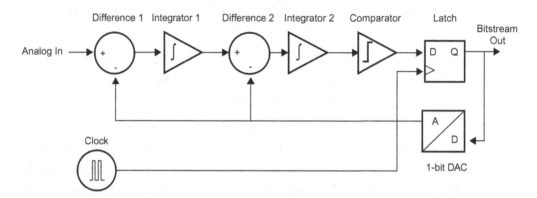

Figure 13.31:
Second-order sigma delta architecture

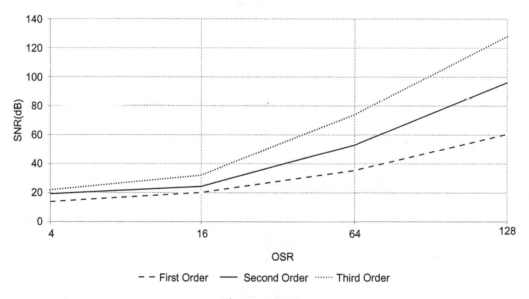

Figure 13.32:
Relationship between signal-to-noise ratio (SNR), order, and oversampling ratio (OSR)

and streaming. It is simple to implement in CMOS, scalable, and very easy to configure both the OSR and the order. One downside is that it can become complex to analyze for higher order converters — a particular problem for use in a sensor interface circuit.

In the consideration of the converter type to be used, the successive approximation is probably the simplest to implement, and, as the system requirements are for relatively low sampling rates, this is possibly the best choice; however, the sigma delta does have excellent noise-shaping characteristics. In any case, the final implementation decision can be made later in the design process. For the time being an idealized configuration can be used where the start and end conversion signals are both used, 16 bits of data transferred to the digital core, and a 3-bit channel select used to request data from a particular channel.

13.3.5 Digital Core Design

13.3.5.1 A Custom Approach

The digital core of the chip design could be implemented using a full custom approach, where the function of the core is defined explicitly and then synthesized to low level gates in the fabrication process, or the functional behavior could also be captured in a processor, which used buffers and memory to store the resulting data, and transfer to and from the RF block.

In this first conceptual level of the design, we do not need to worry about which form the final implementation will take, and so the approach will be to design the main building blocks to manage the digital data, and then these can be implemented as digital hardware or as software running on a processor core.

As we saw in the initial top-level design, the baseband digital block sits between the baseband analog and the RF blocks with two main functions. The first function is to control the baseband analog multiplexer and data converter, receiving the data from the data converter in 16 bit bytes, and the second function is to transmit these data bytes in serial form down the RF channel. In the initial design, the transformation of a bit stream into a more complex coding scheme was to be managed in the RF block itself, and so the output bit stream needs to contain synchronization, identification, data, and, finally, error correction coding.

13.3.6 Summary

So far in this design we have specified the overall top-level structure and have investigated various potential design options for the sensor chip. Models have been created for aspects of the three main areas (baseband analog, digital, and RF), at a system level to enable some basic studies to be undertaken of the overall system behavior, and for us to understand the design and engineering issues, we need to address when the individual elements are combined.

We can now concentrate on the individual blocks and their place in the system as a whole from a very informed perspective. The first stage in that process is an initial design (first cut) where simple models will enable a complete system design to be analyzed.

13.4 Initial Design — First Cut

13.4.1 An Introduction to Design Partitioning and Reuse

As we have seen previously in this book, one of the key concepts that we would like to leverage as engineers is to maximize the material that we can reuse from either standard libraries or previous designs. In this design example we have seen how the first level of hierarchy has been used to divide the initial conceptual design into three main sections: analog and mixed-signal (baseband), digital, and analog (RF). This is a fairly coarse (albeit useful) initial division, but now each section of the design can be investigated in more detail. This is where we can carry out an initial design partition — the concept we introduced in Chapter 12 — and begin to see how the individual elements of the design will take shape.

As we reprised in the previous section, how does this fit into the overall V diagram process? If we look at Figure 13.33, we can see that we are now into the

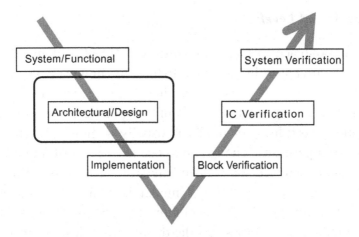

Figure 13.33:
V Diagram — moving into the design phase

design phase "proper", and transitioning from a system level view into decomposition into individual design elements. This means that design partitioning needs to take place in order to accomplish that.

Design Partitioning

MBE facilitates evaluation of partitioning options the designer proposes.

13.4.2 Initial Design Partition

As we have seen already in this chapter, the overall hierarchy is a standard transceiver architecture, as shown in Figure 13.2, with a baseband analog/mixed-signal section for interfacing to sensors and a generic digital core. As in any project, there will be some definitions to consider, which will be effectively prerequisites for the designer and not really a design decision as such. For example, in the case of the digital core in this design, the aim is to use a proprietary processor and this obviously constraints the architecture to operate with that. This will be discussed in more detail later in this section of the chapter. The first step in the process will be to define ideal models for the main blocks in the initial design so that an initial assessment of the design can be made prior to detailed modeling taking place.

13.4.3 Models and Levels

It is quite commonplace to talk about different "levels" of model, and, as we have introduced previously in this book, that refers, in most cases, to the level of abstraction in the model. It is worth refining that somewhat in the context of a real design and clarifying what we mean in the specific context of this example. The initial, most basic level of model is usually a simple linear model that is used for two main purposes. The first purpose is to establish that the connectivity is correct and also to see some basic behavior, such as the gain. This type of model is often referred to as a "level 0" model, and this is the nomenclature that we will use in this chapter. As the level numbers increase (level 1, level 2, etc.), the level of complexity increases and the detail intensifies. For example, a level 1 model would include nonlinear effects and probably power supply connections. A level 2 model would be a very detailed model, including second-order effects, and a level 3 model would be typically used to include physically realistic behavior, such as layout or parasitics. In this example, we will constrain the modeling to level 0 and level 1; however, in many practical situations critical sections of a design may go to level 2 or even level 3 depending on the context.

This section will concentrate on level 0 models to enable that first cut of verification to take place and ensure that the blocks will fit correctly into the overall design strategy. Example blocks will be described that illustrate the approach of not only creating the models, but also establishing the verification framework to ensure that they are correct.

13.4.4 RF System — Level 0 Blocks

13.4.4.1 Level 0 RF Power Amplifier

The power amplifier implemented in this model amplifies the AC component of the voltage, while maintaining the DC bias. The level 0 model consists of a simple ideal amplifier with gain as a parameter. The basic architecture of the level 0 model is shown in Figure 13.34 and shows the basic amplifier blocks used to achieve the power amplifier function.

The basic level 0 model is an idealized amplifier and therefore has gain and not much else, other than the correct connection points.

When the amplifier is simulated by applying a sinusoidal voltage to the input (blue waveform in Figure. 13.35), the output waveform (red) is simply the input multiplied by whatever gain was set as a parameter to the model (in this case 6).

This model enables a basic gain analysis of the system to take place and also to check that the connectivity of the model is correct. As we can see from Figure 13.35, the basic model operates as required.

13.4.4.2 Level 0 RF Digital-to-Analog Converter

The RF digital-to-analog converter (DAC) is defined as an unsigned 12-bit digital-to-analog voltage converter. The digital word is defined using a little

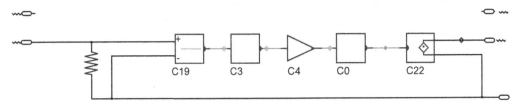

Figure 13.34:
Level 0 power amplifier model architecture

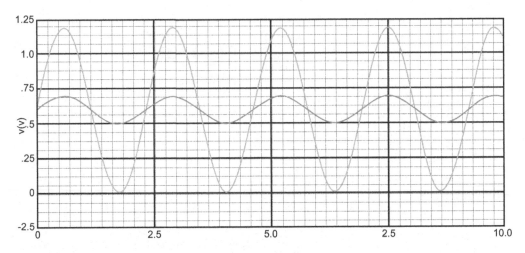

Figure 13.35:
Level 0 radio frequency power amplifier gain test

endian bit designation: 0 (MSB) to 11 (LSB). The analog output voltage range and zero for the level 0 model are defined by parameters of the model.

The topology of the model is given in Figure 13.36 and is an ideal data converter with no nonideal effects at this stage.

In order to test the level 0 DAC, input square waves were applied as the input digital word, with a digital decrementing counter so that the output voltage would ramp from the maximum range to the minimum range. The input signals were set to the nominal VDD of 1.2 V and a reference VSS of 0 V. The resulting output voltage was simulated as shown in Figure 13.37, where the linear transformation of digital word to analog voltage can be seen.

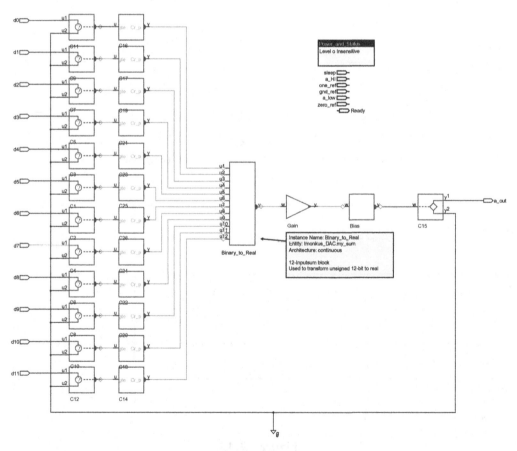

Figure 13.36:
Level 0 radio frequency digital-to-analog converter (DAC) topology

13.4.4.3 Level 0 RF Switch

The level 0 model shows a basic switching function and is not dependent on supply voltages. The enable signal is converted from electrical to variable and passed on to the two comparator blocks (Figure 13.38). The upper comparator will send a high signal to the transmitter line switch if the enable is higher than the `threshold` parameter. The lower comparator will send a high signal to the receiver channel switch if the enable is less than or equal to `threshold`. This action selects either the transmit or receive mode. For the level 0 behavior, no hysteresis is implemented (Figure 13.38).

The simulations to validate this simple behavior are run separately to show each channel. The `threshold` parameter is set to 0.6 V. In both simulations the

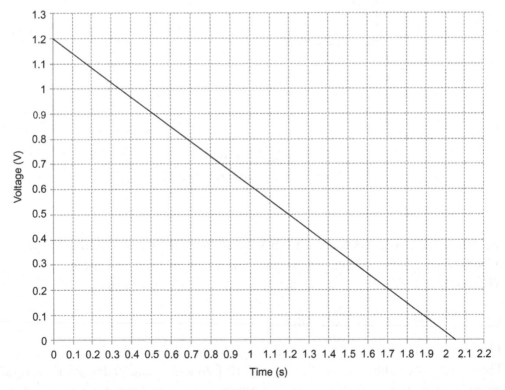

Figure 13.37:
Level 1 radio frequency digital-to-analog converter waveform

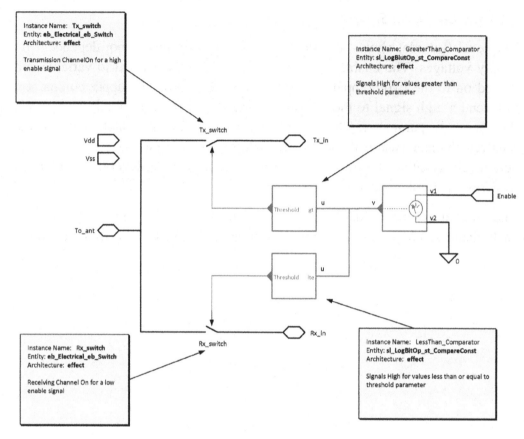

Figure 13.38:
Level 0 radio frequency switch

input is a sinusoidal source with 0.6 V amplitude, 0.6 V bias, and a frequency of 400 GHz. As the enable signal goes from low to high in Figure 13.38, the RF switch changes from receive mode to transmit mode (Figure 13.39).

13.4.4.4 Level 0 RF Low Pass Filter

This entity contains the level 0 and level 1 version of a third-order RC LPF. This filter was tuned for a 2 MHz cutoff frequency and 0 dB gain. Several second-order effects, such as output voltage saturation, power supply control, start-up delay, and group delay, were implemented in the level 1 model (Figures 13.40–13.44).

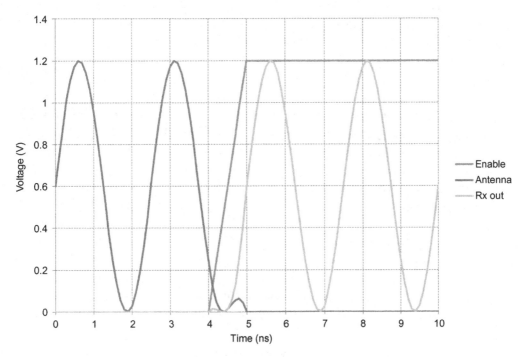

Figure 13.39:
Level 0 radio frequency switch waveforms

13.4.4.5 Level 0 RF Low Noise Amplifier

The low noise amplifier (LNA) is modeled initially as a basic gain element. The input and output to the model are defined as conserved electrical connections to enable them to be connected to other circuit elements using any level of abstraction. The input is converted to a real scalar inside the model and then is converted back to electrical at the output of the model. The input DC voltage level is subtracted from the input and then is sent to an amplifier. The bias of the amplified signal is then set by adding the output DC voltage level before the signal is converted to electrical gain.

The basic model architecture is shown in Figure 13.45, and it consists of linear elements only at this stage to represent the basic gain and offsets in the block.

The model was evaluated using a simple test bench with a time domain sinusoidal signal being applied to the input and the waveforms checked to ensure the basic operation of the LNA. The input to the model consists of a sinusoidal

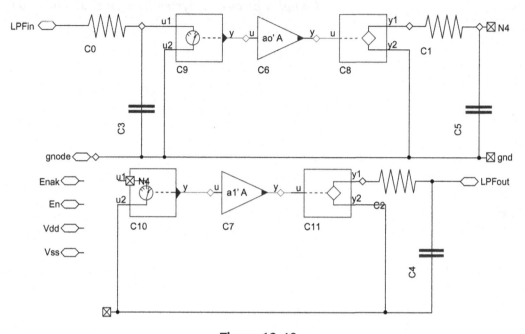

Figure 13.40:
Level 0 radio frequency low-pass filter

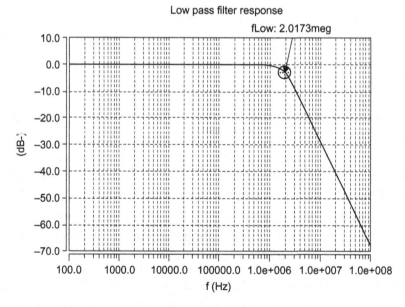

Figure 13.41:
Level 0 low pass filter frequency response

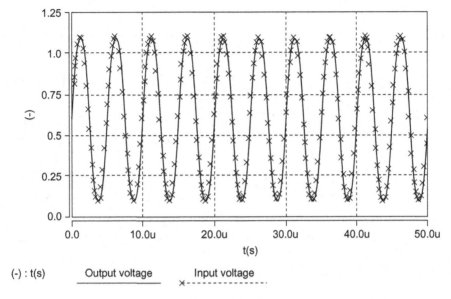

(-) : t(s)   Output voltage   Input voltage

Figure 13.42:
Level 0 radio frequency low pass filter at 200 kHz (∼100% gain)

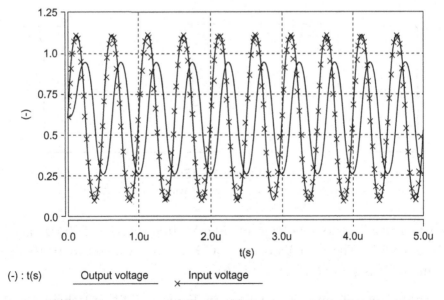

(-) : t(s)   Output voltage   Input voltage

Figure 13.43:
Level 0 RF LPF at 2MHz (reduced gain ∼0.7)

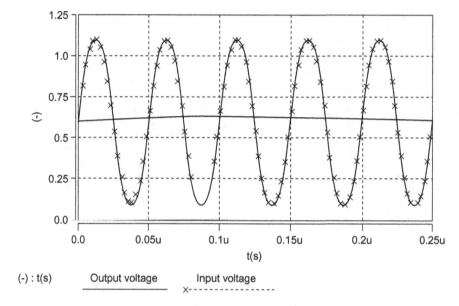

Figure 13.44:
Level 0 radio frequency low pass filter at 20 MHz (~0% gain)

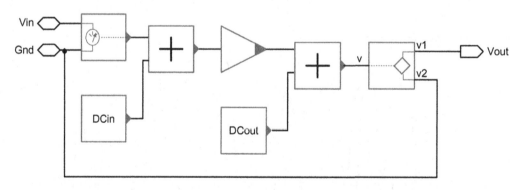

Figure 13.45:
Low noise amplifier level 0 model architecture

input which was set to a voltage of 50 mV, frequency of 500 Hz, and a DC bias level of 5 V. The model's output DC bias level was set to 10 V. The gain of the model was set to 6 (Figure 13.46).

The resulting waveforms can be seen in Figure 13.47 and show clearly the correct biasing on the input and output, and the effect of the linear gain.

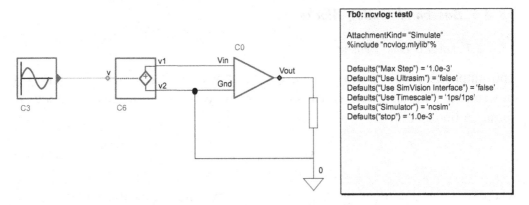

Figure 13.46:
Level 0 low noise amplifier testbench

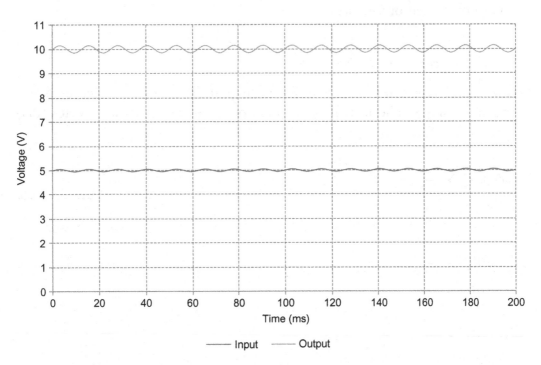

Figure 13.47:
Level 0 low noise amplifier input test signal and output voltage

13.4.5 Baseband Analog Blocks

13.4.5.1 Level 0 Dust Sensor

The output response graph from a datasheet was digitized with Engauge [2] and entered into an Excel spreadsheet for curve fitting. The level 0 model is simply a linear fit

$$V_o = K \cdot x + V_{oc} \qquad (13.22)$$

where x is the dust density, K is the sensitivity, and V_{oc} is the voltage at the "no dust" condition from the datasheet (Figure 13.48).

The datasheet curve, obscured in Figure 13.49 by the level 1 model curve, indicates that as the density of the dust reaches a certain point the output voltage of the sensor saturates.

13.4.5.2 Level 0 Phase Locked Loop

This model represents the control system portion of a linear phase locked loop frequency synthesizer. It uses a multiplication-based phase detector (PD)

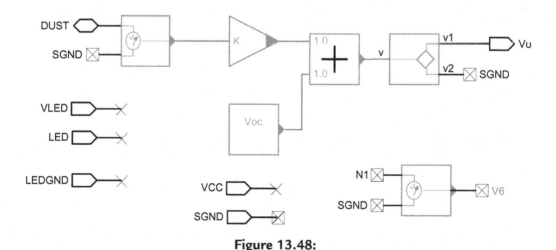

Figure 13.48:
Level 0 dust sensor topology

and a second-order active loop filter (LF). A fractional N divider is also included for frequency multiplication and synthesis. The second order active LF is defined by three parameters (`ka,tau_1, tau_2`). `tau_1` determines the cut-off frequency and `tau_2` determines the stop frequency. `tau_1` should be greater than `tau_2`. The gain, `ka`, is chosen based on damping criteria. The input reference frequency, f_{ref}, can be multiplied up or down by setting the number of divisions, N, of the VCO output signal, f_{vco}. ($f_{vco} = N \cdot f_{ref}$) (Figures 13.50–13.53).

13.4.5.3 Level 0 ADC

The level 0 model of the ADC is shown in Figure. 13.54. The quantizer begins sampling when the input signal is zero, so a step response function is

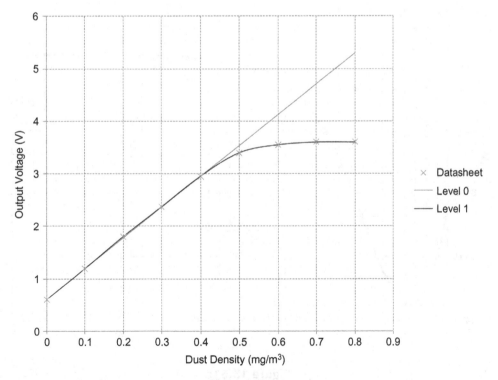

Figure 13.49:
Level 0 dust sensor waveforms

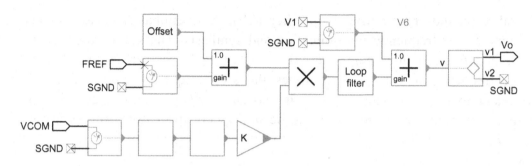

Figure 13.50:
Level 0 phase locked loop topology

logic simulation

—bxs003_pll_pll_tb.inlined_vco_freq_C0

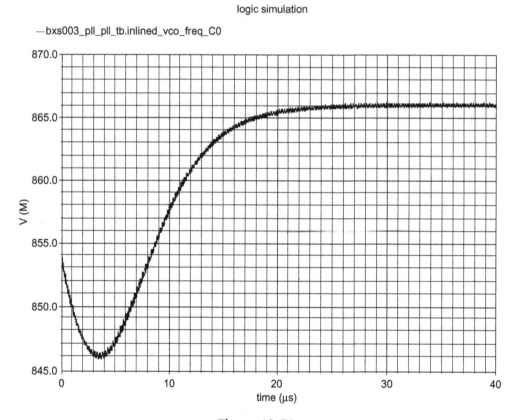

Figure 13.51:
Step response of the system, which settles at 866 MHz

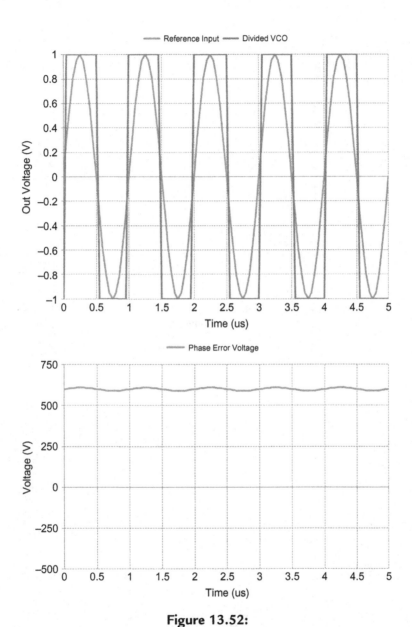

Figure 13.52:
Simulation showing the phase locked loop locked (top graph: reference input vs divided voltage-controlled oscillator (VCO); bottom graph: phase error voltage)

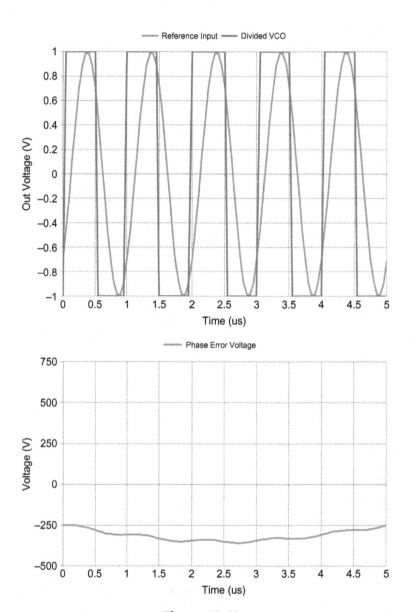

Figure 13.53:
Simulation showing the phase locked loop unlocked (top graph: reference input vs divided voltage-controlled oscillator (VCO) feedback; bottom graph: phase error voltage)

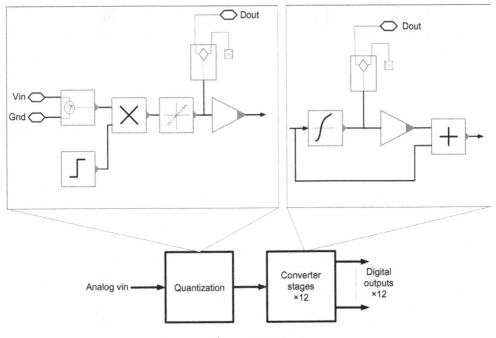

Figure 13.54:
Level 0 analog-to-digital converter

multiplied with the input signal to make the quantizer work correctly. The sampling rate of the quantizer is determined by the `quantizer_interval` parameter. In this model, as the maximum input is 1.2 V the sampling rate is 1.2/4096, which means whenever the input signal changes by 1.2/4096, there is a binary value associated with it. For example, 1.2 V corresponding to 4095 in decimal, 0 corresponding to 0 in decimal, and other values in between 4096 and 0 (Figure 13.54).

Validation simulations are run with a 4 MHz sine wave input. The total simulation running time is 0.6 μs. The 12 output bits are plotted in Figure 13.55. A bus is created to show the output results. The maximum and minimum values are marked in order to verify it is working. Then, a zoomed in version in Figure 13.56 is provided that illustrates that the output results match the input signal with 4095 corresponding to the peak value of the sine wave input (Figures 13.55 and 13.56).

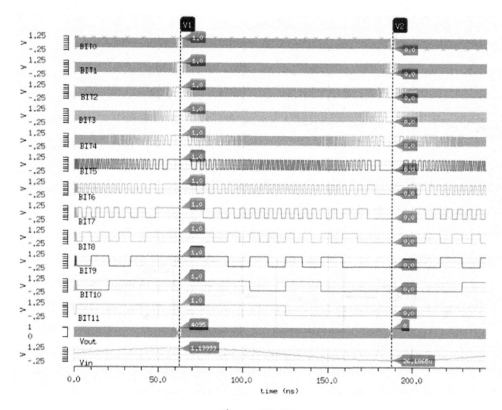

Figure 13.55:
Analog-to-digital converter simulation results over one period

13.4.6 Digital Blocks

13.4.6.1 Level 0 DSP

This DSP model is one comprised of system-level and electrical basic effects. A startup delay of 1 ms is implemented. For the demodulation, the DSP takes in 12-bit streams of data from the two ADCs and determines whether it is above the midpoint of the 12 bit value, 7FF. Once this comparison occurs, it assigns a 1 or 0 to the signal. This bit is then passed to the microcontroller for each stream, I and Q, as a 1 or 0. As for modulation, the DSP is fed individual bits for the two separate streams and relays that value onward such that the 12 separate pins for each stream are either set to all high or all low.

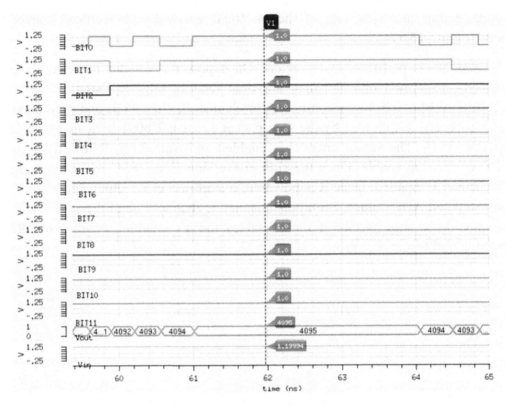

Figure 13.56:
Zoomed in simulation data

This is a very simple, almost digital comparator, approach to converting the received demodulated signal into digital pulses in the IQ plane.

13.4.6.2 Level 0 Microcontroller

This architecture includes all of the necessary power management functions, such as the enabling and power verification of other channels (will not run without these owing to finite state machine (FSM)). The outputs are supply-sensitive. The topology of the microcontroller is fairly simple. It consists of a finite state machine inside the microcontroller. All of the pin outs are converted from the logic domain into the electrical domain and then ported. A variable power-on delay is controlled by a parameter, along with the sampling time and slew rate of the electrical/real and real/electrical conversions (Figure 13.57).

The simulation performs as expected. Once powered on the Rx channel is enabled and waited on. The data are then read in until the start packet is received {2'b11, 2'b11}. A command for channel 3, channel 7, and then channel 6 is received, followed by the stop packet {2'b10, 2'b00} indicated by 8 in the com[35:0]. The sensor channel is enabled and waited on, followed by the ADC conversion enable. Once the start conversion signal FSC is received, the Tx channel is enabled. Once it is powered, a start packet is output, followed by the simulated ADC values. It outputs channel 3, channel 7, and then channel 6, followed by a stop packet. It then disables the Tx channel, enables the Rx channel, and repeats the process (Figure 13.58).

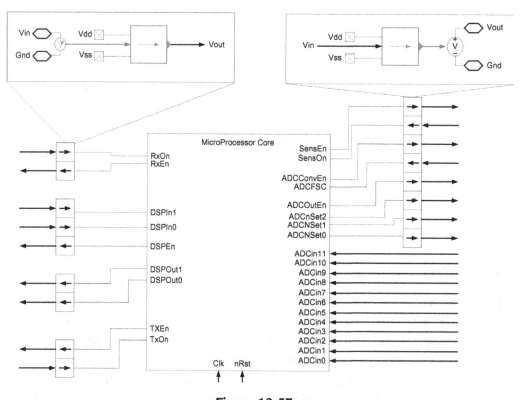

Figure 13.57:
Microcontroller topology

13.4.6.2.1 Finite State Machine Design

The design of the finite state machine is quite simple. The microcontroller waits until it receives power. It then initializes everything and enables the Rx channel. Once the Rx channel is powered, it begins checking the 2-bit input from the DSP. Once it has received a start packet (4'b1111), it starts saving the data in a register until it receives a stop packet (4'b1000). It then enables the sensor block and waits until it has power. Next, it enables the ADC conversion and waits until that is complete. Next, it enables the Tx channel and waits until it is powered. It then transmits a start packet. The ADC output is enabled, and it passes the ADC channel select value from the command register. It outputs the data from the ADC in order from MSB to LSB two bits at a time. If multiple commands were received, it will then repeat until all of the ADC data has been transmitted. It sends a stop packet and then disables the Tx channel. It then repeats this FSM process (Figure 13.59) until its power has been removed.

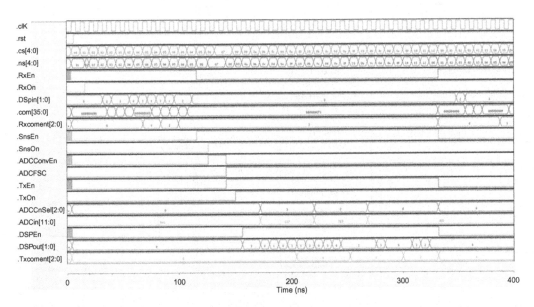

Figure 13.58:
Microcontroller operation

13.4.7 Integration of Level 0 Executable Specifications

Now that all of the constitutive parts of the design have been captured in the form of level 0 models that represent the most basic executable specifications, it is time to gather all of this together to ensure that they work together to accomplish the basic requirements described earlier. Only one particular scenario will be described for brevity, but there are many such analyses that are now poised to be

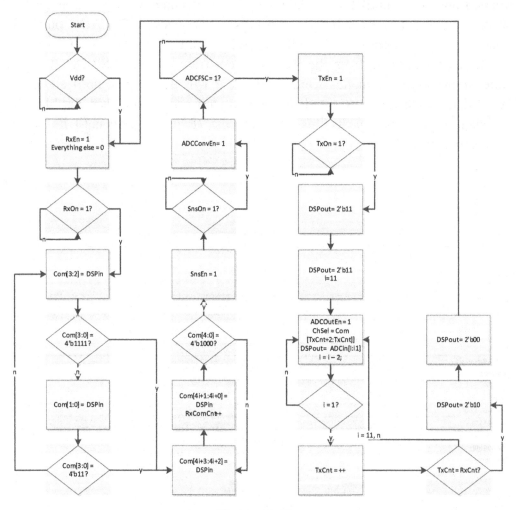

Figure 13.59:
Microcontroller finite state machine process

run. Once the sensor has provided an analog voltage converted to a digital byte to the microprocessor, and it has been coded, then the relevant signal can be transmitted using the RF channel. As we have seen already in this chapter we have level 0 models, the most basic level, already created, and so before embarking on more detailed design it is useful to check that we can actually achieve the basic design requirement of transmitting the signal correctly using these level 0 models.

In this simplified case study, the design focuses on a sensor interface node that receives commands, executes data collection, and transmits that data back to a central computer. That basic scenario will be simulated. In the system, a command set is received from the base station on the receive channel. The chip is assumed to be powered up and in stand-by mode. The first function that occurs is the awakening of the Rx channel and DSP to receive and decode the command. When the DSP has gone from sleep mode to active, a decoded and digitized command sequence instructs the microcontroller which analog channels to activate and process sensor data from. These data are then transmitted through the RF Channel highlighted in Figure 13.60.

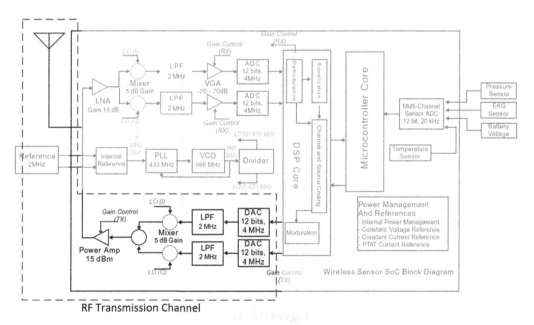

Figure 13.60:
Block diagram of the chip showing the radio frequency transit channel to be tested

In order to test the channel, reference data of a single value are converted to a 1 kHz rotating vector in the IQ channel sent to the I and Q DACs (12 bits, 4 MHz), and the resulting data are then processed through the channel.

When the output of the DAC and filter are observed, for a DSP output signal frequency of 1 kHz, sampled at 4 MHz, we can see that the two signals look coincident, as shown in Figure 13.61, and the effect of the 2 MHz LPF can be seen in Figure 13.62, where the samples are smoothed out somewhat.

The I and Q signals are then mixed with the local oscillator running at 433 MHz and when these signals are then summed together, the resulting combined RF signal is then put through the RF Power Amplifier into a 50 Ω load, to give a power level of 15 dBm, which corresponds to the power required for a typical wireless device, such as a laptop. If the fast fourier transform (FFT) of the output waveform is calculated, we can see the 433 MHz carrier signal clearly in the magnitude response, as shown in Figure 13.63.

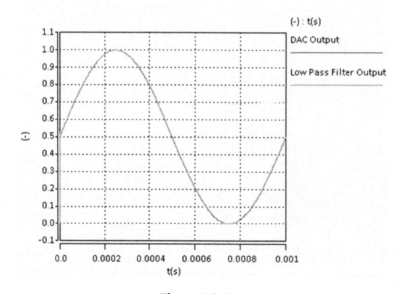

Figure 13.61:
Radio frequency digital-to-analog converter (DAC) and low pass filter output on I channel

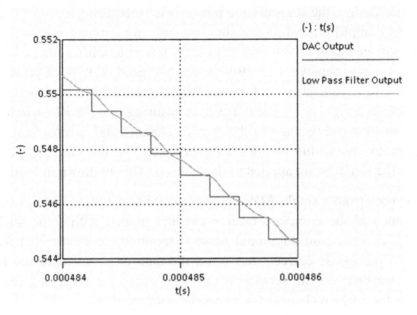

Figure 13.62:
Effect of 2 MHz low pass filter on output

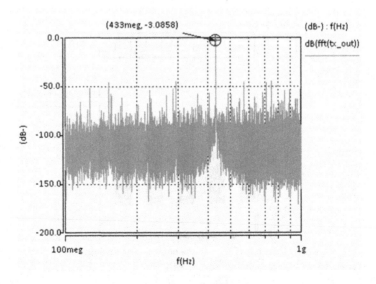

Figure 13.63:
Radio frequency channel carrier FFT

Now, in our design, the transmission power is defined using a model parameter in the power amplifier, and, in the simplest case, we can extrapolate the power from the voltage being driven using this ideal power amplifier into a 50 Ω load, where, in order to achieve a 15 dBm power, we need 32 mW of power, which corresponds to a voltage of 1.26 V into 50 Ω. As we have seen before in the overall system design, one critical aspect is ensuring that the RF switch enables the transmit to avoid damage to the receive channel and so one final test was undertaken to check this by switching between the Rx and Tx modes and observing the signal being applied to the antenna. This is shown in Figure 13.64.

When the data from a single ADC conversion are transmitted through the channel, we can add the complete receive channel model, with some attenuation due to the channel (and additional noise if required) to ensure that the basic modulation process is correct, and when the data are applied to the I and Q channels, we can see the recovered data sequences by observing the output voltage on the receive channel for a specific data input.

For example, Figure 13.65 shows the input digital sequence for the I channel (signal di) and this is then transmitted and received, and after demodulation the

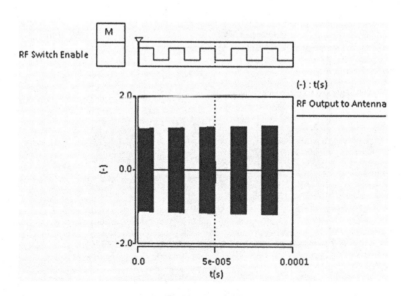

Figure 13.64:
Radio frequency Switch Mode showing Tx power when enable signal is high

voltage output from the LPF of the I channel can be seen in signal rx_i_lpf. Obviously, the amplitude is not being limited (it is a level 0 model), but, clearly, there is enough gain in the channel as a whole to successfully transmit data sequences correctly.

From the description of this single scenario, many others can be conceived, such as data conversion, channel coding, channel selection, power-up sequencing, receiving additional commands before completing the round-trip processing of data and transmission of the previous set (i.e., handling potential collisions), and activation and exercise of the power management functions while these basic scenarios are executed. All of these scenarios can be studied at a level 0 to wring out basic fundamentals of the system design (i.e., capturing executable specs) before proceeding to more detailed models. In this way, the design team stays focused on what they *want* to have happen rather than leaving to chance what *may* happen if key issues go unaddressed. Then, as timing analyses proceed with more detailed models, this design intent is more readily verified and not forgotten, thus leading to expensive rework later in the design cycle (or, sadly, another design iteration).

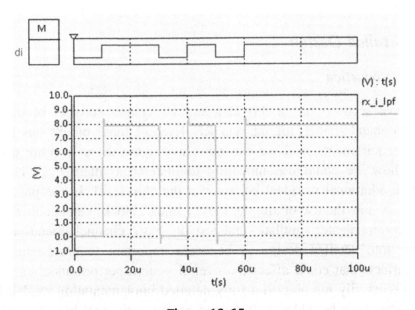

Figure 13.65:
Radio frequency channel data streaming test

13.4.8 Summary of Level 0 Modeling

At this point in the design process, the key question for the design team is "Does this really work at an ideal level?". In other words, has the concept made the transition into instantiated hardware blocks that will ultimately end up being manufactured? This is still more detail than a system-level model as it does have hardware realistic connections, but is proving to be a significantly useful tool to refine the detailed specification for the subsequent design stages.

The next stage of the process, as discussed earlier in this book, is to descend deeper into the details of the design and answer more detailed design questions. A decision at this stage is what level to go to next? It would be possible to go directly to transistor level; however, this would often be prohibitively time consuming, so a useful next stage will be to go to a level 1 behavioral model, which includes the detailed effects that will be directly relevant to circuit designers. The outcome of the next stage of the model will therefore be to not only define the functional specifications of the circuits to be designed, but also the capture of the detailed design specifications — which is leading directly towards verification.

13.5 Detailed Design

13.5.1 Introduction

As we have seen so far in this chapter, the system consists of three main blocks: baseband analog, digital, and RF. Each of these blocks has their own differing requirements, and we can now investigate these in more detail and examine how we could use modeling to understand them even better, and investigate additional nonideal behavior at the chip level. In the previous section, we saw how the use of simple, mostly linear, blocks can enable the design team to communicate, simulate, and test the basic channel operation, but, as we move into detailed design, it becomes important to ensure that all the detailed affects that could affect the overall system performance are included, without necessarily jumping to a fully detailed implementation model. In each case, the details to be added will depend on the individual block requirements and also the context within the system.

We will take each major block in turn and look at it separately, as the requirements are distinctly different in each case. For example, the RF block will be one of the major power loads in our design, and so the trade-off between signal power, range, and lifetime will be critical, whereas the baseband analog design is much more concerned with signal accuracy and noise reduction. The digital block is a clear trade-off between flexibility, functionality, and power consumption. If the sensor node is essentially "dumb" then there may not be much point in having a complex processor in the design, rather a high-quality, low-power digital design may be more appropriate. These issues will be discussed in the remainder of this section.

13.5.2 RF Detailed Design

13.5.2.1 Level 1 RF Power Amplifier

The level 1 model of the power amplifier is where we begin to investigate what might be termed the "second order" effects and behavior in more detail. In the level 0 model, we were only really concerned with the basic amplification of the block. As we look in more detail we have to ensure that the system can cope with supply variations, frequency effects (such as harmonics), and also what happens if the amplifier saturates (under extreme operating conditions).

In an integrated circuit power amplifier, we have to ensure that not only does the circuit have a useful range of operation, but, of course, we are fundamentally limited by the VDD and VSS supplies to the chip. It is ideally the case, therefore, that we bias the amplifier midway between VDD and VSS to ensure the best operating point. The amplifier also has a threshold beyond which the circuit will operate, and obviously this must be less than VDD-VSS. Finally, the amplifier, as in any real circuit, will have its output limited by VDD and VSS, and a dynamic behavior based on a time constants within the circuit.

The amplifier frequency-dependent behavior consists of both small- and large-signal dynamics implemented using a single pole (small-signal) and slew rate limit (large-signal) elements (Figure 13.66).

In order to test the amplifier, a test signal was applied to the amplifier and then the power was turned on, and then off. As can be seen from Figure 13.67, after the amplifier is turned on there is a short delay and then the output waveform

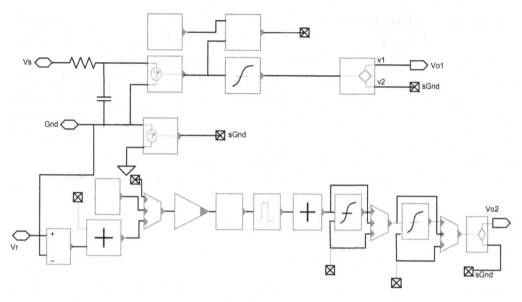

Figure 13.66:
Level 1 radio frequency power amplifier architecture

can be seen to demonstrate the gain, and the transition time is also evident from the start and finish transition.

The small-signal frequency behavior is also implemented using a single pole filter, with a default value of 500 MHz, and the resulting attenuation can be seen in Figure 13.68. The waveforms show the attenuation of the output signal compared with the input signal. This is a linear effect, which operates regardless of the rate of change of the signal and simply attenuates the amplitude of the waveform.

The second frequency-dependent effect is the slew rate limit, and this is manifest as a limit on the rate of change of the output waveform. This model was tested with a slew rate set to 1000 V/μs, and, for an input frequency of 433 MHz, the effect is clear. Figure 13.69 shows how with the addition of a slew rate limit, the amplifier is unable to respond quickly enough to the demanded 433 MHz input signal and this results in distortion of the output voltage. The problem for an RF power amplifier is a loss of quality in the output signal and also extra power required to provide the transmission range required.

When the amplifier is driven to a high level, the effect of the power supply limitations becomes evident. In an ideal model, it is possible to drive the

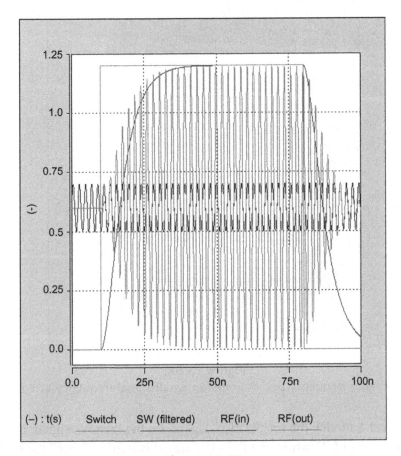

Figure 13.67:
Level 1 radio frequency power amplifier test waveforms

amplifier output to an infinitely high voltage; however, for the real circuit, the output voltage is limited by VDD and VSS, and this is highlighted in Figure 13.70. The output voltage can be seen to be limited to the supply voltage range of 0 V to 1.2 V for this process.

13.5.2.2 Level 1 RF DAC

The RF DAC, when modeled to a level 0, was simply a transformation between a digital word (12 bits) and an analog voltage. In order to extend this to a more detailed level 1 model for analysis, the model was extended to include transport delays and other nonideal effects. The resulting topology is shown in Figure 13.71.

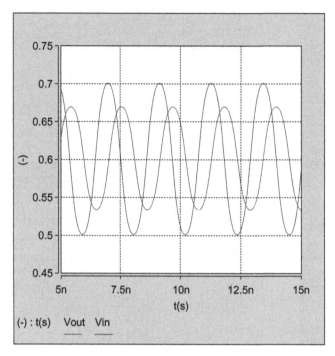

Figure 13.68:
Level 1 radio frequency power amplifier small-signal frequency attenuation

With the level 1 model, the results also depend on the "ready" signal, and this can be seen in Figure 13.72 where the READY signal is used to "gate" the output.

One of the interesting aspects of using a MBE approach is the ability to document graphically the assumptions and features of the model. As can be seen in Figure 13.71, comment boxes were added to the topology of the model to enable another engineer to clearly see and understand the assumptions made for the model, which is one of the most common problems with a nongraphical approach.

In addition to the delay characteristics, the level 1 model includes saturation- and frequency-dependent effects, as we saw in the power amplifier model. This gives a first order approximation of the possible nonlinearities.

13.5.2.3 Level 1 RF Switch

The level 1 model of the RF switch provides switching ability in addition to supply voltage dependence. Additionally, capacitances from the transistors in

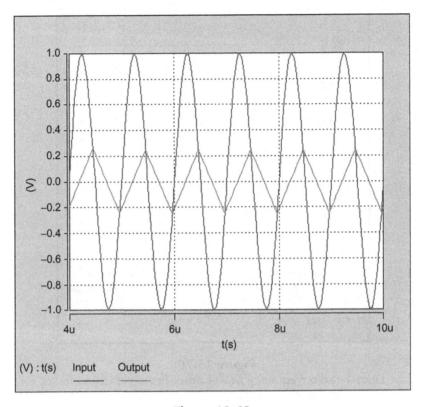

Figure 13.69:
Level 1 radio frequency power amplifier slew rate effects

the switches are implemented using C_{diff} for diffusion capacitances and C_{switch} for drain-source capacitance.

The threshold between the two comparators is obtained by taking the difference between the high supply voltage and the ground. This voltage is then halved to make the threshold the average between the two. A comparator gives a 1 or a 0 if the Enable value compared to ground is above or below the supply voltage, respectively. A slew control block is implemented in the level 1 model to model the delay behavior in the time domain (not just a frequency domain model). A high signal turns on the transmission channel switch, and a low signal turns on the receiver channel switch. Diffusion capacitances from the switches are present on each channel port and the antenna port, with series resistors present to avoid a simulation error.

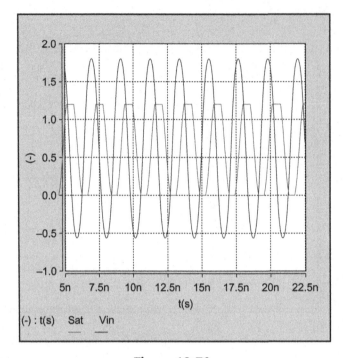

Figure 13.70:
Level 1 radio frequency power amplifier saturation effects

This is a typical modeling fix to avoid the potential problem of "floating" capacitors in the model. As we saw previously in this book, when a circuit is formulated for solution by a simulator, it needs to have a DC path to the reference, otherwise the matrix can become ill-conditioned and difficult to solve. While these internal model resistors need to be in place for convergence, care must be taken with the values used. When testing in isolation the absolute values will probably not matter too much; however, when connecting to other impedances, the values need to be high enough not to cause any attenuation (i.e., a potential divider effect). However, it is not advisable to make the values too high, as some simulators can truncate resistor values. A suitable value in most circumstances would be $1G\Omega$ (Figure 13.73).

The simulations are run separately to show each channel. In both simulations the input is a sinusoidal source with 0.6 V amplitude, 0.6 V bias, and a frequency of 400 MHz (Figure 13.74) — just as when we simulated this test bench

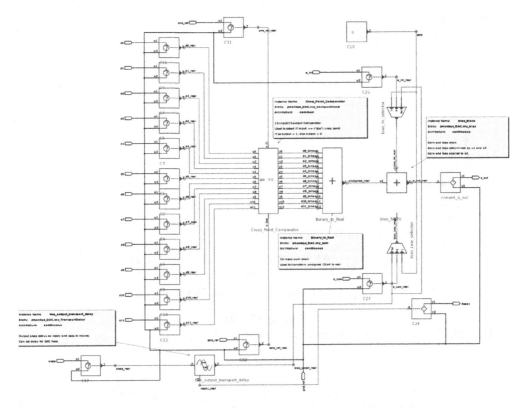

Figure 13.71:
Level 1 radio frequency digital-to-analog converter topology

with the level 0 model. In this case, there is a 100 Ω load, which is consistent with the loading in the final circuit (Figure 13.74).

13.5.2.4 Level 1 RF LPF

The level 1 filter model uses the same basic low frequency response as defined for the level 0 ideal model, with the addition of slew rate and saturation effects to improve the accuracy of the model. The model itself consists of LPF stages, each consisting of an ideal first order filter built using primitive R and C elements, with ideal gain stages (as in the level 0 model). The model is extended to include saturation and slew rate limits, which will introduce realistic distortion for excessive voltage swing demands or unrealistic rates of change (Figure 13.75).

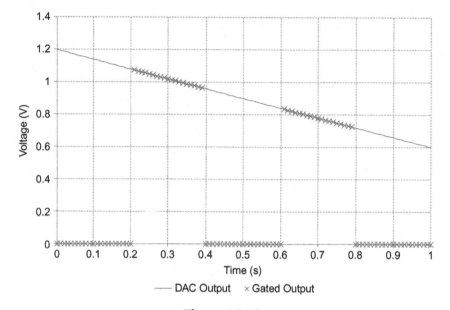

Figure 13.72:
Level 1 radio frequency digital-to-analog converter (DAC) output gated by READY signal

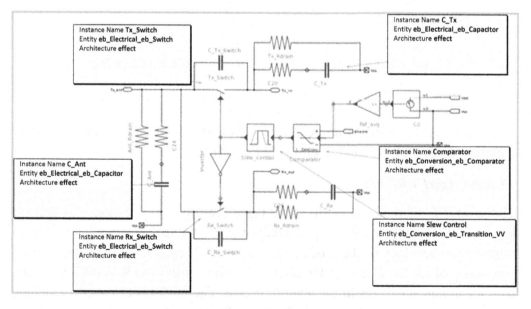

Figure 13.73:
Level 1 radio frequency switch topology

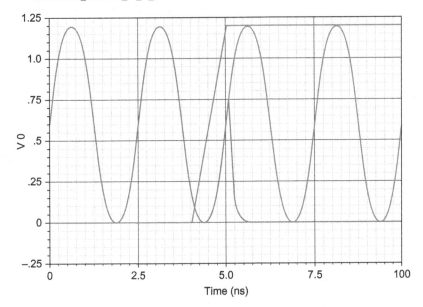

Logic Simulation

— tcbowman_rfswitch_tb_rf_switch.antenna   — tcbowman_rfswitch_tb_rf_switch.Rx_out
— tcbowman_rfswitch_tb_rf_switch.Enable

Figure 13.74:
Level 1 radio frequency switch waveforms

When the model is evaluated, the response of an input voltage that exceeds the VDD value of 1.2 V results in saturation of the output voltage. This can be seen clearly in Figure 13.76, where the output voltage is limited to 1.2 V.

13.5.2.5 Level 1 RF LNA

In the level 1 model of the LNA, saturation, bandwidth, and power dependence effects were added. The saturation effect was modeled to be dependent on VDD using the addition of the model parameters `satmin` and `satmax`. The bandwidth and the quality factor of the LNA were implemented through a transfer function block. The transfer function used represents the behavior of a narrow band filter. The resulting extended model topology can be seen in Figure 13.77.

The input to the model consists of a sinusoidal input of 50 mV, 500 Hz at a DC bias level of 1.5 V. The model's output DC bias level was set to 1.2 V. The gain

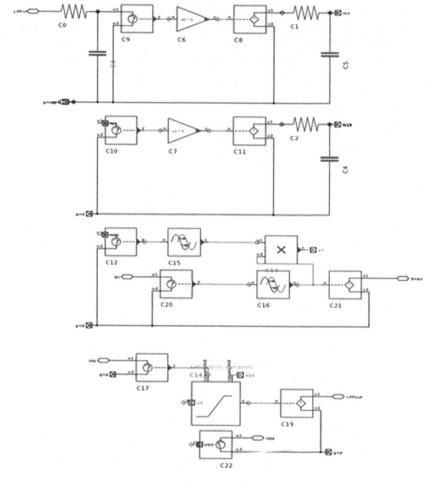

Figure 13.75:
Level 1 radio frequency low pass filter

of the model was set to 2. The rest of the model parameters were set to the default values. The results displayed in Figure 13.78 are the input and output signals.

13.5.3 Baseband Analog

As we have seen, the baseband analog section of the design consists of an eight-channel sensor interface, with a multiplexer and ADC. The section operates off a discrete analog supply and ground, and the requirement is for 16 bits of digital data to be transferred to the digital core. The digital core will request a specific channel, and so the analog baseband has two functions to undertake,

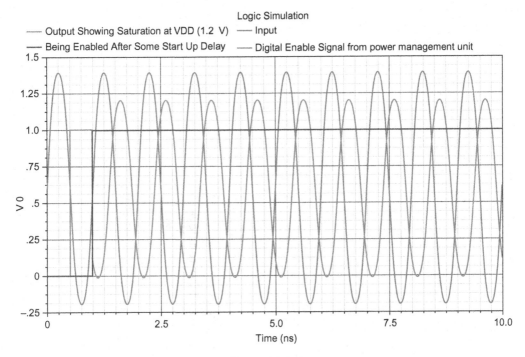

Figure 13.76:
Level 1 radio frequency low pass filter showing saturation effects

the first being to select the appropriate channel and the second is to convert the measured signal into a 16-bit word.

One intended application is measuring the voltage on photovoltaic (PV) cells, and so if the supply voltage is 3.3 V, then the input range is 0–3.3 V. One of the issues in any conversion system is the range of input signal, in this case single polarity, which determines the coding scheme we could use. As we do not care about negative voltages we can simply map 0 V input onto all zeros and 3.3 V input onto all ones. The second significant aspect of the design is that the sensor is a voltage mode system. This means that a high impedance buffer is required; however, as the effect of the multiplexer will be minimal on the system itself, we can have a single buffer amplifier after the multiplexer, which simplifies the design and reduces the size of the final chip.

The second aspect of the design is the rate of change will be low. This means that there is no requirement for either rapid sample rates or changes of channel

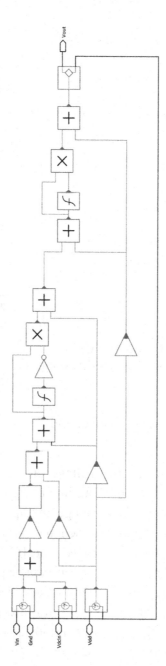

Figure 13.77:

Level 1 radio frequency low pass filter topology

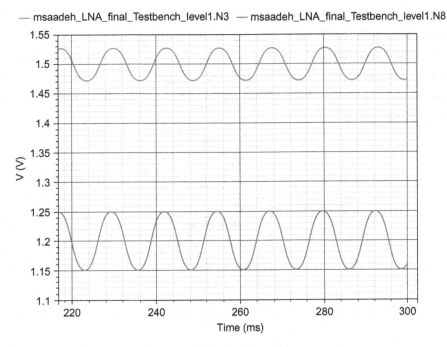

— msaadeh_LNA_final_Testbench_level1.N3   — msaadeh_LNA_final_Testbench_level1.N8

Figure 13.78:
Level 1 radio frequency low noise amplifier waveforms

(the channel may be changed, but in the context of individual sample rates, relatively slowly). This also simplifies the amplifier design, as in each case the amplifier will have time to settle when either a channel or sensor is changed.

13.5.3.1 Level 1 Dust Sensor

The dust sensor is a simple circuit that responds to changes in output as dust is potentially obscuring the light incident on the light-emitting diode (LED). Figure 13.79 shows the output response when the LED is pulsed (from the datasheet). Figure 13.80 shows the level 1 response when the LED is pulsed, and where the sensor is not only representing the basic detection function, but also the time constant of the circuit response.

13.5.3.2 Level 1 PLL

The architecture in Figure 13.81 is the level 1 model of the control system of the linear PLL implemented at a level 0. Transition and transport delays are added to the divider to approximate nonideal behavior of its digital circuitry. The PD output voltage is also limited to the voltage difference between the supplied VDD and VSS. Using these additional effects, the complete PLL model was simulated to evaluate its accuracy. Figure 13.82 shows the PLL step response of the control voltage signal as the PLL circuit locks onto the design frequency of 866 MHz.

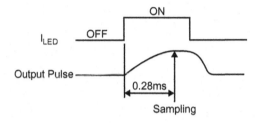

Figure 13.79:
Level 1 dust sensor pulse response

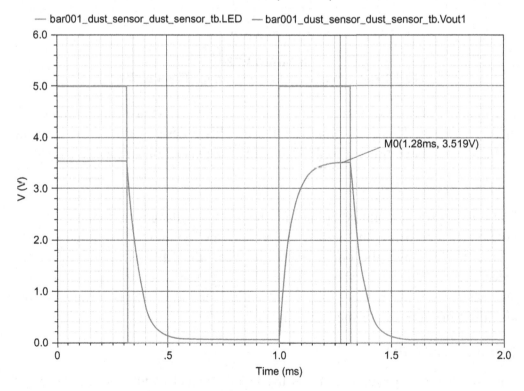

Figure 13.80:
Level 1 dust sensor simulated pulse response

Figure 13.83 shows how the feedback and reference signals are not locked, and the control voltage is "searching" for the correct value but is beginning to move to the approximate region where the PLL will lock. When the control voltage reaches the correct value, the PLL locks and the reference and feedback signals can be seen to be in phase — as shown in Figure 13.84.

13.5.3.3 Level 1 ADC

The baseband ADC model used in the initial level 0 implementation has an ideal data conversion, but does not model the nonlinearities or supply dependence effects in a real device. The level 1 model extends the behavior to include this on the analog input, with the modified topology shown in Figure 13.85.

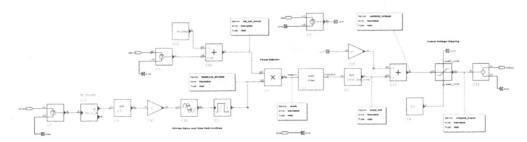

Figure 13.81:
Level 1 phase locked loop topology

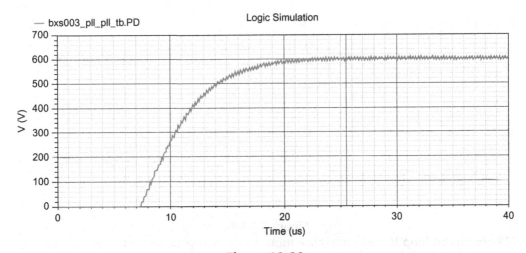

Figure 13.82:
Step response of phase locked loop illustrating lock at 866 MHz

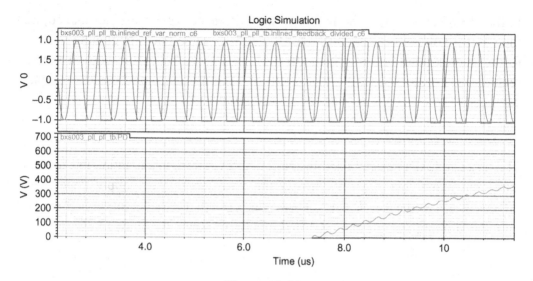

Figure 13.83:
Unlocked phase locked loop detail (reference input vs divided voltage-controlled oscillator; and phase error voltage)

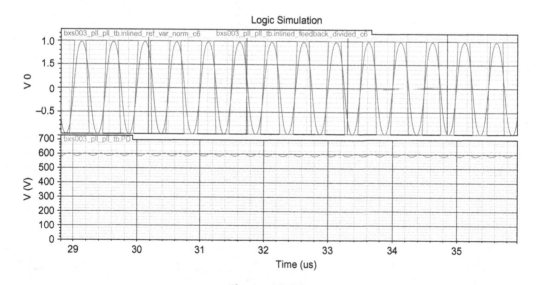

Figure 13.84:
Phase locked loop locked (reference input vs divided voltage-controlled oscillator; and phase error voltage)

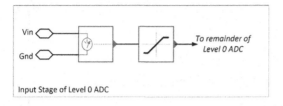

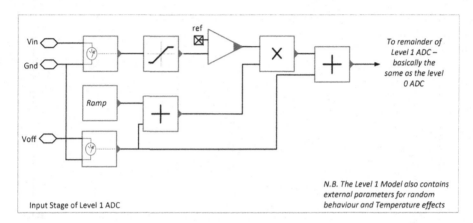

Figure 13.85:

Level 1 analog-to-digital converter

Figure 13.86 shows that the ADC started and worked properly when enabled. After 0.1 ms of delay, the ready signal is sent to inform the power management unit that the ADC is ready to operate. The bit outputs, bus output, and input sine wave are plotted in order to verify its accuracy. Figure 13.87 shows zoomed in results that prove that the ADC works correctly with output data matching the input data.

13.5.4 Digital Blocks

As we have seen so far in this chapter, the extension of baseband analog and RF models from level 0 (ideal and linear) to level 1 (introduction of first-order frequency effects, basic nonlinearities, and supply dependence) is relatively easy to apply. The question is what changes (if any) are required to extend the level 0 digital models to a more detailed level of abstraction?

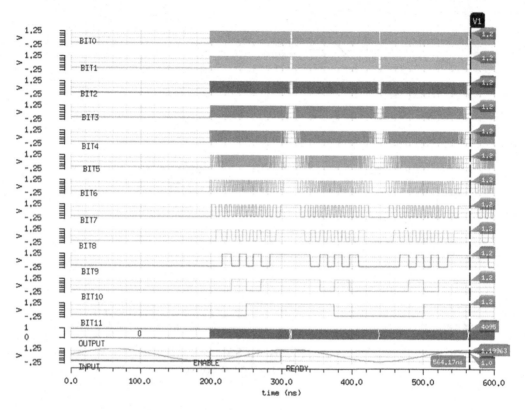

Figure 13.86:
Analog-to-digital converter simulation results over two periods

The main differences between an ideal and more detailed digital model in most cases will be the addition of delays. These could be at a behavioral level or at a gate level, but the main effect will be to reduce the potential operating frequency of the design.

In most cases, post-synthesis extraction of models that are implemented using a cell library for the underlying technology (either ASIC or FPGA) will include those delays, and the synthesis process will include that information when the optimization for speed or area is carried out.

So, how do we as modelers handle this situation? In the digital domain, in fact, this is one situation where manual intervention is generally not helpful. The digital design flow is so automated that manual intervention in the sense of addition of delays would be counter-productive and often a waste of time.

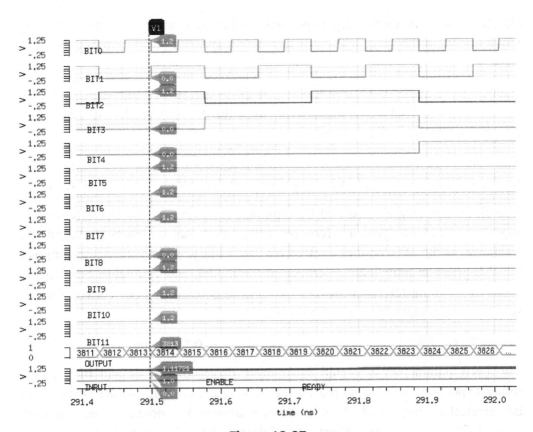

Figure 13.87:
Zoomed in analog-to-digital converter simulation data

Therefore, the good news from a digital designer's perspective is that the development of a synthesized design is very much the preserve of the synthesis flow and the resulting model will be generated automatically during that process, associated with the appropriate model libraries for the particular technology being used.

13.5.5 Integration of Level 1 Executable Specifications

13.5.5.1 Review of the Level 1 Model

At this point in the design process, we know from the level 0 models that the basic system should work, and the question changes from "Does it work?" to "How well does it work?" The more detailed models allow, for example, the channel distortion to be simulated and evaluated against the signal-to-noise requirements in the specification or the overall spectral purity required.

One way of understanding the effect of this MBE approach is to review what changes were made to each individual block as a result of this increased level of detail in the analysis (remembering that the level 0 models demonstrated that the system basically worked).

The variable gain amplifier was modified to include a DC offset as the input range for the ADC is 0–1.2 V and this was effectively ignored in the level 0 model.

The mixer model was modified by replacing an internal ground reference (for model reasons – to give a reference) with a port that connects to the ground reference in the Rx channel itself. This, seemingly, serves the same purpose, but this gives an unrealistic reference in the model, and can cause problems in correctly defining reference paths – leading to potentially poor convergence or sometimes complete failure to find a solution.

The LPF was found to operate correctly and needed no changes.

The LNA parameters DCin and DCout were changed to a default of 0 as the mixer expects a signal centered around 0 V and assuming the RF switch will be centered around 0 V. Unless the LNA is designed to operate with dual rails, level-shifting needs to be addressed and designed into the model and circuit.

The ADC model was modified to include a function that multiplied the input by a unit step function (StepTime = 1e − 9) because the quantizer doesn't start working until the input reaches 0. In this way the quantizer starts up at zero, no matter what the input is. The ADC should have 4096 steps (12 bits), but did not work properly. It tracked well on the rising edge of a signal, but it will get stuck at the highest value when the input falls. It works perfectly using 240 steps. The control sequence of the model therefore needed to be changed (it operates correctly for simple test cases, but needed fixing for more complex real world-type data).

13.5.5.2 Focusing on Interfaces and Design Complexity

The DAC was evaluated in a test bench to verify that the ADC was wired correctly. The parameter avPb was changed to (aV_Hi−Midpoint)/4096, Midpoint to 0, and dV_Hi to 1. Also, k1 in the summer should be 2048 instead of −2048. The DAC and ADC have opposite LSBs — B0 on the ADC connects to B11 on the DAC, B1 to B10, and so on. This is an interesting case where the individual blocks are correct, the system is correct, but only if all the connections are wired correctly. It could be very easy for some other engineer to assume that the DAC and ADC used the same convention for naming and inadvertently connect them together in completely the wrong order. The outcome of this review of the level 1 design against the specification would be to recommend that the individual connection names for either the ADC or the DAC should be renamed (with the recommended change that the DAC naming be reversed such that B0 is the LSB and B11 is the MSB).

13.5.6 Summary of Level 1 Modeling

As we have seen in the integration of the level 1 models into a validation of the design specification, we are now in a position to evaluate in more detail how well the design will potentially meet the specification. This is beginning to link to our concept (used throughout this book) of the Verification "V" diagram, where we can begin to start "verification" of the design from the basic elements upwards.

As we have seen with the revised and more detailed models, we have been able to test individual key blocks, evaluate their basic performance and, ultimately, how they will contribute to the overall system quality. By judiciously replacing level 0 models with level 1 equivalents, we can also see the relative effect of individual blocks on the overall system performance, and the sensitivity of the system as a whole to key elements or even individual block parameters.

Reviewing the level 1 simulations, it was clear that numerous errors were identified, improvements identified, and, finally, one potentially serious source of errors in the ADC/DAC naming conventions located and a solution suggested.

13.6 Bringing It All Together

So, what are our conclusions, both of this chapter and also of MBE in general? It is clear that in the world of modeling and simulation there are numerous pitfalls for the unwary, with false assumptions being a typical cause of problems. There is also a wide variety of techniques to be potentially employed and, as such, the designer can often feel overwhelmed by what they imagine is the amount of work involved in completing an adequate model. This sounds like a negative outcome; however, we hope that by defining many of the techniques and explaining their use and context, we have dispelled some of the myths about modeling — especially the biggest common myth that "designers don't do modeling".

In many ways we could more accurately define the problem as either "designers don't do modeling efficiently" or "designers don't do modeling effectively", but perhaps the most appropriate of all is that, in fact, "designers DO modeling, routinely, every day and just need a bit of help to get better at it". In many ways, that is our goal with this book — to make designers better and more efficient at modeling.

Evaluating Interfaces

MBE is extremely valuable for defining and evaluating interfaces between circuitry — where problems are most likely to occur.

When the task of modeling is understood, especially in the context of design, we can begin to see that it is just one of the tools that we can use as designers to solve problems and make fewer mistakes. As we discussed earlier in this book and, as is well known, the sooner we can identify problems with our design, the better. It is generally cheaper, simpler, and quicker to fix something if we can catch it early (as in most things). One of the key ingredients in systems modeling has to therefore be the ability to evaluate interfaces and establish that we have gotten them right. If we can do this, then we have overcome one of the most potentially destructive areas for error in the whole design process. In our MBE approach, we have described how we can use simple models to ensure that interfaces can be tested and make sure we have them correct — before we spend a long time designing detailed models or circuits.

It is often stated, correctly in our view, that "problems migrate to boundaries" and so much modern design is involved in handling interfaces, data transfer, and signals that if we manage this process effectively good design will probably result.

Managing Complexity

MBE is the key to managing design complexity of many types as we span the V-diagram.

The other serious issue in modern design is the sheer complexity involved. This can be immense for integrated circuits, with literally billions of transistors in single chips, and the low-level transistor simulation of this scale of circuit becomes practically impossible.

What can we do, therefore, to manage this issue? Complexity of design, whether scale or functionality, requires an intelligent approach to design, with a clear understanding of what is necessary to design the system and what is important in assessing its performance.

It is often a complaint made by engineers, "Do I really have to do this task? Is it really necessary? I'm too busy to waste time on tasks not on my critical path....", and often when talking about paperwork, documentation, modeling, and any number of tasks outside the seemingly glamorous world of "design". But is this simply a false economy? In many cases what is really happening is a lack of attention to detail or, perhaps, work done being confused with achievement. This seems a harsh comment to make; however, it is often the case that mistakes are made when individuals are under time pressure, or not enough time is taken to review work properly and make a considered assessment of the true status of a design.

MBE is one method by which designers can improve their ability to understand, document, and explain their design to others, without an excessive addition to workload. In fact, as we have seen in this case study, identifying problems early would probably save more time than doing the modeling itself. Consider the design completed with no modeling. There would be no predictions of performance available other than vague calculations. There would be no checking of connections until the equipment was made. Noise and power would be estimates until hardware was tested, and, finally, the overall system performance would be unconfirmed until the complete system was put together (or the chip fabricated). In the case of a fabricated integrated circuit, one mistake on the connections could cause a complete respin of the chip, and this would involve a minimum of a 3-month delay in most cases *just for the new chip to be fabricated*. Now consider the alternative scenario with an MBE approach. The preliminary designs are evaluated and connections tested with simulation, with any errors fixed at this early stage. Specification refinements are made (such as ADC/DAC pin names) early to make it straightforward for designers to change top-level designs for each block. Detailed modeling enables individual designers to really understand their own designs, not only in isolation, but in partnership with colleagues to ensure the best matching for the benefit of the system. In addition to reports describing the blocks, any member of the design team can see the context of their block in a complete system model. The simulations become an extra layer of documentation to see exactly how the system works. The chip works the first time.

MBE is not the only approach for good design, and it is not to be treated in isolation. However, in conjunction with other good design practice it is one of the most effective tools for the modern designer to get the design completed correctly to specification, on time, and within budget.

References

[1] J.C. Maxwell, A Treatise on Electricity and Magnetism, MacMillan and Co, 1873.
[2] Engauge, Digitization Software, Link correct as of September 2012. <http://digitizer. sourceforge.net/>.

Index

Printed in the United States
By Bookmasters